CRITICAL STABILITY CONSTANTS

Volume 4: Inorganic Complexes

CRITICAL STABILITY CONSTANTS

CRITICAL STABILITY CONSTANTS

Volume 4: Inorganic Complexes

by Robert M. Smith
and Arthur E. Martell

Department of Chemistry
College of Science
Texas A & M University
College Station, Texas

PLENUM PRESS · NEW YORK AND LONDON

Library of Congress Cataloging in Publication Data

Martell, Arthur Earl, 1916-
  Critical stability constants.

  On vol. 2 and 4 Smith's name appears first on t.p.
  Includes bibliographical references.
  CONTENTS: v. 1. Amino acids.—v. 2. Amines.—v. 3. Other organic
ligands—v. 4. Inorganic complexes.
  1. Chemical equilibrium—Tables, etc. 2. Complex compounds—Tables,
etc. I. Smith, Robert Martin, 1927-   joint author. II. Title. [DNLM:
1. Amino acids. 2. Chemistry, Physical. QD503 M376c]
QD503.M37   541'.302   74-10610
ISBN 0-306-35214-1

© 1976 Plenum Press, New York
A Division of Plenum Publishing Corporation
227 West 17th Street, New York, N.Y. 10011

Printed in the United States of America

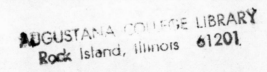

PREFACE

Over the past fifteen years the Commission on Equilibrium Data of the Analytical Division of the International Union of Pure and Applied Chemistry has been sponsoring a noncritical compilation of metal complex formation constants and related equilibrium constants. This work was extensive in scope and resulted in the publication of two large volumes of *Stability Constants* by the Chemical Society (London). The first volume, edited by L. G. Sillen (for inorganic ligands) and by A. E. Martell (for organic ligands), was published in 1964 and covered the literature through 1962. The second volume, subtitled Supplement No. 1, edited by L. G. Sillen and E. Hogfeldt (for inorganic ligands) and by A. E. Martell and R. M. Smith (for organic ligands), was published in 1971 and covered the literature up to 1969. These two large compilations attempted to cover all papers in the field related to metal complex equilibria (heats, entropies, and free energies). Since it was the policy of the Commission during that period to avoid decisions concerning the quality and reliability of the published work, the compilation would frequently contain from ten to twenty values for a single equilibrium constant. In many cases the values would differ by one or even two orders of magnitude, thus frustrating readers who wanted to use the data without doing the extensive literature study necessary to determine the correct value of the constant in question.

Because of difficulties of this nature, and because of the general lack of usefulness of a noncritical compilation for teaching purposes and for scientists who are not sufficiently expert in the field of equilibrium to carry out their own evaluation, we have decided to concentrate our efforts in this area toward the development of a critical and unique compilation of metal complex equilibrium constants. Although it would seem that decisions between available sets of data must sometimes be arbitrary and therefore possibly unfair, we have found the application of reasonable guidelines leads directly to the elimination of a considerable fraction of the published data of doubtful value. Additional criteria and procedures that were worked out to handle the remaining literature are described in the *Introduction* of this book. Many of these methods are quite similar to those used in other compilations of critical data.

In cases where a considerable amount of material has accumulated, it is felt that most of our critical constants will stand the test of time. Many of the data listed, however, are based on only one or a very few literature references and are subject to change when better data come along. It should be fully understood that this compilation is a continually changing and growing body of data, and will be revised from time to time as new results of these systems appear in the literature.

The scope of these tables includes the heats, entropies, and free energies of all reactions involving organic and inorganic ligands. The magnitude of the work is such that far more than a thousand book pages will be required. In order that the material be available in convenient form, the amino acid complexes are presented in Volume 1 and amine complexes (which do not contain carboxylic acid functions) are included in Volume 2. The remaining organic complexes are the subject of Volume 3. Volume 4 comprises the inorganic complexes.

We are grateful to Sten Ahrland, Charles F. Baes, Jr., Gregory R. Choppin, George H. Nancollas, and Reino Näsänen for reviewing portions of the manuscript and making valuable comments. We are also indebted to Charles F. Baes, Jr. for a prepublication copy of his book on the hydrolysis of cations (76BM).

Texas A&M University
College Station, Texas

Robert M. Smith
Arthur E. Martell

CONTENTS

INTRODUCTION

Purpose

This compilation of metal complex equilibrium (formation) constants and the corresponding enthalpy and entropy values represent the authors' selection of the most reliable values among those available in the literature. In many cases wide variations in published constants for the same metal complex equilibrium indicate the presence of one or more errors in ligand purity, in the experimental measurements, or in calculations. Usually, the nature of these errors is not readily apparent in the publication, and the reader is frequently faced with uncertainties concerning the correct values. In the course of developing noncritical compilations of stability constants, the authors have long felt that these wide variations in published work constitute a serious impediment to the use of equilibrium data. Thus these critical tables were developed in order to satisfy what is believed to be an important need in the field of coordination chemistry.

Scope

These tables include all organic and inorganic ligands for which reliable values have been reported in the literature. The present volume is restricted to inorganic ligands.

Values determined in nonaqueous solutions as well as values involving two or more different ligands (i.e., "mixed ligand" complexes) have not been included in this compilation but may be included in a subsequent volume. Mixed ligand complexes containing hydrogen or hydroxide ions are included since these ions are derived from the solvent and are therefore potentially always available. In general, data were compiled for only those systems that involve metal ion equilibria. Data on potentially important ligands for which only acid—base equilibria are presently available are given in a separate table.

Selection Criteria

When several workers are in close agreement on a particular value, the average of their results has been selected for that value. Values showing considerable scatter have been eliminated. In cases where the agreement is poor and few results are available for comparison, more subtle methods were needed to select the best value. This selection was often guided by a comparison with values obtained for other metal ions with the same ligand and with values obtained for the same metal ion with similar ligands.

While established trends among similar metal ions and among similar ligands were valuable in deciding between widely varying values, such guidelines were used cautiously, so as not to overlook occasionally unexpected real examples of specificity or anomalous behavior.

When there was poor agreement between published values and comparison with other metal ions and ligands did not suggest the best value, the results of more experienced research groups who had supplied reliable values for other ligands were selected. When such assurances were lacking, it was sometimes possible to give preference to values reported by an investigator who had published other demonstrably reliable values obtained by the same experimental method.

In some cases the constants reported by several workers for a given group of metal ions would have similar relative values, but would differ considerably in the absolute magnitudes of the constants. Then a set of values from one worker near the median of all values reported were selected as the best constants. By this method it is believed that internal consistency was preserved to a greater extent than would be obtained by averaging reported values for each individual metal ion. When an important constant was missing from the selected set of values, but was available in another set of values not selected for this compilation, the missing constant was obtained by adjusting the nonselected values by a common factor, which was set so as to give the best agreement between the two groups of data.

Values reported by only one investigator are included in these tables unless there was some reason to doubt their validity. It is recognized that some of these values may be in error, and that such errors will probably not be detected until the work is repeated by other investigators, or until more data become available for analogous ligands or other closely related metal ions. Some values involving unusual metal ions have been omitted because of serious questions about the form of their complexes.

Papers deficient in specifying essential reaction conditions (e.g., temperature, ionic strength, nature of supporting electrolyte) were not employed in this compilation. Also used as a basis for disqualification of published data is lack of information on the purity of the ligand. Frequent deficiencies are lack of calibration of potentiometric apparatus, and failure to define the equilibrium quotients reported in the paper. Papers in which both temperature and ionic strength are not controlled have been omitted from the bibliography.

A bibliography for each ligand is included so that the reader may determine the completeness of the literature search employed in the determination of critical values. The reader may also employ these references to make his own evaluation if he has any questions or reservations concerning this compilation.

Arrangement

The arrangement of the tables is based on the periodic table position of the central atom of the ligand except that the hydroxide ion is placed first because of its importance in considering equilibria involving other ligands. This is followed by transition metal ligands and then those of groups III through VII of the periodic table. Within each group of tables involving the same atom, the arrangement is from the lowest oxidation state to higher ones. Next there is a table of protonation constants for ligands for which no stability constants or only questionable metal stability constants are reported. Finally, there is a list of other ligands considered but not included in the tables for various reasons.

Metal Ions

The metal ions within each table are arranged in the following order: hydrogen, alkali metals, alkaline earth metals, lanthanides (including Sc and Y), actinides, transition metals, and posttransition metals. Within each group the arrangement is by increasing oxidation state of the metal, and within each oxidation state the arrangement follows the periodic table from top to bottom and from left to right. An exception is that Cu^+, Ag^+, Pd^{2+}, and Pt^{2+} are included with the posttransition metals.

Equilibrium

An abbreviated equilibrium quotient expression in the order products/ reactants is included for each constant, and periods are used to separate distinct entities. Charges have been omitted as these can be determined from the charge of the metal ion and the abbreviated ligand formulas (such as HL) given after the name. Water has not been included in the equilibrium expressions since all of the values cited are for aqueous solutions. For example, $M_4L_4/M^4 \cdot L^4$ for Mg^{2+} and hydroxide ion would represent the equilibrium: $4Mg^{2+} + 4OH^- \rightleftharpoons Mg_4(OH)_4^{4+}$. The symbol M represents the metal ion given in the first column and may include more than one atom as in the case of Hg_2^{2+}. The symbol H_{-1} (H_{-2}, etc.) is used for the ionization of a proton from the ligand alone at high pH.

Equilibria involving protons are written as stability constants (protonation constants) rather than as ionization constants to be consistent with the metal complex formation constants. Consequently the $\triangle H$ and $\triangle S$ values have signs opposite to those describing ionization constants.

Solids and gases are identified by (s) and (g) respectively and are included for identification purposes even though they are not involved in the equilibrium quotient.

Log K Values

The log K values are the logarithms of the equilibrium quotients given in the second column at the specified conditions of temperature and ionic strength. The selected values are those considered to be the most reliable of the ones available. In some cases the value is the median of several values and in other cases it is the average of two or more values. The range of other values considered reliable is indicated by + or − quantities describing the algebraic difference between the other values and the selected value. The symbol ±0.00 indicates that there are one or more values which agree exactly with the stated value to the number of significant figures given. Values considered to be of questionable validity are enclosed in parentheses. Such values are included when the evidence available is not strong enough to exclude them on the basis of the above criteria. Values concerning which there is considerable doubt have been omitted.

The log K values are given for the more commonly reported ionic strengths. The ionic strengths most used for inorganic ligands are 0.1, 0.5, 1.0, 2.0, 3.0, 4.0, and 0. Zero ionic strength is perhaps more important from a theoretical point of view, but several assumptions are involved in extrapolating or calculating from the measured values. The Davies equation is often used to calculate constants to zero from low-ionic-strength measurements. It was established from results obtained with monovalent and divalent ions and its extension to trivalent ions is extremely questionable.

The temperature of 25°C was given preference in the tables because of its widespread use in equilibrium measurements and reporting other physical properties. When available, enthalphy changes ($\triangle H$) were used to calculate log K at 25°C when only measurements at other temperatures were available.

Other temperatures frequently employed are 20°C, 30°C, and 37°C. These are not included in the tables when there is a lack of column space and $\triangle H$ is available, since they may be calculated using the $\triangle H$ value. Values at other temperatures, especially those at 20°C and 30°C, were converted to 25°C to facilitate quantitative comparisons with the 25°C values listed.

Equilibria involving protons have been expressed as concentration constants in order to be more consistent with the metal ion stability constants which involve only concentration terms. Concentration constants may be determined by calibrating the electrodes with solutions of known hydrogen ion concentrations or by conversion of pH values using the appropriate hydrogen ion activity coefficient. When standard buffers are used, mixed constants (also known as Bronsted or practical constants) are obtained which include both activity and concentration terms. Literature values expressed as mixed constants have been converted to concentration constants by using the hydrogen ion activity coefficients determined in KCl solution before inclusion in the tables. In some cases, papers were omitted because no indication was given as to the use of concentration or mixed constants. Some papers were retained despite this lack of information when it could be ascertained which constant was used by comparing to known values or by personal communication with the authors. For those desiring to convert the listed protonation constants to mixed constants, the following values should be added to the listed values at the appropriate ionic strength (the tabulation applies only to single proton association constants):

Ionic strength	Increase in log K
0.05	0.09
0.10	0.11
0.15	0.12
0.2	0.13
0.5	0.15
1.0	0.14
2.0	0.11
3.0	0.07

The values in the tables have not been corrected for complexation with medium ions for the most part. There are insufficient data to make corrections for most of the ligands, and in order to make values between ligands more comparable, the correction has not been made in the few cases where it could be made. In general the listed formation constants at constant ionic strength include competition by ions from KNO_3 and $NaClO_4$ and are somewhat smaller than they would be if measured in solutions of tetraalkylammonium salts.

Limited comparisons were made between values at different ionic strengths using observed trends. With inorganic ligands the stability constants usually decrease with ionic strength until a minimum is reached and then increase with increasing ionic strength. The minimum is often at about 0.5 ionic strength when hydrogen or hydroxide ion is involved but may change with other ions. With basic ligands such as ammonia there is a continuing increase with ionic strength and no maximum or minimum is generally observed. With phosphorous compounds containing oxygen donors there is a continuing decrease with ionic strength because of the increased competition from sodium ions in the background electrolyte.

The solubility products of precipitates frequently become more negative with longer digestive times. This is apparently due to conversion to less soluble forms of the precipitate with time. Since various digestive times have been used to measure solubility products, their comparison becomes rather tenuous, except for rough approximations. With lanthanide hydroxides, the solubility products have been measured as a function of time at constant temperature in some cases, and the values listed in the table have been corrected to the fresh or active precipitate by using the average change with time of the lanthanide hydroxides as a whole.

The hydrolysis of highly charged metal ions, such as Th^{4+}, apparently leads to different polymerization products in different media. Consequently the species formed may be different with a change in the background electrolyte if the background electrolyte becomes part of the polymeric ions.

Enthalpy Values

The enthalpy of complexation values ($\triangle H$) listed in the tables have the units kcal/mole because of the widespread use of these units by workers in the field. These may be converted to SI units of kj/mole by multiplying the listed values by 4.184.

Calorimetrically determined values and temperature-variation-determined values from cells without liquid junction were considered of equal validity for the tables. Other temperature-variation-determined values were rounded off to the nearest kcal/mole and were enclosed in parentheses because of their reduced accuracy. Other values considered to be reliable but differing from the listed value were indicated by + or − quantities describing the algebraic difference between the other value and the selected value.

The magnitude of $\triangle H$ may vary with temperature and ionic strength, but usually this is less than the variation between different workers and little attempt has been made to show $\triangle H$ variation with changing conditions except for certain carefully measured equilibria such as the protonation of hydroxide ion and of ammonia. These $\triangle H$ values may be used for estimating $\log K$ values at temperatures other than those listed, using the relationship

$$\frac{\triangle H}{2.303RT^2} = \frac{d \log K}{dT}$$

or, at 25°C,

$$\log K_2 = \log K_1 + \triangle H(T_2 - T_1)(0.00246).$$

This assumes that $\triangle C_p = 0$, which is not necessarily the case. The greater the temperature range employed, the greater the uncertainty of the calculated values.

Entropy Values

The entropy of complexation values ($\triangle S$) listed in the tables have the units cal/mole/degree and have been calculated from the listed $\log K$ and $\triangle H$ values, using the expression

$$\triangle G = \triangle H - T\triangle S$$

or, at 25°C,

$$\triangle S = 3.36 \ (1.363 \log K + \triangle H).$$

These entropy values have been rounded off to the nearest cal/mole/degree, except in cases where $\triangle H$ values were quite accurate.

Bibliography

The references considered in preparing each table are given at the end of the table. The more reliable references are listed after the ions for which values are reported. In some tables groups of similar metal ions have been grouped together for the bibliography. The term "Other references" is used for those reporting questionable values, or values at conditions considerably different from those used in the tables, or values for metal ions not included in the tables because of questionable knowledge about the forms of their complexes. These additional references are cited to inform the reader of the extent of the literature search made in arriving at the selected values.

The bibliographical symbols used represent the year of the reference and the first letter of the surnames of the first two listed authors. In cases of duplication, letters a, b, c, etc., or the first letter of the third author's name are employed. The complete reference is given in the bibliography at the end of each volume.

In a work of this magnitude, there will certainly be errors and a few pertinent publications will have been overlooked by the compilers. We should like to request those who believe they have detected errors in the selection process, know of publications that were omitted, or have any suggestions for improvement of the tables, write to:

A. E. Martell, Head
Department of Chemistry
Texas A&M University
College Station, Texas 77843, U.S.A.

It is the intention of the authors to publish more complete and accurate revisions of these tables as demanded by the continually growing body of equilibrium data in the literature.

$$OH^-$$

HO^-			Hydroxide ion			L^-
Metal ion	Equilibrium	Log K 25°, 0.5	Log K 25°, 1.0	Log K 25°, 0	ΔH 25°, 0	ΔS 25°, 0
H^+	HL/H.L	13.74 ±0.02	13.79 ±0.02	13.997±0.003	-13.34 ±0.01	19.3
		13.78[a]±0.01	14.18[e]±0.04		-13.55[b]±0.05	17.7[b]
		13.95[h]			-13.08[e]±0.03	21.0[e]
		13.96[d]±0.01			-12.69[p]	
Li^+	ML/M.L		-0.18[e]	0.36	(0)[r]	(2)
Na^+	ML/M.L			-0.2	(0)[s]	(-1)
K^+	ML/M.L			-0.5		
Be^{2+}	ML/M.L	8.3[h]		8.6		
	$ML_2/M.L^2$	(16.5)	(17.5)[e]	(14.4)		
		(16.7)[h]				
	$ML_3/M.L^3$			18.8		
	$ML_4/M.L^4$			18.6		
	$M_2L/M^2.L$	10.54	10.95[e]	(10.0)	-8.9[e]	20[e]
		10.68[d]				
	$M_3L_3/M^3.L^3$	32.41	33.88[e]	33.1	-24.8[e]	72[e]
		32.98[d]				
	$M_6L_8/M^6.L^8$			(85)	(-58)[t]	(200)
	$M.L^2/ML_2$(s, amorphous)			-21.0		
	$M.L^2/ML_2$(s,α)			-21.31		
	$M.L^2/ML_2$(s,β)			-21.7		
Mg^{2+}	ML/M.L		1.85[e]	2.58 ±0.0		
	$M_4L_4/M^4.L^4$		16.93[e]	16.3		
	$M.L^2/ML_2$(s)			-11.15 ±0.2		
Ca^{2+}	ML/M.L		0.64[e]	1.3 ±0.1	2.0	13
	$M.L^2/ML_2$(s)			-5.19 ±0.2	-4.3	-38
Sr^{2+}	ML/M.L		0.23[e]	0.8 ±0.1	1.2	8

[a] 25°, 0.1; [b] 25°, 0.5; [d] 25°, 2.0; [e] 25°, 3.0; [h] 20°, 0.1; [p] 40°, 0; [r] 15-35°, 0;
[s] 0-50°, 0; [t] 0-60°, 1.0 molal

Hydroxide ion (continued)

Metal ion	Equilibrium	Log K 25°, 0.5	Log K 25°, 1.0	Log K 25°, 0	ΔH 25°, 0	ΔS 25°, 0
Ba^{2+}	ML/M.L	0.4[a]	0.00[e]	0.6 ±0.1	1.1 ±0.1	9
	$M.L^2/ML_2(H_2O)_8(s)$			−3.6	13.7	29
Sc^{3+}	ML/M.L	9.06 9.31[a] ±0.00 9.1[h]	8.63 ±0.02	9.7	(2)[t]	(50)[c]
	$ML_2/M.L^2$	17.4[h]	17.4	18.3		
	$ML_3/M.L^3$	24.9[h]		25.9		
	$ML_4/M.L^4$			30		
	$M_2L_2/M^2.L^2$	21.50 21.53[a]	21.49 ±0.03	22.0	(−12)[t]	(60)[c]
	$M_3L_5/M^3.L^5$	51.69 51.88[a]	51.55 ±0.07	53.8	(−28)[t]	(150)[c]
	$M.L^3/MOL(s)$		−30.9	−32.7		
Y^{3+}	ML/M.L	5.39[u]	5.1[e]	6.3		
	$M_2L_2/M^2.L^2$		14.06[e]	13.8		
	$M_3L_5/M^3.L^5$		37.1[e]	38.4		
	$M.L^3/ML_3(s)$			−23.2		
La^{3+}	ML/M.L	4.67[u]	4.1[e]	5.5		
	$M_2L/M^2.L$		(4.2)[e]			
	$M_5L_9/M^5.L^9$		56.2[e]	54.8		
	$M.L^3/ML_3(s)$			−20.7		
Ce^{3+}	$M_3L_5/M^3.L^5$		35.1[e]	36.5		
	$M.L^3/ML_3(s)$			−21.2		
Pr^{3+}	ML/M.L	5.18[u]				
	$M.L^3/ML_3(s)$			−21.5		
Nd^{3+}	ML/M.L	5.30[u]	4.8[e]	6.0		
	$ML_4/M.L^4$			18.6		
	$M_2L_2/M^2.L^2$		14.43[e]	14.1		
	$M.L^3/ML_3(s)$			−23.1 ±0.0		
Sm^{3+}	ML/M.L	5.39[u]				
	$M.L^3/ML_3(s)$			−25.4		

---

[a] 25°, 0.1; [c] 25°, 1.0; [e] 25°, 3.0; [h] 20°, 0.1; [t] 10–40°, 1.0 molal; [u] 25°, 0.3

A. WATER

Hydroxide ion (continued)

Metal ion	Equilibrium	Log K 25°, 0.5	Log K 25°, 1.0	Log K 25°, 0	ΔH 25°, 0	ΔS 25°, 0
Eu^{3+}	$ML/M.L$	5.42^u				
	$M.L^3/ML_3(s)$			-25.6		
Gd^{3+}	$ML/M.L$	5.38^u	5.0^e			
	$M_2L_2/M^2.L^2$		14.13^e			
	$M.L^3/ML_3(s)$			-25.7		
Tb^{3+}	$ML/M.L$	5.57^u				
	$M.L^3/ML_3(s)$			-25.5		
Dy^{3+}	$ML/M.L$	5.63^u				
	$M.L^3/ML_3(s)$			-25.6		
Ho^{3+}	$ML/M.L$	5.69^u				
	$M.L^3/ML_3(s)$			-25.9		
Er^{3+}	$ML/M.L$	5.74^u				
	$M.L^3/ML_3(s)$			-24.9		
Tm^{3+}	$ML/M.L$	5.78^u				
Yb^{3+}	$ML/M.L$	5.81^u				
	$M.L^3/ML_3(s)$			-25.0		
Lu^{3+}	$ML/M.L$	5.83^u				
	$M.L^3/ML_3(s)$			-26.1		
Pu^{3+}	$ML/M.L$	6.7^a		7.0		
Am^{3+}	$ML/M.L$	7.9^a				
Cm^{3+}	$ML/M.L$	7.9^a -0.1				
Bk^{3+}	$ML/M.L$	8.2^a				
Cf^{3+}	$ML/M.L$	8.2^a				
Ce^{4+}	$ML/M.L$		13.17^k		$(-8)^z$	$(30)^k$
	$M_2L_3/M^2.L^3$		$40.9^{e,n}$			
	$M_2L_4/M^2.L^4$		$54.5^{e,n}$			
	$M_6L_{12}/M^6.L^{12}$		$168.4^{e,n}$			

---

[a] 25°, 0.1; [e] 25°, 3.0; [k] 25°, 1.5, assuming HL/H.L = 13.87; [n] $NaNO_3$ used as background electrolyte; [u] 25°, 0.3; [z] 10-25°, 1.5

Hydroxide ion (continued)

Metal ion	Equilibrium	Log K 25°, 0.5	Log K 25°, 1.0	Log K 25°, 0	ΔH 25°, 0	ΔS 25°, 0
Th^{4+}	$ML/M.L$		9.6	10.8	$(-8)^t$	$(20)^c$
			$9.1^{e,y}$			
	$ML_2/M.L^2$		19.89	21.1	$(-13)^t$	$(50)^c$
	$M_2L_2/M^2.L^2$		22.97	(21.9)	$(-12)^t$	$(60)^c$
			$23.58^{e,y}$			
	$M_2L_3/M^2.L^3$		$33.8^{e,y}$			
	$M_2L_5/M^2.L^5$		$(53.7)^{e,y}$			
	$M_3L/M^3.L$		$(12.7)^{e,y}$			
	$M_3L_3/M^3.L^3$		$(35.7)^{e,y}$			
	$M_4L_8/M^4.L^8$		91.2	(90.9)	$(-51)^t$	$(250)^c$
	$M_6L_{14}/M^6.L^{14}$		$162.1^{e,y}$			
	$M_6L_{15}/M^6.L^{15}$		169.8	173.2	$(-96)^t$	$(470)^c$
	$M.L^4/ML_4(s)$			-44.7^o		
	$M.L^4/MO_2(s)$			-49.7		
Pa^{4+}	$ML/M.L$		14.04^e	14.8		
	$ML_2/M.L^2$		27.84^e	28.0		
	$ML_3/M.L^3$		40.7^e			
	$ML_4/M.L^4$		51.4^e			
U^{4+}	$ML/M.L$	12.24	12.23 −0.01	13.3	$(-2)^r$	(50)
		12.50^a				
		$12.31^d \pm 0.03$	$(12.1)^e$			
	$M_6L_{15}/M^6.L^{15}$		196.1^e	192.8		
	$ML_5/M.L^5$			54.0		
	$M.L^4/MO_2(s)$			-56.2		
Np^{4+}	$ML/M.L$	11.7^d		12.5		
Pu^{4+}	$ML/M.L$	12.14	12.1			
		12.23^d			$(-6)^w$	$(40)^d$
	$M.L^4/ML_4(s)$			-47.3		

---

^{a} 25°, 0.1; ^{c} 25°, 1.0; ^{d} 25°, 2.0; ^{e} 25°, 3.0; ^{o} 22°, 0; ^{r} 10-43°, 0; ^{t} 0-95°, 1.0 molal; ^{w} 15-25°, 2.0; ^{y} NaCl used as background electrolyte.

Hydroxide ion (continued)

Metal ion	Equilibrium	Log K 25°, 0.5	Log K 25°, 1.0	Log K 25°, 0	ΔH 25°, 0	ΔS 25°, 0
Pa(V)	$ML_4/ML_3 \cdot L$		13.13[e]	14.5		
	$ML_5/ML_4 \cdot L$		9.68[e]	9.5		
UO_2^{2+}	$ML/M \cdot L$	8.0 ±0.0 (7.7)[a]	8.1[e]	8.2	(-2)[x]	(30)[b]
	$M_2L/M^2 \cdot L$		9.6			
	$M_2L_2/M^2 \cdot L^2$	21.55 ±0.02 21.73[a]	21.64 ±0.02 22.32[e]	22.4	16.7[e]	46[e]
	$M_3L_4/M^3 \cdot L^4$		42.4 43.5[e]			
	$M_3L_5/M^3 \cdot L^5$	52.4 ±0.1	52.6 ±0.1 54.4[e]	54.4	41.1[e]	111[e]
	$M \cdot L^2/ML_2(s)$			-22.4		
NpO_2^{2+}	$ML/M \cdot L$		8.6	8.9		
	$M_2L_2/M^2 \cdot L^2$		20.9	21.6		
	$M_3L_5/M^3 \cdot L^5$		50.7	52.5		
PuO_2^{2+}	$ML/M \cdot L$		7.8 7.9[e]	8.4		
	$M_2L_2/M^2 \cdot L^2$		19.1 20.1[e]	19.6		
	$M_3L_5/M^3 \cdot L^5$		46.8 49.3[e]	48.4		
Mn^{2+}	$ML/M \cdot L$	2.9[a] +0.1 3.5[d]	3.0	3.4	(0)[r]	(20)
	$ML_4/M \cdot L^4$			7.7		
	$M_2L/M^2 \cdot L$	4.1[d]		(3.4)		
	$M_2L_3/M^2 \cdot L^3$	16.4[d]		18.1		
	$M \cdot L^2/ML_2(s)$			-12.8 +0.1		
Fe^{2+}	$ML/M \cdot L$		4.3	4.5		
	$ML_2/M \cdot L^2$			(7.4)		
	$ML_3/M \cdot L^3$			10.0		
	$ML_4/M \cdot L^4$			9.6		
	$M \cdot L^2/ML_2(s)$			-15.1		

[a] 25°, 0.1; [b] 25°, 0.5; [d] 25°, 2.0; [e] 25°, 3.0; [r] 15-42°, 0; [x] 25-95°, 0.5

Hydroxide ion (continued)

Metal ion	Equilibrium	Log K 25°, 0.5	Log K 25°, 1.0	Log K 25°, 0	ΔH 25°, 0	ΔS 25°, 0
Co^{2+}	ML/M.L	(3.9)	(3.9)	4.3		
			4.2[e]			
	ML$_2$/M.L^2		8.5	8.4		
	ML$_3$/M.L^3		9.7	9.7		
	ML$_4$/M.L^4			10.2		
	M$_2$L/M^2.L		4.7[e]	(2.7)		
	M$_4$L$_4$/M^4.L^4		27.5[e] ±0.0	(25.6)		
	M.L^2/ML$_2$(s)		-14.6	-14.9		
Ni^{2+}	ML/M.L	3.7[a]	3.8 -0.1	4.1	(-1)[r]	(20)
	ML$_2$/M.L^2			8		
	ML$_3$/M.L^3			11		
	M$_2$L/M^2.L		4.2[e]	(3.3)		
	M$_4$L$_4$/M^4.L^4		29.37[e]±0.03	28.3	-12.8[e]	92[e]
	M.L/ML$_2$(s)			-15.2		
Cu^{2+}	ML/M.L			6.3		
	ML$_2$/M.L^2		12.8			
	ML$_3$/M.L^3		14.5			
	ML$_4$/M.L^4		15.6	16.4		
	M$_2$L$_2$/M^2.L^2	17.02[h]±0.03	17.28[j]	17.7		
			17.8[e]		-10.4[e]	47[e]
	M.L^2/ML$_2$(s)	-18.48[q]	-18.9	-19.32	13.2[d]	-40q
	M.L^2/MO(s)	-19.51[q]		-20.35		
Ti^{3+}	ML/M.L	11.8	11.5[e] ±0.1	12.7		
	M$_2$L$_2$/M^2.L^2		24.8[e] ±0.3			
V^{3+}	ML/M.L		11.01[j]	11.7	(-4)[z]	(40)[c]
			11.11[e]			
	M$_2$L$_2$/M^2.L^2		23.8[j]	24.2		
			23.43[e]			
	M$_2$L$_3$/M^2.L^3		34.5[e]			
	M.L^3/ML$_3$(s)			-34.4		

[a] 25°, 0.1; [c] 25°, 1.0; [d] 25°, 2.0; [e] 25°, 3.0; [h] 20°, 0.1; [j] 20°, 1.0; [q] 25°, 0.2
[r] 15-42°, 0; [z] 25-50°, 1.4

A. WATER

Hydroxide ion (continued)

Metal ion	Equilibrium	Log K 25°, 0.5	Log K 25°, 1.0	Log K 25°, 0	ΔH 25°, 0	ΔS 25°, 0
Cr^{3+}	ML/M.L	9.41[i]	9.41	10.07 ±0.02	-1.0	43
		9.77[h]±0.03				
	$ML_2/M.L^2$	17.3[h]			(4.5)	(94)[h]
	$M_2L_2/M^2.L^2$	24.6[d]	24.1			
	$M_4L_4/M^4.L^4$	50.7[d]				
	$M_4L_6/M^4.L^6$	72.8[d]				
	$M.L^3/ML_3(s)$	-29.8[a] ±0.1				
Mn^{3+}	ML/M.L			14.4[f]	(-8)[m]	(40)[f]
Fe^{3+}	ML/M.L	11.01 ±0.07	11.09 ±0.1	11.81 ±0.03		
		11.17[a]±0.09	11.21[e]±0.08		-2[e]	(45)[e]
		11.14[d]				
	$ML_2/M.L^2$		21.9[j]	22.3		
			22.1[e]			
	$ML_4/M.L^4$			34.4		
	$M_2L_2/M^2.L^2$	24.7	24.9 -0.1	25.1		
		24.7[a]	25.6[e] ±0.2		-16.2[e]	(63)[e]
		25.3[d]				
	$M_3L_4/M^3.L^4$		51.0[e]	49.7	-38[e]	(106)[e]
	$M.L^3/ML_3(s)$		-38.6[e]	-38.8 ±0.2		
	$M.L^3/MOOH(s,\alpha)$		-41.1[e]	41.5		
	$M.L^3/(M_2O_3)^{0.5}(s,\alpha)$			-42.7		
Co^{3+}	ML/M.L		13.52[e]			
	$M.L^3/ML_3(s)$			-44.5[o]	(18)[r]	(-140)[o]
Rh^{3+}	ML/M.L		10.67[l]		(-9)[v]	(20)[l]
Ti(IV)	$ML_4/ML_2.L^2$	22.6[a]				
Zr^{4+}	ML/M.L			14.3		
				13.9[f]		
	$ML_5/M.L^5$			54.0		
	$M_3L_4/M^3.L^4$	50.5[d]		55.4		
	$M_4L_8/M^4.L^8$	103.4[d]		106.0		
	$M.L^4/MO_2(s)$		-52.0[j]	-54.1		

[a] 25, 0.1; [d] 25°, 2.0; [e] 25°, 3.0; [h] 20°, 0.1; [i] 20°, 0.5; [j] 20°, 1.0; [l] 25°, 2.5; [m] 1-35°, 4.0; [o] 19°, 0; [r] 19-81°, 0; [v] 25-60°, 2.5

Hydroxide ion (continued)

Metal ion	Equilibrium	Log K 25°, 0.5	Log K 25°, 1.0	Log K 25°, 0	ΔH 25°, 0	ΔS 25°, 0
Hf^{4+}	ML/M.L			13.7		
				13.3[f]		
	$ML_5/M.L^5$			52.8		
	$M.L^4/MO_2(s)$			-54.8		
VO^{2+}	ML/M.L		7.9[e]	8.3		
	$ML_2/M.L^2$		18.31[e]			
	$M_2L_2/M^2.L^2$		28.35[e]	21.3		
	$M.L^2/ML_2(s)$			-23.5		
TcO^{2+}	ML/M.L	12.6[h]				
	$ML_2/M.L^2$	24.2[h]				
OsO_4	ML/M.L		1.8			
	$ML_2/M.L^2$		1.1			
Cu^+	$M.L/(M_2O)^{0.5}(s)$			-14.7 +0.1	-1.5[i]	-72
Ag^+	ML/M.L			2.0		
	$ML_2/M.L^2$		3.55[e]	3.99		
	$M.L/(M_2O)^{0.5}(s)$	-7.37[u]	-7.18	-7.71 ±0.02	8.1	-8
		-7.15[d]	-7.42[e]			
Hg_2^{2+}	ML/M.L	8.7				
CH_3Hg^+	ML/M.L	9.24[a] ±0.02			-8.5[h]	14[a]
	$M_2L/ML.M$	2.37[a]				
Tl^+	ML/M.L	0.30	0.26	0.79	0.4	5
			0.08[e]	-0.10[g]	(1.8)[e]	(6)[e]
	$ML_2/M.L^2$		(-0.8)[e]		(6.8)[e]	(19)[e]
$(CH_3)_2Tl^+$	ML/M.L			1.05		
	$M_2L_2/M^2.L^2$			1.23		
$(CH_3)_3Sn^+$	ML/M.L	7.54[d]	7.58[e]			
	$M_2L/M^2.L$	8.5[d]				
	$M_2L_2/M^2.L^2$	14.1[d]				
$(C_2H_5)_3Sn^+$	ML/M.L		7.37[e]			

---

[a] 25°, 0.1; [d] 25°, 2.0; [e] 25°, 3.0; [f] 25°, 4.0, assuming HL/H.L = 14.4; [g] 25°, 5.0; [h] 20°, 0.1; [i] 19°, 0.25; [u] 25°, 0.3

A. WATER

Hydroxide ion (continued)

Metal ion	Equilibrium	Log K 25°, 0.5	Log K 25°, 1.0	Log K 25°, 0	ΔH 25°, 0	ΔS 25°, 0
$(C_6H_5)_3Sn^+$	$ML/M.L$	9.2^q				
$(C_6H_5)_3Pb^+$	$ML/M.L$	7.7^q				
Pd^{2+}	$ML/M.L$	12.4^a		13.0		
	$ML_2/M.L^2$	25.2^a		25.8		
	$M.L^2/ML_2(s)$			-28.5		
Zn^{2+}	$ML/M.L$			5.0 ± 0.0	$(0)^r$	(25)
	$ML_2/M.L^2$		8.3^e	(11.1)		
	$ML_3/M.L^3$		13.7^e	13.6		
	$ML_4/M.L^4$		18.0^e	(14.8)		
	$M_2L/M^2.L$		5.5	5.0		
	$M.L^2/ML_2(s,\text{amorphous})$		-14.82^q	-15.52	7.5^d	-43
	$M.L^2/ML_2(s,\beta_1)$		-15.54^q	-16.24		
	$M.L^2/ML_2(s,\beta_2)$		-15.50^q	-16.20		
	$M.L^2/ML_2(s,\gamma)$		-15.56^q	-16.26		
	$M.L^2/ML_2(s,\delta)$		-15.45^q	-16.15		
	$M.L^2/ML_2(s,\varepsilon)$		-15.77^q	-16.46 ± 0.04		
	$M.L^2/MO(s)$		-15.96^q	-16.66	5.4^d	-55
Cd^{2+}	$ML/M.L$		$4.1^e \pm 0.2$	3.9	0.0^e	19^e
	$ML_2/M.L^2$		7.7^e	7.7		
	$ML_3/M.L^3$		10.3^e			
	$ML_4/M.L^4$		12.0^e	(8.7)		
	$M_2L/M^2.L$		5.08^e	4.6	-2.2^e	16^e
	$M_4L_4/M^4.L^4$		24.9^e	23.2	-11.8^e	74^e
	$M.L^2/ML_2(s,\beta)$		-14.29^e	-14.35	4.1	-52
	$M.L^2/ML_2(s,\gamma)$		-14.10^e			
Hg^{2+}	$ML/M.L$	10.0	10.1	10.6		
			$10.8^e \pm 0.2$		-5.9^e	30^e
	$ML_2/M.L^2$	21.0 ± 0.2	21.1 ± 0.2	$21.8 +0.1$		
		21.2^a				
		21.6^d	$22.2^e \pm 0.2$		-16.4^e	47^e
	$ML_3/M.L^3$			20.9		
	$M_2L/M^2.L$		11.5^e	10.7	-10.0^e	19^e
	$M_3L_3/M^3.L^3$		36.1^e	35.6		
	$M.L^2/MO(s,\text{red})$		-26.0^e	-25.44	7.9^c	-93^e

^{a} 25°, 0.1; ^{c} 25°, 1.0; ^{d} 25°, 2.0; ^{e} 25°, 3.0; ^{q} 25°, 0.2; ^{r} 15–42°, 0;

Hydroxide ion (continued)

Metal ion	Equilibrium	Log K 25°, 0.5	Log K 25°, 1.0	Log K 25°, 0	ΔH 25°, 0	ΔS 25°, 0
Ge^{2+}	ML_2/MO(s,brown)			-3.7		
Sn^{2+}	$ML/M.L$		10.4[e] ±0.1			
	$M_2L_2/M^2.L^2$		23.9[e]			
	$M_3L_4/M^3.L^4$		49.93[e]±0.03			
	$M.L^2/MO$(s)			-26.2		
$(CH_3)_2Sn^{2+}$	$ML/M.L$	10.53[a]	10.63[e]			
	$ML_2/M.L^2$	19.04[a]	19.36[e]			
	$ML_3/M.L^3$	29.8[a]				
	$ML_4/M.L^4$	43.4[a]				
	$M_2L_2/M^2.L^2$	22.6[a]	23.8[e]			
	$M_4L_6/M^4.L^6$	65.8[a]	68.9[e]			
$(C_2H_5)_2Sn^{2+}$	$ML/M.L$		11.25			
			10.78[e]			
	$ML_2/M.L^2$		20.31			
	$M_2L_2/M^2.L^2$		23.8			
			24.0[e]			
	$M_2L_3/M^2.L^3$		34.1			
$(C_3H_7)_2Sn^{2+}$	$ML/M.L$		11.26[e]			
	$M_2L_2/M^2.L^2$		24.8[e]			
Pb^{2+}	$ML/M.L$	6.0[u]	6.3[e]	6.3		
	$ML_2/M.L^2$	10.3[u]	10.9[e]	10.9		
	$ML_3/M.L^3$	13.3[u]	13.7[e]	13.9		
	$M_2L/M^2.L$		7.9[e]	7.6		
	$M_3L_4/M^3.L^4$	31.7[u]	33.8[e]	32.1	-25.9[e]	63[e]
	$M_4L_4/M^4.L^4$	35.1[u]	37.5[e]	(35.1)	-32.3[e]	68[e]
	$M_6L_8/M^6.L^8$	67.4[u]	71.3[e]	68.4	-55.4[e]	141[e]
	$M.L^2/(M_2OL_2)^{0.5}$(s)		-14.9			
	$M.L^2/MO$(s,yellow)		-15.1			
	$M.L^2/MO$(s,red)		-15.3			
$(CH_3)_2Pb^{2+}$	$ML_2/M.L^2$		12.82[e]			
	$ML_3/M.L^3$		14.0[e]			
	$M_2L_2/M^2.L^2$		17.53[e]			
	$M_3L_4/M^3.L^4$		32.4[e]			

[a] 25°, 0.1; [e] 25°, 3.0; [u] 25°, 0.3

A. WATER

Hydroxide ion (continued)

Metal ion	Equilibrium	Log K $25°, 0.5$	Log K $25°, 1.0$	Log K $25°, 0$	ΔH $25°, 0$	ΔS $25°, 0$
Au(III)	$ML_4/ML_3 \cdot L$			2.3		
	$ML_5/ML_4 \cdot L$			0.6		
	$ML_3/ML_3(s)$			-5.5		
Al^{3+}	$ML/M \cdot L$	$8.48^a + 0.5$	8.31	9.01 ± 0.03	$(-1)^r$	(40)
	$ML_2/M \cdot L^2$	$(17.6)^a$		(18.7)		
	$ML_3/M \cdot L^3$	$(25.7)^a$		(27.0)		
	$ML_4/M \cdot L^4$			33.0		
	$M_2L_2/M^2 \cdot L^2$		20.0	20.3	$(-9)^t$	$(60)^c$
	$M_3L_4/M^3 \cdot L^4$		42.5	42.1	$(-19)^t$	$(130)^c$
	$M_{13}O_4L_{24} \cdot H^8/M^{13} \cdot L^{32}$		336.5	349.2	$(-156)^t$	$(1020)^c$
	$M \cdot L^3/ML_3(s,\alpha)$			-33.5		
Ga^{3+}	$ML/M \cdot L$	10.8		11.4		
		$10.9^a \pm 0.2$				
	$ML_2/M \cdot L^2$	21.5^a		22.1		
	$ML_3/M \cdot L^3$	30.9^a		31.7		
	$ML_4/M \cdot L^4$			39.4		
	$M \cdot L^3/ML_3(s,amorphous)$			-37		
	$M \cdot L^3/MOL(s)$			-39.1		
In^{3+}	$ML/M \cdot L$	$(10.5)^a$	$9.76^e \pm 0.0$	10.0	-8.2^e	17^e
	$ML_2/M \cdot L^2$	$(20.3)^a$	$19.6^e + 0.5$	20.2		
	$ML_3/M \cdot L^3$	$(29.3)^a$		29.6		
	$ML_4/M \cdot L^4$			33.9		
	$M_3L_4/M^3 \cdot L^4$		47.4^e	50.2		
	$M \cdot L^3/ML_3(s)$			-36.9		
	$M \cdot L^3/(M_2O_3)^{0.5}(s)$			-35.9		
Tl^{3+}	$ML/M \cdot L$	$(12.8)^a$	$13.02^e \pm 0.02$	13.4	$(11)^e$	$(100)^e$
	$ML_2/M \cdot L^2$	$(25.3)^a$	$25.75^e \pm 0.02$	26.4	$(23)^e$	$(200)^e$
	$ML_3/M \cdot L^3$	$(37.6)^a$		(38.7)		
	$ML_4/M \cdot L^4$		40.0^e	41.0		
	$M \cdot L^3/(M_2O_3)^{0.5}(s)$		-45.0^e	-45.2		

---

^{a} 25°, 0.1; ^{c} 25°, 1.0; ^{e} 25°, 3.0; ^{r} 0-40°, 0; ^{t} 63-153°, 1.0 molal

Hydroxide ion (continued)

Metal ion	Equilibrium	Log K 25°, 0.5	Log K 25°, 1.0	Log K 25°, 0	ΔH 25°, 0	ΔS 25°, 0
Sb(III)	$ML_2/ML.L$ (in HNO_3)		15.5^g			
	$ML_3/ML_2.L$			12.8		
	$ML_4/ML_3.L$			2.2		
	$M_2L_2/(ML_2)^2.H^2$ (in $HClO_4$)	0.7^g				
	$ML_2/H.(M_2O_3)^{0.5}$ (s,rhombic)	$-3.06g$		-3.09		
	$ML_2/H.(M_2O_3)^{0.5}$ (s,cubic)	-3.18^g				
Bi^{3+}	$ML/M.L$	12.36^a	12.60^e	12.9		
	$ML_3/M.L^3$	31.9^a		33.1		
	$ML_4/ML_3.L$		0.95	1.1		
	$M_6L_{12}/M^6.L^{12}$		164.95			
			170.49^e			
	$M_9L_{20}/(M_6L_{12})^{1.5}.L^2$	$23.9^a \pm 0.2$				
	$M_9L_{21}/M_9L_{20}.L$	$10.6^a \pm 0.0$				
	$M_9L_{22}/M_9L_{21}.L$	$11.1^a \pm 0.1$				
	$ML_2/(M_2O_3)^{0.5}$ (s,α)		-5.34	-5.4		

---

[a] 25°, 0.1; [e] 25°, 3.0; [g] 25°, 5.0, assuming HL/H.L = 14.6

Bibliography:

H^+	17LB,23LR,290,30R,33HH,36W,55A,57AG,	La^{3+}	60AS,61BC,66FK,76BM
	57IL,58A,59L,60CO,60NM,63GL,63HIa,63VS,	Ce^{3+}	59AS,64BN,76BM
	64FB,67A,67AK,67AKa,67EHP,67GL,67VLa,	Pr^{3+},Lu^{3+}	63A,66FK
	68GH,69VL,70GO,72CB,72CK,72JW,72O,73DH,	Nd^{3+}	59TG,63A,66FK,73BL,76BM
	74BA,74SM,74SS	Sm^{3+}	60ASa,66FK
Li^+-K^+	64O,76BM	Eu^{3+},Tb^{3+},Ho^{3+}	61AE,66FK
Be^{2+}	56GG,56KS,60SG,61COa,62COa,63FS,64HS,	Gd^{3+}	63A,66FK,74NBa
	67BT,67MB,67OK,68LC,68PG,69SW,73CG,	Dy^{3+}	60ASb,66FK
	76BM	Er^{3+},Yb^{3+}	60Aa,66FK
Mg^{2+}	29K,48SD,61PB,63FS,63Hb,63L,67YM,76BM	Tm^{3+}	66FK
Ca^{2+}	34K,49BPa,51DH,53BG,54GMa,56BPa,59BBC,	Pu^{3+}	49KD,69DH
	60G,61CO,61PB,63FS,65HW,76BM	Am^{3+}-Cf^{3+}	69DH,69GM
Sr^{2+}	52CM,54GMa,61CO,67H,76BM	Ce^{4+}	51HR,66OS,67D,74AM
Ba^{2+}	49BPa,54GMa,61CO,65HWb,67H,76BM	Th^{4+}	54KH,61KB,65BM,68HS,76BM
Sc^{3+}	56BK,66A,68AN,71AO,76BM	Pa^{4+}	65G,68Ga,76BM
Y^{3+}	60AS,64BC,66FK,76BM	U^{4+}	50KN,55KN,56DZ,59SH,64MW

Hydroxide ion (continued)

Np^{4+} 59SHC,76BM

Pu^{4+} 49H,50KN,60RK,65Pa

Pa(V)   66Ga,68Gb,76BM

UO_2^{2+} 54AHS,56DZ,60GR,62BM,62Sc,63DH,63DS,
         63HR,63RJ,68ASa,69VO,76BM

NpO_2^{2+} 72CMT,76BM

PuO_2^{2+} 71S,72CMP,76BM

Mn^{2+} 41FS,42Na,52CC,62Pa,68FB

Fe^{2+} 53H,53LK,56GW,76BM

Co^{2+} 42Nb,50GG,62BA,63BPa,63SD,67CB,70BZ,
         70GHa,76BM

Ni^{2+} 49GG,52CC,63SD,64P,65BLa,65BLS,66BI,
         68A,73KO,76BM

Cu^{2+} 53SL,56B,60P,64Wa,65SAH,68AP,70GH,76BM

Ti^{3+} 62PF,68PGa,70KB

V^{3+} 50FG,53Ga,63P,68MB,76BM

Cr^{3+} 10B,27BK,54W,58BJ,63TU,64SM,64Wa,67SKV,
         68ML,73MS

Mn^{3+} 67WD

Fe^{3+} 34BH,41Ba,51BB,51SV,52T,52WT,53Ha,54CT,
         55MV,56CH,57BS,57Ma,58La,59P,60RS,62Sd,
         63FS,63PL,63SM,68AS,69F,72S

Co^{3+} 53Se,66CN

Rh^{3+} 66SHa

Ti(IV)   76BM

Zr^{4+} 56ZC,61PM,69NM,73Na

Hf^{4+} 62Pa,71NAa,73Na

VO^{2+} 53Ga,55RR,66Ba

TcO^{2+} 69GK

Cu^{+} 53SL,63FS,76BM

Ag^{+} 27L,37PS,60AD,60AHS,60BH,60BS,60NM

Hg_2^{2+} 52FH

CH_3Hg^{+} 65SS,66GD,73LR,74A

Tl^{+} 53BG,56BP,70KY,73KK,76BM

$(CH_3)_2Tl^{+}$ 74LP

$(CH_3)_3Sn^{+}$ 66TF,68AC

$(C_2H_5)_3Sn^{+}$ 64TY

$(C_6H_5)_3Sn^{+},(C_6H_5)_3Pb^{+}$ 65SM

Pd^{2+} 67IE

Zn^{2+} 53SL,55SL,60DF,62Bb,62Pb,62SA,63PE,
         63SA,64SA,65Sd,72DS,76BM

Cd^{2+} 59Sf,59SLa,62BC,62DL,63Sd,64SM,
         67AKa,76BM

Hg^{2+} 38GH,52HS,54G,58ASP,59SL,61AH,61DT,
         62A,63FS,67AK,70CGM,76BM

Ge^{2+} 52JL

Sn^{2+} 41GH,42GL,59T,74G

$(CH_3)_2Sn^{2+}$ 64TY,64TYa,65TF

$(C_2H_5)_2Sn^{2+}$ 66TF,68AC

$(C_3H_7)_2Sn^{2+}$ 66TF

Pb^{2+} 39GV,60CO,60O,63CO,62PO,76BM

$(CH_3)_2Pb^{2+}$ 66FT

Au(III)   38JL

Al^{3+} 54ST,55Ka,56Kf,62FP,65Ab,69NN,71MB,
         71VP,76BM

Ga^{3+} 52WT,57F,68NA,73BN

In^{3+} 56Ba,56RR,59ASa,61Sc,63FS,69BNR,
         72Fa,76BM

Tl^{3+} 51Sa,53B,58S,61RW,64KY,69BNT,73KK,
         76BM

Sb(III)   52GG,74AB

Bi^{3+} 57O,59O,60Tb,71Ba,72DN,76BM

Other references: 00A,00D,00KH,00L,00N,
         01L,02NK,02S,03AC,03B,05G,05SA,06B,
         06GE,07K,07L,07P,08B,08D,08M,09A,
         09LB,09S,10A,10B,10NK,10W,13K,16V,
         17K,19L,20F,21G,21LF,22AR,23B,23H,
         23K,24JG,24JJ,24RK,24S,24Ba,25G,25H
         25Sa,25W,25WR,26Ba,27Da,27DJ,27S,
         28BV,28FM,28Pa,28Pb,28RS,29B,29BU,
         29S,29T,30E,30NB,30R,31KE,31KK,31P,
         32BR,32E,32IS,32M,32RF,33BW,33FM,33J,
         33KA,34Bb,34BH,34La,34M,35D,36H,
         36HD,36MJ,36SE,36SH,37C,37CB,37P,

Hydroxide ion (continued)

Other references: 37Q,38CF,38LJ,38O,39Ga,
    39GH,39H,39L,40BC,41K,41Ma,42DM,42H,42MR,
    42N,42RS,43B,43N,43P,43SK,44F,44La,44MK,
    45P,46M,47GD,47HK,47M,48F,48GF,48HS,48KA,
    48SP,49A,49B,49EP,49KDa,49KK,49KN,49L,
    59NT,49OS,49S,50A,50AF,50BQ,50BW,50ET,
    50MK,51Ab,51D,51Da,51DB,51DC,51FR,51HD,
    51Mc,51MF,51PN,51Q,51S,51V,52B,52GW,
    52HH,52J,52Ja,52KF,52KH,52KP,52Lb,52Se,
    52VR,53Ea,53IY,53KF,53KP,53M,53MK,53RS,
    53VT,53WS,54AH,54BBS,54D,54F,54FH,54FSa,
    54G,54GL,54H,54KP,54M,54NR,54Ra,55B,55BS,
    55DC,55GL,55Kc,55MS,55PH,55PK,55R,55SC,
    55V,55WW,55ZD,56C,56CK,56DP,56DZd,56GWb,
    56H,56IA,56J,56JP,56Kd,56Ke,56KF,56KSa,
    56OB,56PC,56SP,56SW,57B,57BG,57Ca,57GL,
    57GW,57HW,57Kb,57Kc,57Kd,57MO,57MT,57P,
    57PP,57R,57TE,57TMa,57ZM,58Ab,58AS,58B,
    58BB,58Cb,58GTZ,58K,58KG,58MG,58MP,58TG,
    58VP,58VPa,58VRa,58VS,59A,59ASb,59ASc,,
    59BE,59EG,59GS,59HS,59HSa,59I,59KB,
    59KGa,59KL,59MA,59MV,59OH,59W,60Ba,60BB,
    60BK,60BN,60Cc,60F,60Gb,60GB,60JP,60S,
    60SWa,61A,61B,61BL,61BP,61Ka,61KBa,61KT,
    61MN,61Nc,61P,61PP,61PS,61RK,61RL,61WL,
    62B,62DG,62IN,62KB,62KB,62La,62LG,62M,
    62N,62PPC,62RB,62RF,62BA,62BG,63BF,63BJ,
    63DD,63Hd,63Kb,63KBa,63KS,63LC,63PS,63SA,
    63SB,64Aa,64Bb,63BSa,63Ca,64DSa,64FC,
    64G,64H,64HSa,64KB,64LD,64NK,64NL,64Pa,
    64PH,64Sa,64SAa,64SMa,64ST,64Ta,65AKP,
    65BS,65F,65Ga,65GA,65GCa,65H,65KY,65NT,
    65RD,65Sf,65SAP,65SSa,65ZS,66Ba,66BB,
    66BM,66HF,66KA,66KS,66OP,66Sa,66SI,
    66SWa, 67AKE,67Bc,67C,67GK,67GP,67GS,
    67GSb,67H,67HC,67La,67LK,67MP,67PB,67SI,
    67SSd,68DM,68GS,68HC,68HM,68KA,68MG,
    68RR,68SF,68SM,68SR,68WS,68ZL,68ZP,69BS,

69CR,69FT,69GA,69H,69LS,69M,69MGa,69MKa,69Nb,
69RC,69RS,69SMK,69VB,69WS,70C,70EL,70GH,70HR,
70IE,70Ld,70MS,70NK,70OE,70SB,70SK,70VT,
71BZ,71DB,71GD,71HR,71IB,71KP,71KS,71Ma,
71Mb,71MKK,71NA,72CB,72HH,72MB,72MG,73MV,
72OK,72SK,72US,73AK,73Ba,73BF,73FJ,73G,
73GG,73GT,73HHb,73RR,73SB,74HI,74KY,
74NB,74RN,74VZ

$$\begin{array}{c} O \\ \| \\ HO\text{-}V\text{-}OH \\ | \\ OH \end{array}$$

H_3O_4V <u>Hydrogen vanadate</u>   (<u>vanadic acid</u>) H_3L

Metal ion	Equilibrium	Log K 25°, 0.5	Log K 25°, 1.0	Log K 25°, 0	ΔH 25°, 0	ΔS 25°, 0
H^+	$HL/H.L$	13.0		14.3		
		13.5[h]		13.2[e]		
	$H_2L/HL.H$	7.85	8.04[j]	8.5		
		8.23[h]	8.31[h,r]			
	$H_3L/H_2L.H$	3.78		4.0		
	$VO_2/H_3L.H$	3.20		3.3		
	$V_2O_7/(HL)^2$	0.4		0.6		
	$HV_2O_7/(HL)^2.H$	10.58		11.0		
				10.93[e]		
	$V_3O_9/(HL)^3.H^3$	30.66	31.6[e]	31.8		
	$V_4O_{12}/(HL)^4.H^4$	41.4				
	$V_4O_{12}/V_4O_{13}.H^2$	8.9				
	$V_3O_9/(H_2L)^3$	7.14[s]				
	$V_4O_{12}/(H_2L)^4$	10.10[s]				
	$HV_6O_{17}/(H_2L)^6.H^3$	33.04[s]				
	$V_{10}O_{27}/(H_2L)^{10}.H^6$	61.8[s]				
	$VO_2/H.(V_2O_5)^{0.5}(s)$			−0.68	4.2	11
	$V_{10}O_{27}.H^{14}/(VO_2)^{10}$		−6.8	−5.5[e]		
	$V_{10}O_{27}/HV_{10}O_{28}.H$	4.34[h]	3.6	3.5[e]		
		4.45[h,r]	3.6[j]			
		4.39[s]	4.5[j,r]			
	$HV_{10}O_{28}/V_{10}O_{28}.H$	6.94[h]	5.8			
		7.52[h,r]	6.06[j]			
			7.6[j,r]			
Li^+	$MV_{10}O_{28}/M.V_{10}O_{28}$	1.6[h,r]				
	$M_2V_{10}O_{28}/MV_{10}O_{28}.M$	0.6[h,r]				
	$MHV_{10}O_{28}/M.HV_{10}O_{28}$	0.6[h,r]				

---

[e] 25°, 3.0; [h] 20°, 0.1; [j] 20°, 1.0; [r] $(CH_3)_4NCl$ used as background electrolyte; [s] 40°, 0.5

Hydrogen vanadate (continued)

Metal ion	Equilibrium	Log K 25°, 0.5	Log K 25°, 1.0	Log K 25°, 0	ΔH 25°, 0	ΔS 25°, 0
Na^+	$MHL/M \cdot HL$		$0.30^{j,r}$			
	$MV_{10}O_{29}/M \cdot V_{10}O_{28}$	$1.6^{h,r}$				
	$M_2V_{10}O_{28}/MV_{10}O_{28} \cdot M$	$0.6^{h,r}$				
	$MHV_{10}O_{28}/M \cdot HV_{10}O_{28}$	$0.7^{h,r}$				
K^+	$MHL/M \cdot HL$		$0.04^{j,r}$			
	$MV_{10}O_{28}/M \cdot V_{10}O_{28}$	$2.4^{h,r}$				
	$M_2V_{10}O_{28}/MV_{10}O_{28} \cdot M$	$1.0^{h,r}$				
	$MHV_{10}O_{28}/M \cdot HV_{10}O_{28}$	$1.4^{h,r}$				
Rb^+	$MV_{10}O_{28}/M \cdot V_{10}O_{28}$	$2.8^{h,r}$				
	$MHV_{10}O_{28}/M \cdot HV_{10}O_{28}$	$1.8^{h,r}$				
Cs^+	$MV_{10}O_{28}/M \cdot V_{10}O_{28}$	$3.18^{h,r}$				
	$MHV_{10}O_{28}/M \cdot HV_{10}O_{28}$	$2.20^{h,r}$				
NH_4^+	$M \cdot H_2L/MVO_3(s)$			-3.5	7.2	8

---

^{h} 20°, 0.1; ^{j} 20°, 1.0; ^{r} $(CH_3)_4NCl$ used as background electrolyte.

Bibliography:

H^+ 55LC,56RRa,58NL,59IB,60BI,63SG,64BI,   Other references: 46SC,50SSa,56TS,57La,
         64DS,66B,66BSW,76BM   58SPa,59NQ,60BC,60Ca,60Sa,63SZ,64DG,
Li^+-Cs^+ 63SG   64NH,64YK,65PSZ,66I,67TK,73R,74IG,
NH_4^+ 74VKL   74IGG

$$\begin{array}{c} O \\ \parallel \\ HO-Cr-OH \\ \parallel \\ O \end{array}$$

H_2O_4Cr Hydrogen chromate   (chromic acid) H_2L

Metal ion	Equilibrium	Log K 25°, 0.5	Log K 25°, 1.0	Log K 25°, 0	ΔH 25°, 0	ΔS 25°, 0
H^+	HL/H.L	5.81	5.74	6.51 ±0.02	0.7	32
		6.09^a±0.04	5.90^e		1.1^e	31^e
	$H_2L/HL.H$	-0.7^i	-0.7 ±0.1	-0.2^o	$(9)^r$	$(27)^c$
		-0.6^h	-0.6^e			
	$Cr_2O_7/(HL)^2$	1.84	1.97 ±0.03	1.53 ±0.02	-4.7	-9
		1.72^a±0.02	2.17^e±0.03		-4.8^e	-6^e
K^+	ML/M.L			$(0.57)^n$		
Ba^{2+}	M.L/ML(s)	-8.51	-8.39	-9.67 ±0.01	$(8)^s$	(-17)
		-8.96^a				
Th^{4+}	ML.H/M.HL	0.67^t			$(4)^u$	(15)
NpO_2^{2+}	ML.H/M.HL	1.81^t			$(1)^u$	(12)
Cu^{2+}	M.L/ML(s)			-5.44		
Fe^{3+}	ML.H/M.HL	0.28^t			$(3)^u$	(12)
Ag^+	$M^2.L/M_2L(s)$			-11.92 ±0.03	$(15)^v$	(-4)
Hg_2^{2+}	M.L/ML(s)			-8.70		
Tl^+	$M^2.L/M_2L(s)$			-12.01		

---

^{a} 25°, 0.1; ^{e} 25°, 3.0; ^{h} 20°, 0.1; ^{i} 20°, 0.5; ^{n} 18°, 0; ^{o} 20°, 0; ^{r} 15-35°, 1.0; ^{s} 18-28°, 0; ^{t} 25°, 0.2; ^{u} 1-25°, 0.2; ^{v} 20-40°, 0

Bibliography:

H^+ 34NR,53TK,55DP,58Hb,58HN,58SM,60BC,
          61Tb,62Sa,64HR,64T,65La,66TJ,67LB,
          68HSa,68LJ,70Lc,72AJ,72LL
K^+ 31BR
Ba^{2+} 43BR
Th^{4+},NpO_2^{2+},Fe^{3+} 72BT
Cu^{2+} 51PC
Ag^+ 35CM,41M,54Pa

Hg_2^{2+} 29B
Tl^+ 53Sa

Other references: 05AS,07S,10BS,23B,24B,
                   28H,28Ja,28S,32BR,32HJ,36E,41SW,42D,
                   42KP,43BR,51K,53PH,54TK,56C,57HN,
                   58KGL,60SS,63EK,63K,64MKc,66E,66HN,
                   66MB,67JJ,68PW,73AB

$$\begin{matrix} & O & \\ & \| & \\ HO- & Mo & -OH \\ & \| & \\ & O & \end{matrix}$$

H_2O_4Mo Hydrogen molybdate   (molybdic acid) H_2L

Metal ion	Equilibrium	Log K 20°, 0.1	Log K 25°, 1.0	Log K 25°, 0	ΔH 25°, 0	ΔS 25°, 0
H^+	HL/H.L		3.55	4.24[o]		
			3.89[e]		14[e]	65[e]
	$H_2L/H^2.L$	7.75	7.20	8.24[o]		
			7.50[e]			
	$Mo_7O_{24}/H^8.L^7$		52.81			
			57.74[e]		-56.0[e]	76[e]
	$HMo_7O_{24}/Mo_7O_{24}.H$		4.57			
			4.40[e]		2.6[e]	29[e]
	$H_2Mo_7O_{24}/HMo_7O_{24}.H$		3.63			
			3.54[e]		0.8[e]	19[e]
	$H_3Mo_7O_{24}/H_2Mo_7O_{24}.H$		2.38			
			2.53[e]		-0.6[e]	10[e]
	$M_{19}O_{59}/H^{34}.L^{19}$		196.3[e]			
	$H^2.L/MoO_3(s)$			-12.1		
Ca^{2+}	M.L/ML(s)			(-8.0)	(-0.7)	(-39)
Ag^+	$M^2.L/M_2L(s)$			-11.55	12.6	-11
Pb^{2+}	M.L/ML(s)			(-13.0)	(11.9)	(-20)

[e] 25°, 3.0; [o] 20°, 0

Bibliography:
H^+ 58SM,63RC,63AA,64SS,68ASb,68SS,76BM   Other references: 31N,34BG,53PH,54IK,
Ca^{2+},Pb^{2+} 58MH   56DZc,57DB,58SS,58YA,60Da,61Sd,
Ag^+ 54Pb,56MH   61Ta,63CK,63F,63LZ,63SBa,63YR,65C,
                                                                   65CO,67Ab,67HS,67VDa,68DK,69BW,
                                                                   74JJ

$$\begin{array}{c} O \\ \| \\ HO-W-OH \\ \| \\ O \end{array}$$

H_2O_4W			Hydrogen wolframate	(tungstic acid)		H_2L

Metal ion	Equilibrium	Log K 20°, 0.1	Log K 25°, 1.0	Log K 25°, 0	ΔH 25°, 0	ΔS 25°, 0
H^+	$HL/H.L$	3.5				
	$H_2L/H^2.L$	8.1	11.30^e			
	$HW_6O_{21}/H^7.L^6$				-64.5	
			60.76^e		-62.5^e	68^e
	$HW_6O_{21}/W_6O_{21}.H$		8.30^e		-5^e	21^e
	$W_{12}O_{41}/H^{14}.L^{12}$		123.2^e		127^e	138^e
	$HW_{12}O_{41}/W_{12}O_{41}.H$	6.28	5.0^j			
	$H_2W_{12}O_{41}/HW_{12}O_{41}.H$					
		5.27	4.3^j			
	$H_3W_{12}O_{41}/H_2W_{12}O_{41}.H$					
		3.6				
	$H^2.L/WO_3(s)$			-14.05		
	$H^2.L/H_2WO_4(s)$				13.0	
Ag^+	$M^2.L/M_2L(s)$			-11.26	14.7	-2

e 25°, 3.0; j 20°, 1.0

Bibliography:

H^+ 56DZa,58GH,58SM,62DK,62SG,69A,74AS   Other references: 33LH,34BG,58F,60Cb,64A,

Ag^+ 54Pc,58GH   64YP,65SP,65YR,69NP,71TM

$$ReO_4^-$$

O_4Re^- Rhenate (VII) ion L^-

Metal ion	Equilibrium	Log K 25°, 0
K^+	ML/M.L	(0.72)

Bibliography: 48M

Other references: 60BC,63SK,66OA,70OP

$$H_4Fe(CN)_6$$

$C_6H_4N_6Fe$ <u>Hydrogen hexacyanoferrate (II)</u> H_4L

Metal ion	Equilibrium	Log K 25°, 0.2	Log K 25°, 1.0	Log K 25°, 0	ΔH 25°, 0	ΔS 25°, 0
H^+	HL/H.L	3.25		4.30 ±0.05	0.5	21
	H_2L/HL.H	1.79		2.6 ±0.4	1	15
K^+	ML/M.L	1.5[a]		2.34 ±0.04	1.0	14
Mg^{2+}	ML/M.L			3.8		
Ca^{2+}	ML/M.L			3.8 -0.2	2.1[a]	24
	M_2L/ML.M			1.4		
Ba^{2+}	ML/M.L			3.8		
La^{3+}	ML/M.L			5.1		
Ag^+	M^4.L/M_4L(s)			-44.07		
Tl^+	ML/M.L			3.00	(1)[r]	(17)
			0.82[e]		-1.8[e]	-2[e]
Zn^{2+}	M^2.L/M_2L(s)			-15.68		
Cd^{2+}	M^2.L/M_2L(s)			-17.38		
Pb^{2+}	M^2.L/M_2L(s)			-18.02		

---

[a] 25°, 0.1;   [e] 25°, 3.0;   [r] 25-50°, 0

Bibliography:

H^+ 35KT,41NZ,57HH,67HI

K^+ 37D,57CP,66CL,67EG

Mg^{2+},Ba^{2+} 57CP

Ca^{2+} 49Ja,74HI

La^{3+} 58PW

Ag^+,Zn^{2+}-Pb^{2+} 64RP

Tl^+ 58PW,67MKa

Other references: 34R,38PO,41LK,53BG,56TG,
57BLa,57BP,58BS,58DT,59BBD,59BS,
59BSB,60BR,61BS,62BB,62JE,66MD,66NS,
66SNa,68LM,69NS,70Bb

$$Fe(CN)_6^{3-}$$

$C_6N_6Fe^{3-}$ Hexacyanoferrate (III) ion L^{3-}

Metal ion	Equilibrium	Log K 25°, 0.5	Log K 25°, 1.0	Log K 25°, 0	ΔH 25°, 0	ΔS 25°, 0
Na^+	ML/M.L		-0.3^e			
K^+	ML/M.L	-0.24	0.18^d	1.43 ± 0.03	0.5	8
		0.85^a	0.30^e			
Cs^+	ML/M.L		0.52^e			
Mg^{2+}	ML/M.L		0.04^e	2.79		
Ca^{2+}	ML/M.L		0.15^e	2.83 -0.2	1.6^a	18
Sr^{2+}	ML/M.L		0.23^e	2.85		
Ba^{2+}	ML/M.L			2.88		
La^{3+}	ML/M.L			3.74 ± 0.00	0.9 $+1$	20
Pr^{3+}	ML/M.L			3.6	0.9	19
Nd^{3+}	ML/M.L			3.74	0.8	20
Sm^{3+}	ML/M.L			3.7	0.9	20
Eu^{3+}	ML/M.L			3.7	1.0	20
Gd^{3+}	ML/M.L			3.74	1.0	20
Tb^{3+}	ML/M.L			3.8	0.9	20
Dy^{3+}	ML/M.L			3.7	1.0	20
Ho^{3+}	ML/M.L			3.7	1.1	20
Er^{3+}	ML/M.L			3.7	1.0	20
Tm^{3+}	ML/M.L			3.7	1.1	20
Yb^{3+}	ML/M.L			3.7	1.0	20
Lu^{3+}	ML/M.L			3.7	1.0	20
Fe^{3+}	ML/M.L	1.55 ± 0.01	1.32			

[a] 25°, 0.1; [d] 25°, 2.0; [e] 25°, 3.0

Hexacyanoferrate (III) ion continued)

Bibliography:

Na^+,Cs^+ 67RM

K^+ 49M,66CL,67EG,67RM

$Mg^{2+}-Ba^{2+}$ 52GMa,67RM,74HI

$La^{3+}-Lu^{3+}$ 48DJ,51DJ,63DK,72SC

Fe^{3+} 51ID,67SSb

Other references: 45D,50JMb,53BP,59GR,
                   61PF,62BBa,62JE,63LM,63HPS,65LW,66MR

$$Co(CN)_6^{3-}$$

$C_6N_6Co^{3-}$ Hexacyanocobaltate (III) ion L^{3-}

Metal ion	Equilibrium	Log K 20°, 0.1	Log K 25°, 0	ΔH 25°, 0	ΔS 25°, 0
K^+	ML/M.L		1.22	2	12
La^{3+}	ML/M.L		3.75 ±0.02	1.3	22
Ag^+	$M^3.L/M_3L(s)$		-25.41		
Hg_2^{2+}	$M^3.L^2/M_3L_2(s)$		-36.72		
CH_3Hg^+	ML/M.L	4.15			
	$M_2L/ML.M$	3.50			
Cd^{2+}	ML/M.L		4.17		

Bibliography:

K^+	50JMb	
La^{3+}	50JMb,60M,63DK	
Ag^+,Hg_2^{2+}	65R	
CH_3Hg^+	65SS	
Cd^{2+}	64RSS	

Other references: 59DT,60A,62AY

$$B(OH)_3$$

H_3O_3B Hydrogen borate   (boric acid)   HL

Metal ion	Equilibrium	Log K 25°, 0.5	Log K 25°, 1.0	Log K 25°, 0	ΔH 25°, 0	ΔS 25°, 0
H[+]	HL/H.L	8.97[a]±0.02	8.85[r]	9.236 ±0.001	-3.4 ±0.0	31
		8.94[d]±0.07	8.97[e]		-3.9[d]	
	$B_3O_3(OH)_4/H^2.L^3$	19.62[a]	20.07[e]			
	$B_3O_3(OH)_5/H.L^3$		10.4[e]			
	$B_4O_5(OH)_4/H^2.L^4$		20.9[e]			
	$B_5O_6(OH)_4/H^4.L^5$	38.1[a]	38.2[e]			
Ag[+]	ML/M.L		0.45[e]			
	$M.(HL)^2/H.MHL_2(s)$		4.5[e]			

[a] 25°, 0.1; [d] 25°, 2.0; [e] 25°, 3.0; [r] 25°, 0.7

Bibliography:
H[+]   340,350,430K,44MD,57Ab,62I,63I,67Bb,   Other references: OOWC,07L,09L,20K,22M,
        73BR,73DH   24PW,27KB,30HK,30HKa,31FA,32BR,34BY,
Ag[+]   70HS   38T,49KL,51Sb,53Ea,55L,57Ac,57La,
                                                                     59D,59Sd,61Sb,61SB,63FU,63Sa,65Se,
                                                                     66KG,67MN,68GL,72MBS,74BM

HCN

CHN		Hydrogen cyanide	(hydrocyanic acid)			HL
Metal ion	Equilibrium	Log K 25°, 0.5	Log K 25°, 1.0	Log K 25°, 0	ΔH 25°, 0	ΔS 25°, 0
H^+	HL/H.L	9.01^a	8.95	9.21 ±0.01	-10.43	7.2
		9.14^h	9.48^e		10.9^h	5^a
					-9.57^r	
Fe^{2+}	$ML_6/M.L^6$			35.4	-85.8	-126
Co^{2+}	$ML_5/M.L^5$				-61.5	
	$MHL_5/ML_5.H$				-32.0	
Ni^{2+}	ML/M.L		7.03^e			
	$ML_4/M.L^4$	30.5^a	31.06^e	30.22 ±0.1	-43.2	-7
	$MHL_4/ML_4.H$	5.4^a				
	$MH_2L_4/MHL_4.H$	4.5^a				
	$MH_3L_4/MH_2L_4.H$	2.6^a				
Fe^{3+}	$ML_6/M.L^6$			43.6	-70.1	-36
Cu^+	$ML_2/M.L^2$			16.26	-29.1	-23
	$ML_3/M.L^3$			21.6 ±0.1	-40.2	-36
	$ML_4/M.L^4$			23.1	-51.4	-67
Ag^+	$ML_2/M.L^2$		20.0	20.48	-32.9	-17
	$ML_3/M.L^3$		20.3^s	21.4	-33.5	-13
	$ML_4/M.L^4$		20.8^s			
	MOHL/M.OH.L		12.7	13.2		
	M.L/ML(s)		-15.4	-15.66		
Hg_2^{2+}	$M.L^2/ML_2(s)$			-39.3		
CH_3Hg^+	ML/M.L	13.8^a			-22.1^h	-11^a
Pd^{2+}	$ML_4/M.L^4$			42.4	-92.3	-116
	$ML_5/M.L^5$			45.3	-92.5	-103
Zn^{2+}	ML/M.L		5.3^e			
	$ML_2/M.L^2$		11.7^e	11.07	-11.0	14
	$ML_3/M.L^3$		16.7^e	16.05	-20.2	7
	$ML_4/M.L^4$		21.6^e	19.62	-27.9	-3
	$M.L^2/ML_2(s)$		-15.5^e			

^{a} 25°, 0.1; ^{e} 25°, 3.0; ^{h} 20°, 0.1; ^{r} 40°, 0; ^{s} 30°, 1.0

Hydrogen cyanide (continued)

Metal ion	Equilibrium	Log K 25°, 0.5	Log K 25°, 1.0	Log K 25°, 0	ΔH 25°, 0	ΔS 25°, 0
Cd^{2+}	$ML/M.L$			6.01	-7.3	3
			$5.55^e \pm 0.07$		-7.4^e	1^e
	$ML_2/M.L^2$			11.12	-13.0	7
			$10.7^e \pm 0.1$		-15.1^e	-2^e
	$ML_3/M.L^3$			15.65	-21.6	-1
			$15.5^e \pm 0.3$		-22.2^e	-4^e
	$ML_4/M.L^4$			17.92	-26.7	-8
			$19.0^e \pm 0.2$		-29.3^e	-11^e
Hg^{2+}	$ML/M.L$	18.00^h		17.00	-23.2	0
	$ML_2/M.L^2$	34.71^h	33.9^t	32.75	-46.6	-6
	$ML_3/M.L^3$	38.54^h	38.1^t	36.31	-53.4	-13
	$ML_4/M.L^4$	41.5^h	40.6^t	38.97	-59.7	-22
	$MOHL/M.OH.L$		28.9^t			

---

^{e} 25°, 3.0; ^{h} 20°, 0.1; ^{t} 30°, 2.0

Bibliography:

H^+	57A,59Ab,62IC,65SS,66BK,70CJ,71P
Fe^{2+},Fe^{3+}	65WC
Co^{2+}	68IW
Ni^{2+}	59FSa,63CI,68KM,71IJ,71PH,74P
Cu^+	67IJ,71PK,74KH
Ag^+	56KSa,65ZP,67AD,67IJ,72GC
Hg_2^{2+}	29B
CH_3Hg^+	65SS
Pd^{2+}	67IW
Zn^{2+}	65IC,71IJ,71P
Cd^{2+}	44L,66Gb,71IJ,71P
Hg^{2+}	57A,58NC,59NH,65CI,71IJ

Other references: 00M,00T,00WC,25HW,30RH, 31BDa,32Br,47R,50H,50VK,53S,54W,55F, 56N,56PJ,56SM,57TM,58SWa,60MJ,61Mb, 61MB,63AS,63PB,64GH,64VHa,65FK,67BP, 67ZF,68AD,68EP,69KH,71DG,72CD,72HF, 72P,74Kc

HNCO

CHON   Hydrogen cyanate   (cyanic acid)   HL

Metal ion	Equilibrium	Log K 27°, 1.5	Log K 25°, 0	ΔH 25°, 0	ΔS 25°, 0
H^+	HL/H.L		3.48 ±0.02	-2.0	9
Co^{2+}	$ML_4/M.L^4$	2.67			
Ag^+	$ML_2/M.L^2$		5.00[r]		
	M.L/ML(s)		-6.64[s]		

---

[r] 30°, 0; [s] 19°, 0

Bibliography:

H^+	58C,58J,58M		
Co^{2+}	66CV		

Ag^+ 30BH,54C

Other references: 56Aa,66BK

HNCS

CHNS		Hydrogen thiocyanate		(thiocyanic acid)		HL
Metal ion	Equilibrium	Log K 25°, 0.5	Log K 25°, 1.0	Log K 25°, 0	ΔH 25°, 0	ΔS 25°, 0
H^+	HL/H.L			0.9 ± 0.1		
Be^{2+}	ML/M.L			-0.16[f]		
	$ML_2/M.L^2$			-0.6[f]		
Mg^{2+}	ML/M.L		-0.9[e]			
Sc^{3+}	ML/M.L	0.20[i]		0.8[m]		
Y^{3+}	ML/M.L	(-0.07)[i]				
La^{3+}	ML/M.L		0.12[r]	0.24[g]		
Ce^{3+}	ML/M.L			0.59[g]		
Sm^{3+}	ML/M.L	0.09[i]				
Eu^{3+}	ML/M.L		0.15[r]	0.7		
				0.38[g]± 0.06	(1)[s]	(5)[g]
Gd^{3+}	ML/M.L	0.21[i]				
Tb^{3+}	ML/M.L		0.23[r]			
Dy^{3+}	ML/M.L	0.12[i]				
Er^{3+}	ML/M.L	0.16[i]				
Lu^{3+}	ML/M.L		0.20[r]	0.45[g]		
Ac^{3+}	ML/M.L		0.05			
Pu^{3+}	ML/M.L		0.46			
Am^{3+}	ML/M.L		0.43 ± 0.07	0.59[g]± 0.2	(3)[s]	(12)[g]
Cm^{3+}	ML/M.L		0.44 ± 0.01	0.61[g]	(3)[s]	(12)[g]
Bk^{3+}	ML/M.L		0.49	(0.86)[g]	(1)[s]	(6)[g]
Cf^{3+}	ML/M.L		0.57			
Es^{3+}	ML/M.L		0.56			
Th^{4+}	ML/M.L		1.08			
	$ML_2/M.L^2$		1.78			

[e] 25°, 3.0; [f] 25°, 4.0; [g] 25°, 4.0; [i] 20°, 0.6; [m] 20°, 4.0; [r] 30°, 1.0; [s] 10-55°, 5.0

Hydrogen thiocyanate (continued)

Metal ion	Equilibrium	Log K 25°, 0.5	Log K 25°, 1.0	Log K 25°, 0	ΔH 25°, 0	ΔS 25°, 0
U^{4+}	ML/M.L	1.49[d]	1.49[j]		(-6)[t]	(-13)[d]
	$ML_2/M.L^2$	2.11[d]	1.95[j]		(-8)[t]	(-17)[d]
	$ML_3/M.L^3$		2.2[j]			
UO_2^{2+}	ML/M.L		0.75	0.93	-0.8[c]	1[c]
			0.72[u]	0.71[f]		
	$ML_2/M.L^2$		0.72		-2.1[c]	-4[c]
			0.70[u]	0.72[f]		
	$ML_3/M.L^3$		0.18		-1.4[c]	0[c]
V^{2+}	ML/M.L		1.43		(-5)[v]	(-10)[c]
Cr^{2+}	ML/M.L			1.09		
	$ML_2/M.L^2$			0.77		
Mn^{2+}	ML/M.L	0.80	0.65	1.23	-0.9	3
		1.30[k]				
Fe^{2+}	ML/M.L		0.81[e]	1.31		
Co^{2+}	ML/M.L	1.11 ±0.04	0.98 ±0.03	1.72 -0.2	-1.6	3
			1.27[e]			
	$ML_2/M.L^2$		1.32			
Ni^{2+}	ML/M.L	1.23 ±0.02	1.13 ±0.01	1.76 -0.09	-2.3	0
			1.34[e]		-2.9[c]	-4[c]
	$ML_2/M.L^2$		1.58 -0.01		-5.0[c]	-10[c]
	$ML_3/M.L^3$		1.5 ±0.2		-7[c]	-16[c]
Cu^{2+}	ML/M.L	1.90[a]	1.74 +0.02	2.33	-3.0	1
			1.91[e]		-3.0[c]	-2[c]
	$ML_2/M.L^2$	3.00[a]	2.74	3.65	-3.1[c]	2[c]
V^{3+}	ML/M.L		2.13 ±0.06		(-4)[w]	(-4)[c]
Cr^{3+}	ML/M.L		(1.87)	3.08	(-2)[x]	(7)
	$ML_2/M.L^2$		(2.98)			

---

[a] 25°, 0.1; [c] 25°, 1.0; [d] 25°, 2.0; [e] 25°, 3.0; [f] 23°, 4.0; [j] 20°, 1.0; [k] 20°, 1.5; [t] 10-40°, 2.0; [u] 23°, 2.5; [v] 11-45°, 0.8; [w] 5-37°, 1.0; [x] 10-40°, 0

Hydrogen thiocyanate (continued)

Metal ion	Equilibrium	Log K 25°, 0.5	Log K 25°, 1.0	Log K 25°, 0	ΔH 25°, 0	ΔS 25°, 0
Fe^{3+}	$ML/M.L$	2.14 ± 0.03	2.10 ± 0.03	3.02 ± 0.01	-1.3^c	5^c
			$2.21^e \pm 0.03$		-0.6^d	
	$ML_2/M.L^2$	3.3 ± 0.2	3.2 ± 0.1	4.64^n		
			$3.64^e \pm 0.04$			
	$ML_3/M.L^3$		5.0^e			
	$ML_4/M.L^4$		6.3^e			
	$ML_5/M.L^5$		6.2^e			
	$ML_6/M.L^6$		6.1^e			
VO^{2+}	$ML/M.L$	0.92		3.32	$(-4)^x$	(-4)
	$ML_2/M.L^2$			3.68	$(-1)^x$	(14)
Pd^{2+}	$ML_4/M.L^4$		27.2 ± 1			
Cu^+	$ML_2/M.L^2$			11.00^g		
	$ML_3/M.L^3$			10.9^g		
	$ML_4/M.L^4$			10.4^g		
	$M.L/ML(s)$			-13.40^g		
Ag^+	$ML/M.L$	4.6^a		4.8		
				4.6^f		
	$ML_2/M.L^2$	8.06^a		8.23		
		8.18^d		8.29^f		
	$ML_3/M.L^3$	9.6^a		9.5		
		9.3^d		10.0^f		
	$ML_4/M.L^4$	10.5^a		9.7		
		10.0^d		11.3^f		
	$M.L/M.(s)$	-11.80^a		-11.97 ± 0.03	22.6	21
Au^+	$ML/M.L$		15.27^e			
	$ML_2/M.L^2$		16.98^e			
Hg_2^{2+}	$M.L^2/ML_2(s)$		-19.00	-19.52		
CH_3Hg^+	$ML/M.L$	6.05^h				
	$M_2L/ML.M$	1.65^h				
$C_2H_5Hg^+$	$ML_2/ML.L$		-0.10			
	$ML_3/ML_2.L$		0.20			

[a] 25°, 0.1; [c] 25°, 1.0; [d] 25°, 2.0; [e] 25°, 3.0; [f] 25°, 4.0; [g] 25°, 5.0; [h] 20°, 0.1; [n] 18°, 0; [x] 10-40°, 0

Hydrogen thiocyanate (continued)

Metal ion	Equilibrium	Log K 25°, 0.5	Log K 25°, 1.0	Log K 25°, 0	ΔH 25°, 0	ΔS 25°, 0
Tl^+	ML/M.L	0.24	0.17	0.58		
		0.12^d	$0.15^e \pm 0.05$	$0.17^f \pm 0.03$		
	$ML_2/M.L^2$	0.00	-0.05			
		-0.12^d	$-0.12^e \pm 0.08$	$-0.03^f \pm 0.03$		
	$ML_3/M.L^3$		-0.4			
		-0.5^d	$-0.5^e \pm 0.1$	$-0.5^f \pm 0.1$		
	M.L/ML(s)			-3.79 ± 0.02		
				-3.16^f		
Zn^{2+}	ML/M.L		0.71	1.33	-1.4^c	-1^c
		0.74^d		1.11^f		
	$ML_2/M.L^2$		1.04	1.91	-1.8^c	-1^c
		1.15^d		1.81^f		
	$ML_3/M.L^3$		1.2	2.0	-2^c	-1^c
		1.3^d		2.8^f		
	$ML_4/M.L^4$		1.5	1.6	-4^c	-6^c
		1.7^d		2.8^f		
Cd^{2+}	ML/M.L	1.35 -0.02	1.32 ±0.02	1.89	-2.3^c	-2^c
		1.53^a				
		1.34^d	$1.41^e -0.05$		-1.9^e	0^e
	$ML_2/M.L^2$	2.04	1.99	2.78	-4.2^c	-5^c
		2.05^d	$2.24^e -0.2$		-3.7^e	-2^e
	$ML_3/M.L^3$	2.1	2.0	2.8	-6^c	-12^c
		2.2^d	$2.5^e \pm 0.1$		-5.2^e	-6^e
	$ML_4/M.L^4$	2.0	1.9	2.3		
		2.0^d	$2.5^e \pm 0.0$		-6.2^e	-10^e
Hg^{2+}	ML/M.L		9.08		-11.9^c	2^c
	$ML_2/M.L^2$	16.43^a	16.86	17.26	-24.0^c	-3^c
	$ML_3/M.L^3$	19.14^a	19.70	19.97	-29.0^c	-7^c
	$ML_4/M.L^4$	$21.2^a \pm 0.1$	21.7	21.8 ±0.1	$-34.4^c \pm 0.4$	-16^c
	$M.L^2/ML_2(s)$		-19.56			
Sn^{2+}	ML/M.L	1.17^k				
	$ML_2/M.L^2$	1.77^k				
	$ML_3/M.L^3$	1.74^k				

[a] 25°, 0.1; [c] 25°, 1.0; [d] 25°, 2.0; [e] 25°, 3.0; [f] 25°, 4.0; [k] 20°, 2.2

Hydrogen thiocyanate (continued)

Metal ion	Equilibrium	Log K 25°, 0.5	Log K 25°, 1.0	Log K 25°, 0	ΔH 25°, 0	ΔS 25°, 0
$(CH_3)_2Sn^{2+}$	$ML/M.L$		0.43			
	$ML_2/M.L^2$		1.0			
Pb^{2+}	$ML/M.L$	0.54[d]	0.78[e]	1.08[f]	0.3	3[d]
	$ML_2/M.L^2$	0.87[d]	0.99[e]	1.48[f]		
	$ML_3/M.L^3$		1.0[e]			
Al^{3+}	$ML/M.L$			0.42[o]		
Ga^{3+}	$ML/M.L$	1.18[i]		2.15[y]		
In^{3+}	$ML/M.L$	2.34[i]		3.15[y]		
		2.56[d]			-1.7[d]	6[d]
	$ML_2/M.L^2$	3.53[d]			-5.5[d]	-2[d]
	$ML_3/M.L^3$	4.6[d]			-3.1[d]	11[d]
CH_3Sn^{3+}	$ML/M.L$		1.48			
	$ML_2/M.L^2$		2.20			
	$ML_3/M.L^3$		3.3			
Bi^{3+}	$ML/M.L$	1.67[a]	1.32	2.21		
			(1.28)[e]	2.02[f]		
	$ML_2/M.L^2$	3.0[a]	2.1	2.7		
			2.7[e]	3.5[f]		
	$ML_3/M.L^3$	4.0[a]	3.0	4.4		
			3.8[e]	4.8[f]		
	$ML_4/M.L^4$	4.8[a]	(2.9)	5.2		
			5.3[e]	6.3[f]		
	$ML_5/M.L^5$	5.5[a]	3.6	5.8		
			6.0[e]	6.8[f]		
	$ML_6/M.L^6$	6.1[a]	4.0	5.4		
			6.9[e]	8.3[f]		

---

[a] 25°, 0.1; [c] 25°, 1.0; [d] 25°, 2.0; [e] 25°, 3.0; [f] 25°, 4.0; [k] 20°, 2.2; [y] 30°, 0

Bibliography:

H^+ 42SH,66BK Y^{3+}-Lu^{3+} 64B,64KS,65Sc,74KC,74KM
Be^{2+} 71SK Ac^{3+} 68RS
Mg^{2+} 73HHa Pu^{3+}-Es^{3+} 65CK,65Sc,72HPa,74KC
Sc^{3+} 64KS,67KG Th^{4+} 50WS

Hydrogen thiocyanate (continued)

U^{4+} 54AL,55DW

UO_2^{2+} 49Aa,57DM,64VM,71AKa

V^{2+} 68MSa,68OP

Cr^{2+} 58YF

Mn^{2+} 58YK,63TC,64TCa,67NT,73HH

Fe^{2+} 58YK,67CS

Co^{2+} 51SS,58SP,58YK,62TZ,62W,63TC,64KS,
                 67NT,70MM,71SM,73HH

Ni^{2+} 58YK,62W,63TC,67NT,68MT,70MM,73HH,
                 74K

Cu^{2+} 59TT,62W,67NT,70MM,73HH,74K

V^{3+} 51FG,67BSW,68KT

Cr^{3+} 54PB,55PK

Fe^{3+} 51MM,51SS,53BD,55LR,56L,57YT,58BCC,
                 58P,61Y,64KS,65MR,67CS,68Ma,68PC,69VM

VO^{2+} 51FG,68SW

Cu^+ 59FS

Ag^+ 12K,54KT,54LN,55LN,56VS

Au^+ 66K

Hg_2^{2+} 29B,70CG

CH_3Hg^+ 65SS

$C_2H_5Hg^+$ 65BB

Tl^+ 52Sb,57N,58BC,58PD,60KM,62KC,65KM,
                 66OL,71FR

Pd^{2+} 63GKG,65FK,66BSA

Zn^{2+} 66MK,70DSa,71AKb,73RS

Cd^{2+} 41L,57TH,63TC,64KS,66Gb,68G,68GJa,
                 73HH,73RS

Hg^{2+} 56T,62TE,70CG,71AKb,74Kb

Sn^{2+} 61GO

$(CH_3)_2Sn^{2+},CH_3Sn^{3+}$ 68PC

Pb^{2+} 56LS,59TH,63MKT,67NT

Al^{3+} 63VM

Ga^{3+} 64KS,68DD

In^{3+} 54Sa,64KS,68DD,69R

Bi^3 71FKa

Other references: 26B,49KH,51HDC,52AP,
                    52YV,53BG,53CH,53FH,53J,56LS,56Ta,
                    57GS,57I,57TS,57YT,58HT,58PD,58SW,
                    60NA,61BT,61GS,61MD,61SN,62LY,62S,
                    62Va,63RS,64K,64VM,65HS,65MSW,65NH,
                    66CM,66SD,66VV,68P,69SMT,69SS,70FS,
                    70SGK,71BSB,71DD,71KN,71MO,71MS,71PT,
                    72HPB,72L,73CDa,73RT,74RB,74TM

NCSe⁻

CNSe⁻   Selenocyanate ion   L⁻

Metal ion	Equilibrium	Log K 25°, 0.3	Log K 25°, 1.0	Log K 19°, 0	ΔH 25°, 1.0	ΔS 25°, 1.0
Ni^{2+}	$ML/M.L$		0.99		-3.1	-6
	$ML_2/M.L^2$		1.26		-6	-14
Ag^+	$ML_3/M.L^3$	13.90				
	$M.L/ML(s)$			-15.40		
Zn^{2+}	$ML/M.L$		0.44		-1.4	-3
	$ML_2/M.L^2$		0.64		(-1.5)	(-2)
Cd^{2+}	$ML/M.L$		1.47		-2.4	-1
	$ML_2/M.L^2$		2.30		-6.3	-11
	$ML_3/M.L^3$		2.8		(-1)	(9)
	$ML_4/M.L^4$		4.04		-10	-15
Hg^{2+}	$ML_3/M.L^3$	26.4				
	$ML_4/M.L^4$	28.9	28.7[r]		-46.1	-23[r]

---

[r] 25°, 0.8

Bibliography:
Ni^{2+} 74Ka
Ag^+ 30BH,56T   Other references: 59G,59GK,60GSa,60LC,
Hg^{2+} 56T,74Kb   61GSa,61GSb,62GA,66BK,67HB,72MT,73RT
Zn^{2+},Cd^{2+} 74AA

$$C(CN)_3^-$$

$C_4N_3^-$ Tricyanomethane ion L^-

Metal ion	Equilibrium	Log K 19°, 0
Ag^+	M.L/ML(s)	-8.34

Bibliography: 30BH

$$N(CN_2)^-$$

$C_2N_3^-$ Dicyanimide ion L^-

Metal ion	Equilibrium	Log K 19°, 0
Ag^+	M.L/ML(s)	-8.85

Bibliography: 30BH

$$\underset{\text{HO-C-OH}}{\overset{\displaystyle O}{\overset{\displaystyle \|}{}}}$$

CH_2O_3		Hydrogen carbonate		(carbonic acid)		H_2L
Metal ion	Equilibrium	Log K 25°, 0.5	Log K 25°, 1.0	Log K 25°, 0	ΔH 25°, 0	ΔS 25°, 0
H^+	HL/H.L	10.00[a]	9.57	10.329±0.01	-3.5 -0.1	36
			9.56[e]			
	H_2L/HL.H	6.16[a]	6.02 ±0.03	6.352+0.01	-2.0 ±0.1	22
		5.97[d]	6.33[e]			
	H_2L/CO_2(g)		-1.51	-1.464±0.01		
			-1.55[e]			
Mg^{2+}	ML/M.L	(2.37)[r]		2.88	(3)[s]	(20)
	MHL/M.HL	0.77[r]		0.95		
	M.L/ML$(H_2O)_5$(s)			-4.54		
	M.L/ML$(H_2O)_3$(s)			-4.67		
	M.L/ML(s)			-7.46	(5)[t]	(-20)
Ca^{2+}	ML/M.L	3.00[r]		3.15 ±0.05	(4)[s]	(30)
	MHL/M.HL	0.81[r]		1.0 ±0.0		
	M.L/ML(s,calcite)	-8.01[r]		-8.35 ±0.1	(2)[u]	(-30)
	M.L/ML(s,aragonite)			-8.22		
Sr^{2+}	M.L/ML(s)			-9.03		
Ba^{2+}	ML/M.L			2.78		
	M.L/ML(s)			-8.30 ±0.01		
Y^{3+}	$M^2.L^3/M_2L_3$(s)			-30.6		
La^{3+}	$M^2.L^3/M_2L_3$(s)			-33.4		
Nd^{3+}	$M^2.L^3/M_2L_3$(s)			-33.0		
Sm^{3+}	$M^2.L^3/M_2L_3$(s)			-32.5		
Gd^{3+}	$M^2.L^3/M_2L_3$(s)			-32.2		
Dy^{3+}	$M^2.L^3/M_2L_3$(s)			-31.5		
Yb^{3+}	$M^2.L^3/M_2L_3$(s)			-31.1		
Mn^{2+}	MHL/M.HL		0.45[e]	1.8		
	M.L/ML(s)		-9.68[e]	-9.30		
Fe^{2+}	M.L/ML(s)			-10.68		

[a] 25°, 0.1; [d] 25°, 2.0; [e] 25°, 3.0; [r] 22°, 0.15; [s] 10-50°, 0; [t] 25-40°, 0; [u] 0-70°, 0

Hydrogen carbonate (continued)

Metal ion	Equilibrium	Log K 25°, 0.5	Log K 25°, 1.0	Log K 25°, 0	ΔH 25°, 0	ΔS 25°, 0
Co^{2+}	M.L/ML(s)			-9.98		
Ni^{2+}	M.L/ML(s)			-6.87		
Cu^{2+}	ML/M.L			6.75 ±0.02		
	$ML_2/M.L^2$	8.6^v		9.92 ±0.09		
	M.L/ML(s)			-9.63		
	$M.(OH)^2.L/M(OH)_2L(s)$			-15		
	$M^2.(OH)^2L.L/M_2(OH)_2L(s,malachite)$			-33.78		
	$M^3.(OH)^2.L^2/M_3(OH)_2L_2(s,azurite)$			-45.96		
Ag^+	$M^2.L/M_2L(s)$			-11.09		
Hg_2^{2+}	M.L/ML(s)			-16.05		
Zn^{2+}	M.L/ML(s)			-10.00		
Cd^{2+}	M.L/ML(s)			-13.74		
Pb^{2+}	$ML_2/M.L^2$	8.2^v				
	M.L/ML(s)		-11.01	-13.13	-0.01	
	$(M(OH)_3)^3.L^2/(OH)^7.M_3(OH)_2L_2(s)$			-5.10		

---

^{v} 18°, 1.7

Bibliography:

H^+ 14MR,25MM,28SH,35MB,35SM,35W,37P,38CH,
       41HS,43HD,46N,57MS,58FN,58NR,59Ea,60GL,
       61NM,71Nb,73DH,73MC

Mg^{2+} 41G,62H,63Ha,74RL

Ca^{2+} 29FJ,35HR,37BH,41G,46Na,62GT,68La,70Lb,
       74JL,74RL

Sr^{2+} 35KA

Ba^{2+} 35KA,46Na,73BS

$Y^{3+}-Yb^{3+}$ 66JH

Mn^{2+} 35KA,63H,70GK

$Fe^{2+},Ni^{2+},Zn^{2+},Cd^{2+}$ 35KA

Co^{2+} 67B

Cu^{2+} 35KA,58Sb,59FB,68SRG

Ag^+ 27WB

Hg_2^{2+} 29B

Pb^{2+} 28RS,35KA,59FB,61NM

Other references: 00Bc,03AC,07P,09SL,11AV,
    13AP,14W,15J,17SL,220,23M,26HB,29K,
    30M,39HJ,42N,42Na,46KD,50M,51M,52Lb,
    52Sf,55MB,56BC,57Sa,57Sb,57Sc,58LG,
    58MG,58ML,58VG,59E,59Kb,59KS,59U,
    60BG,60BK,60R,61GM,61GT,61PK,62GM,
    62PNN,62SH,62WS,63E,63MG,63R,63SR,
    64FD,65BBa,65GSS,66BT,67Ba,68B,68BB,
    68Na,69Bc,69H,69NR,70HKS,71Na,72MV,
    72OS

D. GROUP IV LIGANDS

$$\begin{array}{c} OH \\ \| \\ HO-Si-OH \\ \| \\ OH \end{array}$$

H_4O_4Si <u>Hydrogen silicate</u>   (silicic acid) H_2L

Metal ion	Equilibrium	Log K 25°, 0.5	Log K 25°, 1.0	Log K 25°, 0	ΔH 25°, 0	ΔS 25°, 0
H^+	HL/H.L	12.56	12.71[e]	13.1	(-10)[r]	(24)[b]
	H_2L/HL.H	9.46 ±0.00	9.47	9.86	(-5)[r]	(27)[b]
			9.43[e]			
	$H_2L_2/H^2.L^2$		26.16[e]			
	$H_4L_4/H^4.L^4$		56.08[e]	55.9		
	$H_6L_4/H^6.L^4$	75.51		78.2	(-20)[r]	(280)[b]
	H_2L/SiO_2(s,amorphous)		-2.89	-2.74		
Mg^{2+}	ML/M.L		4.17			
	MHL/M.HL		0.64			
	$M(HL)_2/M.(HL)^2$		3.82			
	$M^2.(HL)^3/M_2(HL)_3(H_2O)_4$(s)			-38.8[u]		
Ca^{2+}	ML/M.L		3.09			
	MHL/M.HL		0.39			
	$M(HL)_2/M.(HL)^2$		2.89			
	M.L/ML(s)			-7.2		
UO_2^{2+}	MHL/M.HL	7.5[s]				
Fe^{3+}	MHL/M.HL	8.9[a] ±0.4			(4)[t]	(50)[a]

---

[a] 25°, 0.1;   [b] 25°, 0.5;   [e] 25°, 3.0 molal; [r] 25-50°, 0.5;   [s] 25°, 0.2;   [t] 18-32°, 0.1; [u] 51, 0

Bibliography:

H^+ 57GP,58G,58SMa,59L,67BI,68J,74SS,76BM   Other references: 27H, 27H,28RL,30HK,34FW,

Mg^{2+} 73CH,74SS   40RE,43OKa,54Ma,54AHI,57TK,58AK,60GC,

Ca^{2+} 65GC,74SS   61KT,62FR,62MF,65Ac,67R,73PE

UO_2^{2+} 71PW

Fe^{3+} 65WS,71PW,7300

$$NH_3$$

H_3N			Ammonia			L

Metal ion	Equilibrium	Log K 25°, 0.5	Log K 25°, 1.0	Log K 25°, 0	ΔH 25°, 0	ΔS 25°, 0
H^+	HL/H.L	9.32 ±0.03	9.40 ±0.04	9.244±0.005	-12.45 ±0.05	0.5
		9.29[a]-0.01	9.49[d]±0.03	9.80[g]	-13.51[e]	
					-12.75[r]	
	HL/H.L(g)		11.11		(-11)[s]	(11)[c]
Li^+	ML/M.L	-0.3[t]				
Mg^{2+}	ML/M.L	0.23[t]				
	$ML_2/M.L^2$	0.08[t]		0.2	-1.2	-3
	$ML_3/M.L^3$	-0.3[t]				
Ca^{2+}	ML/M.L	-0.2[t]				
	$ML_2/M.L^2$	-0.8[t]				
Mn^{2+}	ML/M.L	1.00[k]				
	$ML_2/M.L^2$	1.54[k]				
	$ML_3/M.L^3$	1.70[k]				
	$ML_4/M.L^4$	1.3[k]				
Co^{2+}	ML/M.L	2.10[u]±0.02		1.99[o]	-3.2[v]	-1[u]
	$ML_2/M.L^2$	3.67[u]±0.07		3.50[o]		
	$ML_3/M.L^3$	4.78[u]±0.01		4.43[o]		
	$ML_4/M.L^4$	5.53[u]±0.02		5.07[o]		
	$ML_5/M.L^5$	5.75[u]±0.02		5.13[o]		
	$ML_6/M.L^6$	5.14[u]±0.03		4.39[o]		
Ni^{2+}	ML/M.L	2.81[d]±0.04		2.72	-4.0[d]±0.0	-1[d]
	$ML_2/M.L^2$	5.08[d]±0.06		4.89	-7.8[d]±0.2	-3[d]
	$ML_3/M.L^3$	6.85[d]±0.07		6.55	-12.1[d]+0.1	-9[d]
	$ML_4/M.L^4$	8.12[d]±0.05		7.67	-15.6[d]±0.4	-15[d]
	$ML_5/M.L^5$	8.93[d]±0.02		8.34	-19[d] ±1	-22[d]
	$ML_6/M.L^6$	9.08[d]±0.04		8.31	-24[d] ±1	-39[d]

[a] 25°, 0.1; [c] 25°, 1.0; [d] 25°, 2.0; [e] 25°, 3.0; [g] 25°, 5.0; [k] 20°, 2.0; [o] 20°, 0; [r] 40°, 0; [s] 25-40°, 1.0; [t] 23°, 2.0; [u] 30°, 2.0; [v] 27°, 2.0

Ammonia (continued)

Metal ion	Equilibrium	Log K 25°, 0.5	Log K 25°, 1.0	Log K 25°, 0	ΔH 25°, 0	ΔS 25°, 0
Cu^{2+}	ML/M.L	$4.24^d \pm 0.03$	4.12 ± 0.03	4.04 ± 0.03	$-5.5^d \pm 0.1$	1^d
	$ML_2/M.L^2$	$7.83^d \pm 0.04$	7.63 ± 0.02	7.47	$-11.1^d +0.1$	-1^d
	$ML_3/M.L^3$	$10.80^d \pm 0.07$	10.51 ± 0.03	10.27	$-16.6^d \pm 0.1$	-6^d
	$ML_4/M.L^4$	$13.00^d \pm 0.05$	12.6 ± 0.1	11.75	$-22.0^d \pm 0.1$	-14^d
	$ML_5/M.L^5$	$12.43^d \pm 0.04$			$-26^d \pm 1$	-30^d
Co^{3+}	$ML_5/ML_4.L$	5.07^d			(-1.5)	$(18)^d$
	$ML_6/ML_5.L$	4.50^d	4.33	4.15	-6.9	-3^d
	$ML_6/M.L^6$	35.21^u	34.36^w			
Cu^+	ML/M.L	5.93^x				
	$ML_2/M.L^2$	10.58^d			$(-16)^y$	$(-5)^d$
Ag^+	ML/M.L	3.30		3.31 ± 0.06	-4.9	-1
		3.26^d	3.20^g			
	$ML_2/M.L^2$		7.21	7.22 ± 0.01	-13.4 ± 0.1	-12
		7.20^d	7.13^g			
CH_3Hg^+	ML/M.L	7.25				
		7.60^h				
Tl^+	ML/M.L	-0.9^t				
Pd^{2+}	ML/M.L		9.6			
	$ML_2/M.L^2$		18.5			
	$ML_3/M.L^3$		26.0			
	$ML_4/M.L^4$		32.8			
Zn^{2+}	ML/M.L	$2.38^d \pm 0.03$	2.32	2.21	-2.6^v	2^d
	$ML_2/M.L^2$	$4.88^d \pm 0.2$	4.81	4.50	-5.7^v	3^d
	$ML_3/M.L^3$	$7.43^d -0.3$	7.11	6.86	-9.6^v	2^d
	$ML_4/M.L^4$	$9.65^d \pm 0.1$	9.32	8.89	-14.8^v	-6^d
Cd^{2+}	ML/M.L	$2.72^d \pm 0.03$	$2.62 -0.08$	2.55	-3.5^v	1^d
	$ML_2/M.L^2$	$4.90^d \pm 0.05$	4.79 ± 0.01	4.56	-7.0^v	-1^d
	$ML_3/M.L^3$	$6.32^d \pm 0.00$	6.16 ± 0.08	5.90	-10.5^v	-6^d
	$ML_4/M.L^4$	$7.38^d \pm 0.08$	7.1 ± 0.1	6.74	-14.0^v	-13^d
	$ML_5/M.L^5$	7.02^d	6.9		-17.5^v	-27^d
	$ML_6/M.L^6$	5.41^d			-21.0^v	-46^d

---

[d] 25°, 2.0; [g] 25°, 5.0; [h] 20°, 0.1; [t] 23°, 2.0; [u] 30°, 2.0; [v] 27°, 2.0; [w] 30°, 1.0; [x] 18°, 2.0; [y] 18–25°, 2.0

Ammonia (continued)

Metal ion	Equilibrium	Log K 25°, 0.5	Log K 25°, 1.0	Log K 25°, 0	ΔH 25°, 0	ΔS 25°, 0
Hg^{2+}	$ML/M.L$	8.8^t				
	$ML_2/M.L^2$	17.4^d	(17.8)		-24.7^v	-3^d
	$ML_3/M.L^3$	18.4^d			-28.0^v	-10^d
	$ML_4/M.L^4$	$(19.3)^a$				
		19.1^d			-31.6^v	-19^d
Au(III)	$ML_4/ML_3.L$		10.3			
	$ML_4/MH_{-1}L_4.H$		7.48			

---

^{a} 25°, 0.1; ^{d} 25°, 2.0; ^{t} 23°, 2.0; ^{v} 27°, 2.0

Bibliography:

H^+ 30HO,34Oa,37P,41B,49BP,50BL,50BP,
       53Sd,54EL,57KD,65PSV,67KZ,68RJ,68VK,
       69ES,72VK,73CP,74RO

Li^+-Ca^{2+} 41B,54W

Mn^{2+} 72KB

Co^{2+} 41B,58YM,66LM

Ni^{2+} 41B,43DV,57YMa,59Sb,66LM

Cu^{2+} 31B,32B,34B,41B,44B,44Na,53SPa,57YMa,
          58E,59Sc,69ES,73CP

Co^{3+} 41B,49YP

Cu^+ 34B

Ag^+ 00BD,37SBP,41B,41DS,43VM,44KN,47N,
        55Fa,66Ja

CH_3Hg^+ 65SS,74RO

Tl^+ 41B

Pd^{2+} 68RJ

Zn^{2+} 41B,53SPa,57YM,66LM,72BP

Cd^{2+} 41B,43L,43DV,53SP,57YM,58E,72BPa

Hg^{2+} 41B,57YM,62TR,64WD

Au^{3+} 74SB

Other references: 01B,02BF,03E,07K,07L,
          10NK,20LL,24Ka,25B,25W,28J,30K,30RH,
          33AT,33BW,33T,34LS,35BW,36C,36SE,38EW,
          40SF,44C,49J,49SB,51KL,51S,52Fb,52YG,
          53LK,53LU,53Ya,54L,54LP,55SG,57TSa,
          58CPC,58L,59Pa,60MT,61F,61KY,61LP,
          61ML,61Sa,61WL,62B,64SA,65FK,65MB,
          65RP,66FL,66GC,66Mb,67FH,67HL,67LK,
          68GS,68QM,69BL,70GH,70GHa,70La,70MA,
          71SS,73LG,73SB,74FS

$$H_2NNH_2$$

H_4N_2			Hydrazine			L

Metal ion	Equilibrium	Log K 25°, 0.5	Log K 25°, 1.0	Log K 25°, 0	ΔH 18°, ~0	ΔS 25°, 0
H^+	HL/H.L	8.06 ±0.02	8.18	7.98 ±0.01	-9.7	4
		8.48[d]				
	H_2L/HL.H			-0.9[o]		
Co^{2+}	ML/M.L		1.57			
	$ML_2/M.L^2$		2.15			
	$ML_3/M.L^3$		3.09			
Ni^{2+}	ML/M.L	2.76[i]				
	$ML_2/M.L^2$	5.20[i]				
	$ML_3/M.L^3$	7.35[i]				
	$ML_4/M.L^4$	9.20[i]				
	$ML_5/M.L^5$	10.75[i]				
	$ML_6/M.L^6$	11.99[i]				
Zn^{2+}	ML/M.L	2.4[i]				
	$ML_2/M.L^2$	4.2[i]				
	$ML_3/M.L^3$	5.5[i]				

[d] 25°, 2.3; [i] 20°, 0.5 $N_2H_5BF_4$; [o] 20°, 0

Bibliography:
H^+ 00Ba,36S,36WS,67SL,70AB,73KN,74KN Ni^{2+},Zn^{2+} 52SZ
Co^{2+} 73SS   Other references: 00Bb,28H,29G,41Y,53RL,
                                                      54R,54Sd,67BS,72AG,72AK

HONH$_2$

H$_3$ON   Hydroxylamine   L

Metal ion	Equilibrium	Log K 25°, 0.5	Log K 25°, 1.0	Log K 25°, 0	ΔH 25°, 0	ΔS 25°, 0
H$^+$	HL/H.L	6.00	6.06	5.96 ±0.02	-9.3 +0.1	-4
		6.21[d]				
Mn^{2+}	ML/M.L	0.53[i]				
Co^{2+}	ML/M.L	0.93[i]				
Ni^{2+}	ML/M.L	1.47[i]				
Cu^{2+}	ML/M.L	2.42[i]				
	ML$_2$/M.L^2	4.1[i]				
VO$_2^+$	ML/M.L		1.10			
Ag$^+$	ML/M.L	1.85[i]				
Zn^{2+}	ML/M.L	0.48[i]	0.40			
	ML$_2$/M.L^2		1.01			

---

[d] 25°, 2.2; [i] 20°, 0.5

Bibliography:

H$^+$ 00B,00T,270a,41H,61RB,63S,65LL   Zn^{2+} 55N, 63S

Mn^{2+}-Cu^{2+},Ag$^+$ 63S   Other references: 01W,40IA,57MR,57MRH,

VO$_2^+$ 73B   61Kb,65Fa,66FPS,66GS,68JD,68S,74IS

HN_3

HN_3		Hydrogen azide	(hydrazoic acid)			HL
Metal ion	Equilibrium	Log K 25°, 0.5	Log K 25°, 1.0	Log K 25°, 0	ΔH 25°, 0	ΔS 25°, 0
H^+	HL/H.L	4.38	4.44 ±0.00	4.65 ±0.02	-3.6	9
		4.45[a]	4.78[e]	4.99[f]	-3.1[c]	10[c]
	HL/HL(g)			1.08		
Co^{2+}	ML/M.L		0.72			
Ni^{2+}	ML/M.L		0.86 ±0.02		-0.2[c]	3[c]
			1.04[e]			
	$ML_2/M.L^2$		1.3		-1.0[c]	2[c]
	$ML_3/M.L^3$		1.3		-3.5[c]	-6[c]
Cu^{2+}	ML/M.L	2.44[h]		2.86[o]		
				2.56[f]		
	$ML_2/M.L^2$			4.48[f]		
	$ML_3/M.L^3$			6.11[f]		
	$ML_4/M.L^4$			7.82[f]-0.01		
	$M.L^2/ML_2(s)$				3.6	
Fe^{3+}	ML/M.L	4.49[a]±0.07	4.20 ±0.1	4.85 ±0.04	(-2)[r]	(16)
Cu^+	$ML_3/M.L^3$			7.76[f]		
	M.L/ML(s)			-8.31		
Ag^+	ML/M.L			2.49[f]		
	$ML_2/M.L^2$			4.2[f]		
	$ML_3/M.L^3$			4.2[f]		
	$ML_4/M.L^4$			3.7[f]		
	M.L/ML(s)			-8.56 ±0.02	16.7	20
				-8.80[f]		
Hg_2^{2+}	$M.L^2/ML_2(s)$			-9.15	29.9	58
Tl^+	ML/M.L			0.39	-1.3	-3
	M.L/ML(s)			-3.66	11.1	21
Zn^{2+}	ML/M.L	0.78[d]	0.76		0.6[c]	6[c]
	$ML_2/M.L^2$	1.34[d]	1.3		1.2[c]	10[c]
	$ML_3/M.L^3$	2.3[d]	2.2		3.0[c]	20[c]
	$ML_4/M.L^4$	2.9[d]	2.4		-1.8[c]	5[c]

[a] 25°, 0.1; [c] 25°, 1.0; [d] 25°, 2.0; [e] 25°, 3.0; [f] 25°, 4.0; [h] 20°, 0.1; [o] 20°, 0; [r] 20-58°,0

Hydrogen azide (continued)

Metal ion	Equilibrium	Log K 25°, 0.5	Log K 25°, 1.0	Log K 25°, 0	ΔH 25°, 0	ΔS 25°, 0
Cd^{2+}	$ML/M.L$	1.4^d	1.61^e		-1.2^e	3^e
	$ML_2/M.L^2$	2.6^d	2.78^e		-2.6^e	4^e
	$ML_3/M.L^3$	2.9^d	3.2^e		-4.3^e	0^e
	$ML_4/M.L^4$	3.0^d	3.9^e		-5.6^e	-1^e
Hg^{2+}	$ML/M.L$	7.42^a	6.98	7.80	-7.3^c	8^c
	$ML_2/M.L^2$	14.63^a	14.39	15.36	-16.1^c	12^c
Pd^{2+}	$M.L^2/ML_2(s,\alpha)$			-8.57 ± 0.03	15.9	14
Tl^{3+}	$ML/M.L$	3.00^k			$(-2)^s$	$(7)^k$
	$ML_2/M.L^2$	5.38^k			$(-5)^s$	$(8)^k$
	$ML_3/M.L^3$	6.90^k			$(-13)^s$	$(-12)^k$

---

[a] 25°, 0.1; [c] 25°, 1.0; [d] 25°, 2.0; [e] 25°, 3.0; [k] 20°, 2.0; [s] 13-50°, 2.0

Bibliography:

H^+	41Ya,56GW,59BC,61BD,63DW,66BK,67MR, 72NS,76AA,	Tl^+	52Sa,56GW,57NN
Co^{2+}	70SG	Zn^{2+}	70DSa,76AA
Ni^{2+}	67MRa,70SG,76AA	Cd^{2+}	43L,61SN,66Gb,76AA
Cu^{2+}	56GW,58SO,71N,72NS,72SN	Hg^{2+}	65MK,76AA
Fe^{3+}	61BD,61WD,65MK,67CE,76AA	Pb^{2+}	52Sc,54FS,56GW
Cu^+	53Sb,72SN	Tl^{3+}	65V
Ag^+	38TN,52Sa,54LS,56GW		
Hg_2^{2+}	52Sc,56GW		

Other references: 00H,00W,16O,28H,32BR,40Q, 57BPa,59ES,61NP,61SA,62BS,68DS,73AA, 74Pb

$$HNO_2$$

| HO$_2$N | | Hydrogen nitrite | (nitrous acid) | | | HL |

Metal ion	Equilibrium	Log K 25°, 0.5	Log K 25°, 1.0	Log K 25°, 0	ΔH 25°, 0	ΔS 25°, 0
H$^+$	HL/H.L	2.94 3.24[d]	3.00	3.15	(-2)[r]	(7)
Cu^{2+}	ML/M.L	1.34 1.26[d]	1.19 +0.01 1.36[e]	2.02		
	ML$_2$/M.L^2	1.68 1.45[d]	1.43 ±0.01 1.54[e]	3.03		
Ag$^+$	ML/M.L	1.70 1.31[k]	1.56[j] 1.14[l]	2.32	(-7)[r]	(-14)
	ML$_2$/M.L^2	2.07 2.06[k]	2.15[j] 2.00[l]	2.51	(-11)[r]	(-26)
	M.L/ML(s)	-3.72[i] -3.36[k]	-3.57[j] -3.22[l]	-4.13	(15)[r]	(30)
Tl$^+$	ML/M.L			0.83		
Pd^{2+}	ML$_4$/M.L^4	20.3	21.6 +0.3			
Cd^{2+}	ML/M.L	 1.78[d]	1.7 1.80[e]	2.4	 -2.1[e]	 1[e]
	ML$_2$/M.L^2	2.9[d]	3.0[e]		-4.2[e]	-1[e]
	ML$_3$/M.L^3	 3.5[d]	3.1[t] 3.8[e]		 -5.8[e]	 -2[e]
Pb^{2+}	ML/M.L	1.89 1.91[d]	1.90	2.51		
	ML$_2$/M.L^2	2.7[u]	2.4[t]	3.0[v]		
	ML$_3$/M.L^3	3.0[u]		3.2[v]		

[d] 25°, 2.0; [e] 25°, 3.0; [i] 20°, 0.5; [j] 20°, 1.0; [k] 20°, 2.0; [l] 20°, 3.0; [r] 15-35°, 0; [t] 30°, 1.0; [u] 25°, 2.5; [v] 20°, 3.8

Bibliography:

H$^+$ 65LL,65LT,68TL   Cd^{2+} 43L,58VE,61TB,65JG,65SGa,66Gb

Cu^{2+} 46KS,51F,71T   Pb^{2+} 61TB,64GA,67JG,71TL

Ag$^+$ 72TL   Other references: 00S,02B,06Ba,29KH,37SM,

Tl$^+$ 57NBC   57H,58TW,59VK,60SW,60Ta,63BW,64PS,66SNb,

Pd^{2+} 65FK,70SS   67G,67SK,71Ta,71Tb,73CZ,73T

$$NO_3^-$$

O_3N^- <u>Nitrate ion</u> L^-

Metal ion	Equilibrium	Log K 25°, 0.5	Log K 25°, 1.0	Log K 25°, 0	ΔH 25°, 0	ΔS 25°, 0
Na^+	ML/M.L			$(-0.6)^n$		
K^+	ML/M.L	-0.37^a		$(-0.15)\pm0.05$	$(-3)^r$	(-12)
Cs^+	ML/M.L			$(0.01)^n$		
Be^{2+}	ML/M.L			-0.6^f		
Ca^{2+}	ML/M.L	0.06 -0.02^d	-0.06 0.04^e	0.7 -0.5 0.08^f	$(-6)^s$	(-17)
	$ML_2/M.L^2$	-0.3 -0.4^d	-0.5 -0.4^e	0.6 -0.4^f		
Sr^{2+}	ML/M.L	(0.06) 0.06^d	0.05 0.08^e	0.8 ±0.0 0.10^f	-2.4^b	-8^b
	$ML_2/M.L^2$	(-0.5) -0.2^d	-0.3 $(-0.4)^e$	0.8 -0.2^f		
Ba^{2+}	ML/M.L	0.21 0.14^d	0.16 0.20^e	0.9 ±0.1 0.24^f	-3.2^b	-10^b
	$ML_2/M.L^2$	0.1 0.0^d	0.0 $(-0.1)^e$	1.0 0.0^f		
Sc^{3+}	ML/M.L			0.28^f		
	$ML_2/M.L^2$			-0.3^f		
La^{3+}	ML/M.L		0.1 $+0.1$			
Ce^{3+}	ML/M.L		0.2 -0.1			
Pr^{3+}	ML/M.L		0.2			
Nd^{3+}	ML/M.L		0.3			
Pm^{3+}	ML/M.L		0.39 -0.1			
Sm^{3+}	ML/M.L		0.3			
Eu^{3+}	ML/M.L	0.44 0.26^d	0.31 ±0.0	1.23 0.2^f	$(-1)^t$	$(-1)^c$
	$ML_2/M.L^2$			-0.6^f		
Gd^{3+}	ML/M.L		0.0			

---

^{a} 25°, 0.1; ^{b} 25°, 0.5; ^{c} 25°, 1.0; ^{d} 25°, 2.0; ^{e} 25°, 3.0; ^{f} 25°, 4.0; ^{n} 18°, 0;
^{r} 25-39°, 0; ^{s} 18-25°, 0; ^{t} 0-55°, 1.0

Nitrate ion (continued)

Metal ion	Equilibrium	Log K 25°, 0.5	Log K 25°, 1.0	Log K 25°, 0	ΔH 25°, 0	ΔS 25°, 0
Tb^{3+}	ML/M.L	0.10	0.05 -0.1	0.88		
Dy^{3+}	ML/M.L		-0.3			
Ho^{3+}	ML/M.L		-0.2			
Er^{3+}	ML/M.L		-0.3			
Tm^{3+}	ML/M.L		-0.25 -0.1			
Yb^{3+}	ML/M.L		-0.2			
Lu^{3+}	ML/M.L		-0.2			
Ac^{3+}	ML/M.L		0.1			
	$ML_2/M.L^2$		0.0			
Am^{3+}	ML/M.L	0.20[d]	0.26 ±0.00			
Ce^{4+}	ML/M.L			0.32[u]		
Th^{4+}	ML/M.L	0.67		0.45[g]		
	$ML_2/M.L^2$			0.15[g]		
U^{4+}	ML/M.L	0.20[d]	0.28[e]	1.6		
	$ML_2/M.L^2$	0.2[d]	0.3[e]			
	$ML_3/M.L^3$	0.0[d]	0.2[e]			
	$ML_4/M.L^4$	-0.5[d]	-0.2[e]			
Np^{4+}	ML/M.L	0.34[d]	0.38	1.7[o]		
	$ML_2/M.L^2$	0.2[d]	0.1[j]			
	$ML_3/M.L^3$		-0.3[j]			
Pu^{4+}	ML/M.L	0.46[d]	0.54	1.8		
				0.74[m]		
	$ML_2/M.L^2$			1.37[m]		
	$ML_3/M.L^3$			1.2[m]		
NpO_2^+	ML/M.L	-0.25[d]				
UO_2^{2+}	ML/M.L	-0.6[d]	-0.3[j]		(-3)[v]	(-13)[d]
NpO_2^{2+}	ML/M.L	-0.9				
		-0.4[d]				

---

[d] 25°, 2.0; [e] 25°, 3.0; [g] 25°, 6.0; [j] 20°, 1.0; [m] 20°, 4.0; [o] 20°, 0; [u] 23°, 3.5; [v] 10-40°, 2.0

Nitrate ion (continued)

Metal ion	Equilibrium	Log K 25°, 0.5	Log K 25°, 1.0	Log K 25°, 0	ΔH 25°, 0	ΔS 25°, 0
Mn^{2+}	ML/M.L	-0.38	-0.43	0.2		
		-0.41^d	-0.24^e	-0.14^f		
	$ML_2/M.L^2$	-0.3	-0.6	0.6		
		-0.9^d	-0.8^e	-0.7^f		
Co^{2+}	ML/M.L	(-0.46)	-0.46	0.2		
		-0.48^d	-0.60^e	-0.38^f		
	$ML_2/M.L^2$	-0.3	-0.4			
		-0.6^d	-0.6^e	-0.4^f		
Ni^{2+}	ML/M.L		-0.22	0.4		
		-0.44^d	-0.55^e	-0.30^f		
	$ML_2/M.L^2$	-0.5^d	-0.9^e	-0.6^f		
Cu^{2+}	ML/M.L	(-0.13)	-0.01	0.5		
		-0.06^d	-0.02^e	0.11^f		
	$ML_2/M.L^2$		-0.6	-0.4		
		-0.6^d	-0.5^e	-0.4^f		
	$M.(OH)^{1.5}.L^{0.5}/M(OH)_{1.5}L_{0.5}(s)$			-16.37		
Fe^{3+}	ML/M.L	-0.22^q	-0.5	1.00	$(-9)^y$	$(-30)^c$
Zr^{4+}	ML/M.L	0.3^d		0.34^m		
	$ML_2/M.L^2$			0.1^m		
	$ML_3/M.L^3$			-0.3^m		
	$ML_4/M.L^4$			-0.8^m		
Hf^{4+}	ML/M.L	0.34^d		0.40^m		
	$ML_2/M.L^2$	0.0^d		0.1^m		
	$ML_3/M.L^3$	-0.7^d				
VO_2^+	ML/M.L		-0.5^j			
Ag^+	ML/M.L	-0.34^d		(-0.2) ±0.1		
Hg_2^{2+}	ML/M.L	0.08	0.02^e			
	$ML_2/M.L^2$		-0.3^e			
Tl^+	ML/M.L		-0.48^e	0.33 ±0.00	-0.7	-1

---

^{c} 25°, 1.0; ^{d} 25°, 2.0; ^{e} 25°, 3.0; ^{f} 25°, 4.0; ^{j} 20°, 1.0; ^{m} 20°, 4.0; ^{q} 20°, 0.6; ^{y} 10-40°, 1.0

Nitrate ion (continued)

Metal ion	Equilibrium	Log K 25°, 0.5	Log K 25°, 1.0	Log K 25°, 0	ΔH 25°, 0	ΔS 25°, 0
Zn^{2+}	$ML/M.L$	-0.18	-0.19	0.4		
		-0.14^d	0.01^e	0.11^f		
	$ML_2/M.L^2$		-0.6	-0.3		
		-0.8^d	-1.1^e	-0.8^f		
Cd^{2+}	$ML/M.L$	-0.11	-0.05	0.5 -0.1	-5.2	-15
		0.02^d	$0.04^e \pm 0.07$	0.08^f		
	$ML_2/M.L^2$		-0.8	0.2		
		-0.4^d	$(-0.6)^e$	0.0^f		
Hg^{2+}	$ML/M.L$		0.11^e			
	$ML_2/M.L^2$		0.0^e			
Pb^{2+}	$ML/M.L$	0.25	0.33 ± 0.02	1.17 ± 0.02	-0.6	3
		$0.40^d \pm 0.1$	$0.51^e \pm 0.06$		-1.3^e	-2^e
	$ML_2/M.L^2$	0.4	0.4 ± 0.1	1.4		
		$0.4^d \pm 0.2$	$0.4^e \pm 0.1$		-1.6^e	-4^e
	$ML_3/M.L^3$	$0.1^d \pm 0.1$	$0.2^e \pm 0.1$		$(-2)^w$	$(-6)^e$
	$ML_4/M.L^4$		$-0.3^e \pm 0.2$		$(-8)^w$	$(-28)^e$
In^{3+}	$ML/M.L$	0.18^q				
	$ML_2/M.L^2$	-0.3^q				
Tl^{3+}	$ML/M.L$		0.90^e		0.0^e	4^e
	$ML_2/M.L^2$		0.1^e			
	$ML_3/M.L^3$		1.1^e			
Bi^{3+}	$ML/M.L$	0.72	0.81	1.7		
		$(0.72)^d$	$(0.72)^e$	0.92^f	$(3)^x$	$(13)^d$
	$ML_2/M.L^2$	(0.94)	0.90	2.5		
		0.98^d	0.96^e	1.23^f		
	$ML_3/M.L^3$		0.7			
		$(0.2)^d$	$(0.1)^e$	1.1^f		
	$ML_4/M.L^4$	$(0.6)^d$	$(-0.2)^e$	0.4^f		
	$M.L/MOL(s).H^2$			-2.55		

^{d} 25°, 2.0; ^{e} 25°, 3.0; ^{f} 25°, 4.0; ^{q} 20°, 0.7; ^{w} 2-64°, 3.0; ^{x} 5-55°, 2.0

Nitrate ion (continued)

Bibliography:

Na^+ 27D

K^+ 27D,27O,37RD,66CL

Cs^+ 31BR

Be^{2+} 71SK

$Ca^{2+}-Ba^{2+}$ 30RD,63VV,64V,66MB,74FRa

Sc^{3+} 66SH

$La^{3+}-Lu^{3+},Am^{3+}$ 62PM,65CS,67K,67SS,69MK

Ac^{3+} 68SMR

Ce^{4+} 65PF

Th^{4+} 50DS,51ZA,56FM

U^{4+} 62EK,66SN

Np^{4+} 58SPS,66RY,66SN

Pu^{4+} 49Ha,51RL,60GN,66SN

NpO_2^+ 64GS

UO_2^{2+} 51Aa,54DP,59VN

NpO_2^{2+} 66RY,70AW

Mn^{2+} 74FR

Co^{2+},Ni^{2+} 73FS

Cu^{2+} 49NT,73FR

Fe^{3+} 51ID,52S,59M,69MS

Zr^{4+} 49CMa,57S

Hf^{4+} 63PA,69HS

VO_2^+ 66Ia

Ag^+ 27D,31BR,37RD,46OA

Tl^+ 37RD,57NN,65KMa

Zn^{2+} 73FR

Cd^{2+} 30RD,41L,61V,62V,74FRP

Hg_2^{2+},Hg^{2+} 46IS

Pb^{2+} 30RD,53HS,55BPR,55Na,56BD,63MK,63MKc,
          65Ha,67FR,69FRa,72FR

In^{3+} 68FD

Tl^{3+} 65KY,67MK

Bi^{3+} 51SG,71FKM

Other references: 01E,28HE,36RR,37R,38R,
          43RB,45Na,49BM,49ZN,51CM,51Mc,53Y,
          54Pd,55Kb,55Ra,56HS,56M,57BW,57MP,
          58FK,58MF,58PS,59CH,59ST,59T,59TC,
          59TS,60D,60HR,60LP,60PB,60PN,61Kc,
          61Ma,61NR,61TJ,62Hb,62MR,62NP,62PB,
          62SK,62ST,63Hc,63KB,63LK,63M,63NPa,
          63PF,63RSa,63SI,64BP,64DB,64FW,64HMF,
          64HP,64LP,64MW,64NKa,64S,64SD,64DK,
          65HD,64HS,65MS,66BA,66CK,66DO,66Gc,
          66R,67AS,67EME,67KR,67VD,67VG,68DF,
          68DP,68TR,69FR,69KM,69MF,69RP,69SGM,
          70AS,70AW,70HK,70KS,70LK,70MM,70PH,
          71GF,71M,71PJ,73Ab,73CDa,73HH,74FG,
          74M,74MS

$$H_2N_2O_2$$

$H_2O_2N_2$ <u>Hydrogen hyponitrite</u>   (<u>hyponitrous acid</u>) H_2L

Metal ion	Equilibrium	Log K 25°, 1.0	Log K 25°, 0	ΔH 25°, 0	ΔS 25°, 0
H^+	HL/M.L	10.85	11.54	$(-8)^r$	(26)
	H_2L/HL.H	6.75	7.18 ±0.03	$(-5)^s$	(16)
Ag^+	M^2.L/M_2L(s)		-18.89		

---

[r] 25-45°, 0; [s] 15-45°, 0

Bibliography:
H^+ 63BPa,63HS,63Pa Ag^+ 61PY

                                                Other references: 39LZ,59Pc

$$\begin{array}{c} O \\ \parallel \\ H-P-OH \\ \mid \\ H \end{array}$$

H_3O_2P <u>Hydrogen hypophosphite</u>   (<u>hypophosphorous acid</u>)   HL

Metal ion	Equilibrium	Log K 20°, 0.2	Log K 25°, 1.0	Log K 25°, 0	ΔH 25°, 0	ΔS 25°, 0
H^+	HL/H.L			1.23 ±0.2	1.6	11
Eu^{3+}	ML/M.L			2.27		
Cr^{3+}	ML.H/M.HL		1.32[r]			
Fe^{3+}	ML/M.L	4.01				
	$ML_2/M.L^2$	6.79				
	$ML_3/M.L^3$	8.96				
Zn^{2+}	ML/M.L		0.54[f]			
	$ML_2/M.L^2$		0.2[f]			

[f] 25°, 4.0; [r] 45°, 1.0

Bibliography:

H^+ 50Fa,54PM,59D
Eu^{3+} 64B
Cr^{3+} 66EB

Fe^{3+} 67MAN
Zn^{2+} 68HG

Other references: 20M,30Ma,34GM,37N,64NM,
                66CT,67MG,68LN

$$\begin{array}{c} O \\ \| \\ HO\!-\!P\!-\!OH \\ | \\ H \end{array}$$

H_3O_3P <u>Hydrogen phosphite</u>   <u>(phosphorous acid)</u> H_2L

Metal ion	Equilibrium	Log K 25°, 0.5	Log K 25°, 1.0	Log K 25°, 0	ΔH 25°, 0	ΔS 25°, 0
H^+	HL/H.L	6.08	6.01	6.79 +0.01	2.2	39
		6.34[a]	6.00[d]	6.06[e]		
	H_2L/HL.H	1.1[r]		1.5 ±0.1	2.2	14
Na^+	ML/M.L			1.05[o]		
	MHL/M.HL			0.96[o]		
K^+	ML/M.L			0.80[o]		
	MHL/M.HL			0.74[o]		
Ni^{2+}	MHL/M.HL	3.60[s]				
	MH_2L/M.H_2L	1.45[s]				
Cu^{2+}	ML_2/M.L^2		4.57[t]			
	M.L/ML(s)		-6.72[t]			
Fe^{3+}	MHL/M.HL			4.92		
	$M(HL)_2$/M.$(HL)^2$			7.84		
CH_3Hg^+	ML/M.L	4.67[h]				

[a] 25°, 0.1;   [d] 25°, 2.0;   [e] 25°, 3.0;   [h] 20°, 0.1;   [o] 20°, 0;   [r] 25°, 0.6;   [s] 25°, 0.2;   [t] 25°, 3.5

Bibliography:

H^+	30N,40GM,41TY,59D,68MS	Fe^{3+}	66MA
Na^+,K^+	64FP	CH_3Hg^+	65SS
Ni^{2+}	70EE	Other references:	20B,27K,30Ma,37N,50Fa,
Cu^{2+}	64N		65FP,66P,67PS,68HG

```
            O
            ||
      HO-P-OH
            |
            OH
```

H_3O_4P <u>Hydrogen phosphate</u>   (<u>phosphoric acid</u>) H_3L

Metal ion	Equilibrium	Log K 25°, 0.5	Log K 25°, 1.0	Log K 25°, 0	ΔH 25°, 0	ΔS 25°, 0
H^+	HL/H.L	11.74[a]±0.08	10.79[e]±0.07	12.35 ±0.02	-3.5 ±0.9	45
	H_2L/HL.H	6.57 ±0.05	6.46 ±0.02	7.199±0.002	-0.8 ±0.2	30
		5.72[a]±0.05	6.36[d]		-1.2[a]	
		6.79[r,s]	6.26[e]±0.02			
	H_3L/H_2L.H	1.72	1.70 ±0.02	2.148+0.001	1.9 ±0.1	16
		2.0[a] ±0.1	1.86[e]±0.03			
Li^+	MHL/M.HL	0.72[r,s]			(6)[r,t]	(23)[r,s]
Na^+	MHL/M.HL	0.60[r,s]			(8)[r,t]	(30)[r,s]
K^+	MHL/M.HL	0.49[r,s]			(6)[r,t]	(22)[r,s]
Mg^{2+}	ML/M.L		3.4[u]			
	MHL/M.HL	1.7[a]	1.8[u] ±0.0	2.91	3	23
		1.88[r,s]	1.42[e]			
	MH_2L/$M.H_2L$		0.7[u]			
			0.16[e]			
	$M^3.L^2$/$M_3L_2(H_2O)_8$(s)			-25.20		
	$M.HL$/MHL$(H_2O)_3$(s)			-5.82		
Ca^{2+}	ML/M.L			6.46	(3)[v]	(40)
	MHL/M.HL	1.50[s]	1.3[u]	2.74 -0.06	(3)[v]	(23)
		1.70[r,s]				
	MH_2L/$M.H_2L$		0.6[u]	1.4 -0.6	(3)[v]	(17)
	$M.HL$/MHL$(H_2O)_2$(s)			-6.58 ±0.03	(1)[w]	(-28)
Sr^{2+}	ML/M.L	(4.2)[h]				
	MHL/M.HL	1.2[h]				
		1.52[r,s]				
	MH_2L/$M.H_2L$	0.3[h]				
	$M.HL$/MHL(s)			-6.92[o]		
Ba^{2+}	$M.HL$/MHL(s)			-7.40[o]		
Y^{3+}	MH_2L/$M.H_2L$	1.84[x]		2.65		

[a] 25°, 0.1; [d] 25°, 2.0; [e] 25°, 3.0; [h] 20°, 0.1; [o] 20°, 0; [r] $(C_3H_7)_4$NCl used as background electrolyte; [s] 25°, 0.2; [t] 0-25°, 0.2; [u] 37°, 0.15; [v] 25-37°, 0; [w] 18-37°, 0; [x] 20°, 0.2

Hydrogen phosphate (continued)

Metal ion	Equilibrium	Log K 25°, 0.5	Log K 25°, 1.0	Log K 25°, 0	ΔH 25°, 0	ΔS 25°, 0
La^{3+}	$MH_2L/M.H_2L$	1.61				
	$M.L/ML(s)$	-22.43				
Ce^{3+}	$ML/M.L$			(18.52)		
	$MH_2L/M.H_2L$	1.52[x]		2.33		
Pm^{3+}	$MH_2L/M.H_2L$	1.69[x]		2.51		
Gd^{3+}	$M.L/ML(s)$	-22.26				
Ac^{3+}	$MH_2L/M.H_2L$	1.59				
Am^{3+}	$MH_2L/M.H_2L$	1.69[x]		2.51		
Th^{4+}	$MH_2L/M.H_2L$	3.96[d]				
	$M(H_2L)_2/M.(H_2L)^2$	7.5[d]				
UO_2^{2+}	$M^3.L^2/M_3L_2(s)$	-49.7[i]				
	$M.HL/MHL(s)$	-12.17[i]				
Fe^{2+}	$MHL/M.HL$			3.6		
	$MH_2L/M.H_2L$			2.7		
	$M^3.L^2/M_3L_2(H_2O)_8(s)$			-36.0		
Co^{2+}	$MHL/M.HL$	2.18[a]				
Ni^{2+}	$MHL/M.HL$	2.08[a]	2.00[y]			
Cu^{2+}	$MHL/M.HL$	3.2[a]	3.3[u]			
	$MH_2L/M.H_2L$		1.3[u]			
Fe^{3+}	$MHL/M.HL$	8.30				
	$MH_2L/M.H_2L$	3.47				
	$M.L/ML(H_2O)_2(s)$			-26.4		
VO^{2+}	$M^3.L^2/M_3L_2(s)$	-24.01[a]		-25.1		
Ag^+	$M^3.L/M_3L(s)$			-17.55		
Hg_2^{2+}	$M.HL/MHL(s)$			-12.40		
CH_3Hg^+	$MHL/M.HL$	5.03[h]				
Zn^{2+}	$MHL/M.HL$	2.4[a]	2.4[u]			
	$MH_2L/M.H_2L$		1.2[u]			
	$M^3.L^2/M_3L_2(H_2O)_4(s)$			-35.3		

[a] 25°, 0.1; [d] 25°, 2.0; [h] 20°, 0.1; [i] 20°, 0.3; [u] 37°, 0.15; [x] 20°, 0.2; [y] 15°, 0.1

Hydrogen phosphate (continued)

Metal ion	Equilibrium	Log K 25°, 0.5	Log K 25°, 1.0	Log K 25°, 0	ΔH 25°, 0	ΔS 25°, 0
Pb^{2+}	MHL/M.HL			3.1		
	$MH_2L/M.H_2L$			1.5		
	$M^3.L^2/M_3L_2(s)$			-43.53^z		
	M.HL/MHL(s)			-11.43		
Ga^{3+}	M.L/ML(s)		-21.0			
In^{3+}	$MH_2L/M.H_2L$		1.43^j			
	M.L/ML(s)		-21.63			

_____

^{j} 20°, 0.9; ^{z} 38°, 0

Bibliography:

H^+	29BU,32JP,37P,43BA,51B,55CD,56Bb,56MS, 56SA,57DS,57ET,57Ta,58Ga,59D,61DK,61VQ, 62CI,63GS,63Sb,63SG,64SSL,65HC,66CI, 66IT,67SB,69C,69MKb,69PN,70BS,71Pa, 72MR,74HH,74MB
Li^+-K^+	56SA
Mg^{2+}	54CC,54HP,56SAa,63TF,70C,74HH
Ca^{2+}	40GR,49Bb,53DH,56SAa,57SN,60M,66MG,68CM, 70C
Sr^{2+}	56SAa,62GG,66SM
Ba^{2+}	66SM
Y^{3+},Pm^{3+},Am^{3+}	66BE
La^{3+}	63TV,70RS
Ce^{3+}	50MS,66BE
Gd^{3+}	67TP
Ac^{3+}	70RS
UO_2^{2+}	65VP
Th^{4+}	51ZA
Fe^{2+}	72N
Co^{2+}	67SB
Ni^{2+}	67SB,72FS
Cu^{2+}	67SB,70C
Fe^{3+}	63GS,72Na
VO^{2+}	56ZK

Ag^+	70Bd
Hg_2^{2+}	49DC
CH_3Hg^+	65SS
Zn^{2+}	67SB,70C,73N
Pb^{2+}	32JP,72Nb
Ga^{3+}	64TC
In^{3+}	68DT

Other references: 03B,03RD,09AB,14MG,17B, 20B,20K,24PWa,25DS,25HL,25MM,26SN,27B, 27C,27K,27SH,29B,29JM,29Ka,29MJ,30HKa, 30Ma,31BD,31L,32BR,33N,34GS,34N,36SE, 40GR,42Ha,42LK,42TL,42W,44A,56G,45M, 46H,49Ka,50CJ,50Fa,51HM,51Z,51ZA,52TM, 53BS,53BSL,53GC,54BR,54HP,54TO,55C, 55KE,55KJ,56CS,57CJ,57D,57DS,57TV, 58ES,58KB,58KC,58Mb,58Mc,59LP,59SV, 60DM,60FSA,60GL,60MM,61BM,61BN,61CA, 61CAa,61EA,61ICa,61K,61KZ,61TG,61WL, 62AM,62F,62FE,62L,62ML,62RD,63G,63MG, 63PG,63UK,64DRC,64LA,64MP,64WE,65HSE, 65PE,65PT,66DM,66GM,66LA,67DS,67KPb, 67ME,67MSP,67WW,68Ba,68Ca,68MD,69BPa, 69IVa,70GM,70GS,70IV,70LS,71MM,73FA, 73IV,73NM,73RM,73SZ,74Fa,74FGA,74IK, 74RM

```
     O   O
     ||  ||
HO-P-O-P-OH
     |   |
    OH  OH
```

$H_4O_7P_2$		Hydrogen diphosphate	(pyrophosphoric acid)			H_4L

Metal ion	Equilibrium	Log K 25°, 0.1	Log K 25°, 1.0	Log K 25°, 0	ΔH 25°, 0	ΔS 25°, 0
H^+	HL/H.L	8.37 ±0.08	7.43 +0.07	9.40 ±0.1	-0.39 ±0.0	42
		9.00[r] ±0.05	8.74[r] ±0.04	7.17[e]		
	H_2L/HL.H	6.04 ±0.04	5.41 ±0.05	6.70 ±0.1	-0.13 ±0.0	30
		6.19[r] +0.07	5.98[r] ±0.00	5.29[e]		
	H_3L/H_2L.H	1.8	1.4 +0.1	2.2 ±0.1	1.0	13
		2.0[r] ±0.1	1.7[r] ±0.0	1.4[e]		
	H_4L/H_3L.H		0.8	0.8 ±0.1	1.5	9
		0.8[r]	0.7[r] ±0.1			
Li^+	ML/M.L			3.4 ±0.3	1.0	19
			2.39[r]		0.3[c,r]	12[c,r]
	MHL/M.HL			2.0	0.3	10
			1.03[r]		-0.2[c,r]	4[c,r]
Na^+	ML/M.L	0.21[d]		2.29 ±0.07	1.4	15
			1.00[r]		0.5[c,r]	6[c,r]
	M_2L/ML.M	-0.8[d]		1.9 ±0.6		
	MHL/M.HL	-0.5[d]		1.4 ±0.1		
K^+	ML/M.L			2.1	1.7	15
			0.80[r]		0.7[c,r]	6[c,r]
Mg^{2+}	ML/M.L	5.45	5.42[r] -0.01	7.2	3[t]	43
	$ML_2/M.L^2$		7.80[r] ±0.05			
	MHL/M.HL	3.18[s]	3.06[r] -0.01			
	MOHL/M.OH			2.1		
Ca^{2+}	ML/M.L	5.4[r]	4.9[r] ±0.0	6.8		
	MHL/M.HL	3.3[r]	2.3[r] -0.1			
	MOHL/M.OH			2.1		
	$ML.M/M_2L(s)$			-7.9		

---

[c] 25°, 1.0; [d] 25°, 2.0; [e] 25°, 3.0 Na^+; [r] $(CH_3)_4N$ salt used as background electrolyte; [s] 15°, 0.1; [t] 25°, var.

Hydrogen diphosphate (continued)

Metal ion	Equilibrium	Log K 25°, 0.1	Log K 25°, 1.0	Log K 25°, 0	ΔH 25°, 0	ΔS 25°, 0
Sr^{2+}	ML/M.L	3.26[h]		5.4		
	MOHL/ML.OH			2.3		
	$ML.M/M_2L(s)$			-7.5		
La^{3+}	ML/M.L			16.72		
	$ML_2/M.L^2$			18.57		
	$M_2L/M^2.L$			19.59 ±0.06		
Ce^{3+}	ML/M.L			17.15		
Nd^{3+}	$M_2L/M^2.L$			19.98		
Sm^{3+}	$M_2L/M^2.L$			20.16		
Eu^{3+}	$M_2L/M^2.L$			20.27		
Gd^{3+}	$M_2L/M^2.L$			20.45		
Tb^{3+}	$M_2L/M^2.L$			20.50		
Dy^{3+}	$M_2L/M^2.L$			20.64		
Ho^{3+}	$M_2L/M^2.L$			20.9		
Er^{3+}	$M_2L/M^2.L$			21.29		
Yb^{3+}	ML/M.L			17.5		
	$ML_2/M.L^2$			19.4		
	$M_2L/M^2.L$			21.88		
	$M^4.L^3/M_4L_3(s)$			(-75.0)		
Lu^{3+}	$M_2L/M^2.L$			22.23		
Ce^{4+}	ML/M.L	(18.41)				
	M.L/ML(s)	-23.36				
Co^{2+}	ML/M.L	6.1				
		7.36[r]				
	MHL/M.HL	3.4				
		4.07[r]				
Ni^+	ML/M.L	5.94			4.2[t]	41[a]
		7.01[r]				
	$ML_2/M.L^2$				2.0[t]	
	MHL/M.HL	(3.71)				
		3.81[r]				

[h] 20°, 0.1; [r] $(CH_3)_4N$ salt used as background electrolyte; [t] 25°, var

Hydrogen diphosphate (continued)

Metal ion	Equilibrium	Log K 25°, 0.1	Log K 25°, 1.0	Log K 25°, 0	ΔH 25°, 0	ΔS 25°, 0
Cu^{2+}	ML/M.L		7.6			
			9.07[r]			
	$ML_2/M.L^2$		12.45		(-0.7)[t]	(55)[c]
			16.65[r]			
	MHL/M.HL		4.45			
			5.37[r]			
	$MH_2L/M.H_2L$		1.99			
			2.55[r]			
	$MHL_2/ML_2.H$		4.9			
			6.61[r]			
	$MH_2L_2/MHL_2.H$		4.7			
			5.63[r]			
	$MH_3L_2/MH_2L_2.H$		3.7			
			4.25[r]			
	$MH_4L_2/MH_3L_2.H$		2.7			
			3.06[r]			
	$MH_5L_2/MH_4L_2.H$		1.7			
Mn^{3+}	ML/M.L	16.68[u]				
	$ML_2/M.L^2$	31.85[u]				
	$MH_2L/M.H_2L$	5.11[u]				
	$M(H_2L)_2/M.(H_2L)^2$	8.41[u]				
	$M(H_2L)_3/M.(H_2L)^3$	11.24[u]				
Hg_2^{2+}	$ML_2/M.L^2$		12.38[v]			
	MOHL/M.OH.L		15.64[v]			
Tl^+	ML/M.L	1.69[w]				
	$ML_2/M.L^2$	1.9[w]				
Zn^{2+}	ML/M.L			8.7		
	$ML_2/M.L^2$			11.0	2.6[t]	59
	MOHL/ML.OH			4.4		
Cd^{2+}	ML/M.L			(8.7)		
	MOHL/ML.OH			3.1		
Hg^{2+}	MOHL/M.OH.L		17.45[v]			

[c] 25°, 1.0; [r] $(CH_3)_4N$ salt used as background electrolyte; [t] 25°, var.; [u] 25°, 0.3; [v] 27°, 0.75; [w] 35°, 2.0

Hydrogen diphosphate (continued)

Metal ion	Equilibrium	Log K 25°, 0.1	Log K 25°, 1.0	Log K 25°, 0	ΔH 25°, 0	ΔS 25°, 0
Pb^{2+}	ML/M.L		7.3			
	$ML_2/M.L^2$		10.15		$(-1.0)^t$	$(43)^c$

[c] 25°, 1.0; [t] 25°, var.

Bibliography:

H^+	28KB,49M,50SZ,54BR,55D,57LW,59WO,60N, 61I,61ICa,63JW,64HM,64WS,66IT,66MM,68BC, 68MHB,70VA,70VAa,72FS,73PS,73VAK
Li^+-K^+	49M,55D,57LW,59WO,73VA
Mg^{2+}	57LWa,57V,59WO,61I,61IC,72FS
Ca^{2+}	59WL,59WO,60IC
Sr^{2+}	59WO,62GG
$La^{3+},Nd^{3+}-Lu^{3+}$	66SS,67SSc
Ce^{3+}	50MS
Ce^{4+}	67MSK
Co^{2+}	63JW,64HM
Ni^{2+}	56YVa,64HM,73PS
Cu^{2+}	56YVa,63SS,68BC
Mn^{3+}	70GSM
Hg_2^{2+}	59YD,60YD
Tl^+	52SD
Zn^{2+}	56YVa,59WO
Cd^{2+}	59WO
Hg^{2+}	60YD
Pb^{2+}	56YVa,68CFa

Other references: 09AB,28M,30Ma,32Ma,47SF, 49E,49RR,50Ha,50LO,50VC,53GC,53LU, 53WA,54GC,54UL,56UL,56YV,56YVb,58O, 58PT,58VR,60FSA,60FT,60Oa,62AM,62NM, 64GL,64SSa,64WE,65BCY,65HS,65PE,65SMc, 66ASS,66GL,66MI,66VV,67MNU,67SA,68MSb, 68PV,69SA,71BSb,71MM,72BPb,72LG,73RM

$$\begin{array}{c} \text{O} \quad\; \text{O} \quad\; \text{O} \\ \| \quad\; \| \quad\; \| \\ \text{HO--P--O--P--O--P--OH} \\ | \qquad | \qquad | \\ \text{OH} \quad \text{OH} \quad \text{OH} \end{array}$$

| $H_5O_{10}P_3$ | | Hydrogen triphosphate | (triphosphoric acid) | | H_5L |

Metal ion	Equilibrium	Log K 25°, 0.1	Log K 25°, 1.0	Log K 25°, 0	ΔH 25°, 0	ΔS 25°, 0
H^+	HL/H.L	8.00 ±0.1		9.25 ±0.01	-0.1	42
		8.70[r] ±0.1	8.61[r]±0.05		-0.1[h,r]	40[a,r]
	H_2L/HL.H	5.50 ±0.1		9.54 ±0.07	1.4 ±0.1	35
		5.90[r] ±0.1	5.69[r]-0.01			
	H_3L/H_2L.H	(2.6)		2.5 ±0.3		
		2.2[r] -0.1	2.0[r] ±0.0			
	H_4L/H_3L.H		1.0[r] ±0.1			
Li^+	ML/M.L		2.87[r]	3.9		
	MHL/M.HL		1.88[r]			
Na^+	ML/M.L		1.64[r]	2.7 ±0.1		
	MHL/M.HL		0.77[r]			
K^+	ML/M.L		1.39[r]			
Be^{2+}	ML/M.L				4,7[h,r]	
	MHL/ML.H	5.35[h,r]				
Mg^{2+}	ML/M.L	5.76 ±0.1		8.6		
		7.11[r]	5.82[r]±0.02		4.3[h,r]	47[a,r]
	MHL/M.HL	3.5 ±0.2				
		4.45[h,r]	3.35[r]±0.01			
	MOHL/ML.OH			2.4		
Ca^{2+}	ML/M.L	5.20 ±0.2		8.1		
		6.38[r] ±0.03	5.40[r]±0.04		3.3[h,r]	40[a,r]
	MHL/M.HL	3.04 +0.1				
		4.02[h,r]-0.2	2.9[r] ±0.1			
	MOHL/ML.OH			2.3		
Sr^{2+}	ML/M.L	4.10 ±0.2		7.2		
		5.50[r]			3.3[h,r]	36[a,r]
	MHL/M.HL	2.53 ±0.3				
		3.56[h,r]				
	MOHL/ML.OH			2.1		

[a] 25°, 0.1; [h] 20°, 0.1; [r] $(CH_3)_4N$ salt used as background electrolyte.

Hydrogen triphosphate (continued)

Metal ion	Equilibrium	Log K 25°, 0.1	Log K 25°, 1.0	Log K 25°, 0	ΔH 25°, 0	ΔS 25°, 0
Ba^{2+}	ML/M.L	3.3		6.3		
	MHL/M.HL	1.8				
	ML.M/M_2L(s)			-9.8		
Mn^{2+}	ML/M.L	7.15				
		8.08[r]			2.8[h,r]	46[a,r]
	MHL/M.HL	3.77				
		5.08[h,r]				
Co^{2+}	ML/M.L	6.94				
		8.01[r]+0.1			4.5[h,r]	52[a,r]
	MHL/M.HL	3.81				
		4.93[h,r]				
Ni^{2+}	ML/M.L	6.75				
		7.86[r]+0.04			5.0[h,r]	53[a,r]
	MHL/M.HL	3.65				
		4.9[h,r]				
Cu^{2+}	ML/M.L	8.3				
		9.36[r]			4.9[h,r]	59[a,r]
	MHL/M.HL	4.34				
		6.1[h,r]				
Hg_2^{2+}	ML_2/M.L^2		11.23[t]			
	MOHL/M.OH.L		15.00[t]			
Tl^+	ML/M.L		1.34[u]			
	ML_2/M.L^2		2.26[u]			
Zn^{2+}	M./M.L	7.5		(9.7)		
		8.43[r]			6.3[h,r]	60[a,r]
	MHL/M.HL	3.92				
		5.13[h,r]				
	MOHL/ML.OH			3.3		
Cd^{2+}	ML/M.L	6.58		9.8		
		8.13[r]			2.7[h,r]	46[a,r]
	MHL/M.HL	3.60				
		4.97[h,r]				
	MOHL/ML.OH			2.8		

[a] 25°, 0.1; [h] 20°, 0.1; [r] $(CH_3)_4N$ salt used as background electrolyte; [t] 27°, 0.75; [u] 20°, 2.4

Hydrogen triphosphate (continued)

Bibliography:

H^+ 54BR,56MS,56WL,57Lb,61I,61ICa,62CI,
           63JW,64EM,64HM,65Aa,65PE,66IT,68MHB,
           71TR,72FS,73TR

Li^+-K^+ 49Ma,57WL,59WO

Be^{2+} 65Aa

Mg^{2+}-Ba^{2+} 56MS,57LWa,59WL,59WO,60IC,61I,
           61ICa,62GG,63JW,64EM,65Aa,72FS

Mn^{2+}-Cu^{2+},Zn^{2+},Cd^{2+} 64EM,64HM,65Aa

Hg_2^{2+} 60YD

Tl^+ 74F

Other references: 49T,50VC,53GC,55Lb,56Kc,
           57Ka,57Mb,57PL,60G,62RK,62RKa,62SLW,
           63Ka,64SSa,64SSG,64WS,65KS,66MI,67ASa,
           68ASc,68SA,68WSa,71SB,74TR

$$
\begin{array}{cccc}
O & O & O & O \\
\parallel & \parallel & \parallel & \parallel \\
HO-P-O-P-O-P-O-P-OH \\
\mid & \mid & \mid & \mid \\
OH & OH & OH & OH
\end{array}
$$

$H_6O_{13}P_4$ <u>Hydrogen tetraphosphate</u>   (<u>tetraphosphoric acid</u>) H_6L

Metal ion	Equilibrium	Log K 25°, 1.0
H^+	$HL/H.L$	8.34^r
	$H_2L/HL.H$	$(6.63)^r$
	$H_3L/H_2L.H$	2.23^r
	$H_4L/H_3L.H$	1.4^r
Li^+	$ML/M.L$	$(2.64)^r$
	$MHL/M.HL$	$(1.59)^r$
Na^+	$ML/M.L$	1.79^r
	$MHL/M.HL$	1.10^r
K^+	$ML/M.L$	1.71^r
	$MHL/M.HL$	$(1.11)^r$
Mg^{2+}	$ML/M.L$	6.04^r
	$MHL/M.HL$	3.74^r
	$M_2L/ML.M$	2.19^r
Ca^{2+}	$ML/M.L$	$5.46^r-0.01$
	$MHL/M.HL$	$3.54^r-0.2$
	$M_2L/ML.M$	$3.07^r+0.3$
Sr^{2+}	$ML/M.L$	4.82^r
	$MHL/M.HL$	3.49^r
	$M_2L/ML.M$	3.42^r
Cu^{2+}	$ML/M.L$	9.44^r
	$ML_2/M.L^2$	10.60^r
	$MHL/M.HL$	6.66^r
	$MH_2L/M.H_2L$	3.48^r
	$MOHL/ML.OH$	3.86^r
	$MHL_2/ML_2.H$	8.40^r
	$MH_2L_2/MHL_2.H$	7.28^r
	$MH_3L_2/MH_2L_2.H$	4.52^r
	$MH_4L_2/MH_3L_2.H$	3.55^r

---

[r] $(CH_3)_4NCl$ used as background electrolyte.

Hydrogen tetraphosphate (continued)

Bibliography:

H^+ 63WS

Li^+-K^+ 67WM

$Mg^{2+}-Sr^{2+}$ 68WM,69WKa

Cu^{2+} 68WM

Other references: 68WS,68MHB

$H_3O_9P_3$ <u>Hydrogen trimetaphosphate</u>   (<u>trimetaphosphoric acid</u>) H_3L

Metal ion	Equilibrium	Log K 20°, 0.1	Log K 25°, 1.0	Log K 25°, 0
H^+	HL/H.L	1.35	0.65^r	2.05
Na^+	ML/M.L			1.40 -0.2
Mg^{2+}	ML/M.L	1.11		3.31
Ca^{2+}	ML/M.L	2.06	1.64^r	3.47 ±0.02
Sr^{2+}	ML/M.L	1.99^s±0.04		3.35
Ba^{2+}	ML/M.L			3.35
La^{3+}	ML/M.L			5.70
Pm^{3+}	ML/M.L	3.80^s		5.74
Am^{3+}	ML/M.L	3.48^s		6.06
Cm^{3+}	ML/M.L	3.64^s		5.92
Mn^{2+}	ML/M.L			3.57
Ni^{2+}	ML/M.L			3.22
Cu^{2+}	ML/M.L		1.58^r	
	$ML_2/H.L^2$		2.2^r	
Zn^{2+}	ML/M.L	1.94^s		

---

^{r} $(CH_3)_4NNO_3$ used as background electrolyte; ^{s} 20°, 0.2

Bibliography:
H^+ 49DM,49Z,69WK La^{3+} 52Ma
Na^+ 49DM,69GN Pm^{3+}-Cm^{3+} 67EL
Mg^{2+} 49JM,49Z Cu^{2+} 69WK
Ca^{2+} 49DM,49JM,69WKa,74KO Zn^{2+} 74KO
Sr^{2+} 52M,62GG,74KO   Other references: 53GC,58I,62RK,65IM,68WSa
Ba^{2+},Mn^{2+},Ni^{2+} 49JM

$$
\begin{array}{c}
\text{OH} \quad \text{O} \\
| \qquad \| \\
\text{O=P-O-P-OH} \\
| \qquad | \\
\text{O} \qquad \text{O} \\
| \qquad | \\
\text{HO-P-O-P=O} \\
\| \qquad | \\
\text{O} \qquad \text{OH}
\end{array}
$$

$H_4O_{12}P_4$ <u>Hydrogen tetrametaphosphate</u>   (<u>tetrametaphosphoric acid</u>) H_4L

Metal ion	Equilibrium	Log K 20°, 0.1	Log K 25°, 1.0	Log K 25°, 0
H^+	HL/H.L		1.53[r]	2.76 ±0.02
Na^+	ML/M.L		0.81[r,u]	2.10 ±0.05
Mg^{2+}	ML/M.L			5.17
Ca^{2+}	ML/M.L	3.28 2.90[s]	3.1[r]	5.37 ±0.05
	$ML_2/M.L^2$			8.02 ±0.1
Sr^{2+}	ML/M.L	2.80[t]		5.12
	$ML_2/M.L^2$			7.54
Ba^{2+}	ML/M.L			4.99
La^{3+}	ML/M.L			6.66
Mn^{2+}	ML/M.L			5.74
Co^{2+}	ML/M.L	2.62[s]		
Ni^{2+}	ML/M.L		2.63[r,u]	4.95
	$ML_2/M.L^2$		3.48[r,u]	
Cu^{2+}	ML/M.L		3.04[r] 3.18[r,u]	
	$ML_2/M.L^2$		4.28[r] 4.64[r,u]	
Zn^{2+}	ML/M.L	2.86[s]		

[r] $(CH_3)_4NNO_3$ used as background electrolyte; [s] 20°, 0.2; [t] 20°, 0.15; [u] 30°, 1.0

Bibliography:

H^+ 49DM,54BR,69WK La^{3+} 52Ma

Na^+ 49DM,55GG,69GN,72KO Co^{2+},Zn^{2+} 74KO

Mg^{2+},Ba^{2+},Mn^{2+} 50JMa Ni^{2+} 50JMa,55GG

Ca^{2+} 49DM,50JMa,69WKa,74KO Cu^{2+} 55GG,69WK

Sr^{2+} 52M,62GG   Other references: 53GC,62RK,65IM

```
              HO   O OH O
               |   \|/  ||
             O=P-O-P-O-P-OH
               |       |
               O       O
               |       |
          HO-P-O-P-O-P=O
             ||  /|\  ||
             O  HO O   OH
```

$H_6O_{18}P_6$ <u>Hydrogen hexametaphosphate</u>   (<u>hexametaphosphoric acid</u>) H_6L

Metal ion	Equilibrium	Log K 20°, 0.1	Log K 25°, 0.2	Log K 25°, 0
Na^+	ML/M.L			4.3
Ca^{2+}	ML/M.L	4.59	4.11	6.9[o]
Sr^{2+}	ML/M.L		3.73	
Co^{2+}	ML/M.L		3.65	
Zn^{2+}	ML/M.L		3.95	

[o] 20°, 0

Bibliography: Na^+ 72KO Ca^{2+}-Zn^{2+} 74KO

```
            HO   O OHO OH O
             |   \|/ \|/ ||
           O=P-O-P-O-P-O-P-OH
             |           |
             O           O
             |           |
        HO-P-O-P-O-P-O-P=O
           ||  /|\ /|\  ||
           O  HO OHO O   OH
```

$H_8O_{24}P_8$ <u>Hydrogen octametaphosphate</u>   (<u>octametaphosphoric acid</u>) H_8L

Metal ion	Equilibrium	Log K 20°, 0.1	Log K 25°, 0.2	Log K 25°, 0
Na^+	ML/M.L			4.6
Ca^{2+}	ML/M.L	5.18	4.62	8.1[o]
Sr^{2+}	ML/M.L		4.30	
Co^{2+}	ML/M.L		4.80	
Zn^{2+}	ML/M.L		5.02	

[o] 20°, 0

Bibliography: Na^+ 72KO Ca^{2+}-Zn^{2+} 74KO

$$
\begin{array}{ccc}
& O & & O \\
& \parallel & & \parallel \\
HO-&P-&NH-P-&OH \\
& \mid & & \mid \\
& OH & & OH
\end{array}
$$

$H_5O_6NP_2$		Hydrogen imidodiphosphate		(imidodiphosphoric acid)		H_4L

Metal ion	Equilibrium	Log K 25°, 0.1	Log K 25°, 1.0	Log K 25°, 0	ΔH 25°, 0	ΔS 25°, 0
H^+	HL/H.L	10.22	10.36			
	$H_2L/HL.H$	7.3	7.6			
	$H_3L/H_2L.H$	2.66	3.1			
	$H_4L/H_3L.H$	1.5	1.5			
Ca^{2+}	ML/M.L	5.59	4.59	6.1	(-6)[r]	(0)
	MHL/M.HL	3.33	3.15	3.4		

[r] 25-50°, 0

Bibliography: H^+ 61ICa Ca^{2+} 61IC

$$
\begin{array}{cccc}
& O & O & O \\
& \parallel & \parallel & \parallel \\
HO-&P-&NH-P-&NH-P-OH \\
& \mid & \mid & \mid \\
& OH & OH & OH
\end{array}
$$

$H_7O_8N_2P_3$		Hydrogen diimidotriphosphate		(diimidotriphosphoric acid)		H_5L

Metal ion	Equilibrium	Log K 25°, 0.1	Log K 25°, 1.0	Log K 25°, 0	ΔH 25°, 0	ΔS 25°, 0
H^+	HL/H.L	9.84	10.00			
	$H_2L/HL.H$	6.61	6.86			
	$H_3L/H_2L.H$	3.03	3.36			
	$H_4L/H_3L.H$	2	2.0			
	$H_5L/H_4L.H$	1	1.0			
Ca^{2+}	ML/M.L	6.7	5.7	7.1	(-9)[r]	(-6)
	MHL/M.HL	4.44	4.16	4.6		

[r] 25-50°, 0

Bibliography: H^+ 61ICa Ca^{2+} 61IC

$H_6O_6N_3P_3$ Cyclo-tri-μ-imidotris(dioxophosphate)   (trimetaphosphimic acid) H_3L

Metal ion	Equilibrium	Log K 20°, 0.1
H^+	L/H$_{-1}$L.H	12.19
	HL/H.L	3.23
Mg^{2+}	ML/M.L	1.28

Bibliography: 49Z

$H_4O_6P_2$ Hydrogen diphosphate (III,V)   (isohypophosphoric acid) H_3L

Metal ion	Equilibrium	Log K 25°, 0.5	Log K 25°, 1.0	Log K 25°, 0
H^+	HL/H.L	6.10[r]	6.19[r]	6.26
		6.14[a,r]		
	H_2L/HL.H	1.35[r]	1.57[r]	1.67
		1.56[a,r]		
Li^+	ML/M.L	0.82[r]		
Na^+	ML/M.L	0.50[r]		
K^+	ML/M.L	0.36[r]		
Mg^{2+}	ML/M.L	2.65[r]		
Ca^{2+}	ML/M.L	2.27[r]		

[a] 25°, 0.1; [r] $(CH_3)_4NCl$ used as background electrolyte.

Bibliography: 67CM

```
             O  O
             ||  ||
        HO-P-P-OH
             |  |
             HO  OH
```

$H_4O_6P_2$ <u>Hydrogen hypophosphate</u>   (<u>hypophosphoric acid</u>) H_4L

Metal ion	Equilibrium	Log K 25°, 0.1	Log K 25°, 0
H^+	HL/H.L	9.48	
	$H_2L/HL.H$	6.77	
	$H_3L/H_2L.H$	2.1	
Na^+	ML/M.L		2.31
	MHL/M.HL		1.32

Bibliography:

H^+ 50SZ Na^+ 67NS

```
             O      O
             ||      ||
        HO-P-O-O-P-OH
             |          |
             HO        OH
```

$H_4O_8P_2$ <u>Hydrogen peroxodiphosphate</u> H_4L

Metal ion	Equilibrium	Log K 25°, 1.0	Log K 25°, 0
H^+	HL/H.L	7.19[r]	7.68
	$H_2L/HL.H$	4.64[r]	5.18
Li^+	ML/M.L	1.34[r]	
	MHL/M.HL	0.70[r]	
Na^+	ML/M.L	1.02[r]	
	MHL/M.HL	0.25[r]	
K^+	ML/M.L	1.01[r]	
Mg^{2+}	ML/M.L	3.33[r]	
	MHL/M.HL	1.76[r]	
	$M_2L/M^2.L$	1.32[r]	

---

[r] $(CH_3)_4NCl$ used as background electrolyte.

Bibliography: 60CE

$$PF_6^-$$

F_6P^-		Hexafluorophosphate ion			L^-

Metal ion	Equilibrium	Log K 25°, 0	ΔH 25°, 0	ΔS 25°, 0
K^+	ML/M.L	0.38	$(-3)^r$	(-8)
	$M_2L_2/(ML)^2$	0.5		

^{r} 25-50°, 0

Bibliography: 61RSS

$$AsF_6^-$$

F_6As^-		Hexafluoroarsenate ion	L^-

Metal ion	Equilibrium	Log K 25°, 0
K^+	ML/M.L	(0.25)

Bibliography: 60AH

$$H-O-O-H$$

H_2O_2			Hydrogen peroxide			H_2L
Metal ion	Equilibrium	Log K 25°, 0.5	Log K 25°, 1.0	Log K 25°, 0	ΔH 25°, 0	ΔS 25°, 0
H^+	$H_2L/HL.H$		12.13^r	11.65 ± 0.03	-7.4^a	29
	$H_4L_2/(H_2L)^2$		-0.96^r			
	$H_4L_2/H_3L_2.H$		10.89^r			
	$H_3L_2/H_2L_2.H$		13.86^r			
Pu^{4+}	$M_2OHL.H^3/M^2.H_2L$	6.94				
	$M_2L_2.H^4/M^2.(H_2L)^2$	8.80				
UO_2^{2+}	$MHL/H.ML(s)$			-1.4		
	$MOHL/OH.ML(s)$			-2.0		
Fe^{3+}	$MHL/M.HL$	9.31^h			$(2)^s$	$(50)^h$
TiO^{2+}	$MH_2L/M.H_2L$		3.86 ± 0.0		-10.5^c	-18^c
		3.94^d	4.01^e			
	$M(H_2L)_2/M.(H_2L)^2$		6.3			
VO_2^+	$VOL/M.H_2L$		4.53^u			
$HCrO_4^-$	$CrO_5/M.(H_2L)^2$	7.73^v				
$Mo_2O_7^{2-}$	$MH_2L/M.H_2L$	3.42				
	$M(H_2L)_2.H^2/M.(H_2L)^2$	3.30				
$B(OH)_4^-$	$B(OH)_3HL/M.H_2L$	1.52		1.32		
	$B(OH)_2(HL)_2/M.(H_2L)^2$			1.53		
$GeO(OH)_3^-$	$M(H_2L)_2/M.(H_2L)^2$	1.68^a				
$H_5TeO_6^-$	$MH_2L/M.H_2L$	-0.15^a		-0.17		

^{a} 25°, 0.1; ^{c} 25°, 1.0; ^{d} 25°, 2.1; ^{e} 25°, 3.2; ^{h} 20°, 0.1; ^{r} 5°, 3.0; ^{s} 8-40°, 0.1; ^{u} 22°, 1.0; ^{v} 10°, 0.09

Bibliography:
H^+ 49EU,57MRa,57SM,64CH
Pu^{4+} 49CM
UO_2^{2+} 58GT
Fe^{3+} 49EG
TiO^{2+} 60Gb,68VV,70VV
VO_2^+ 61Da
$HCrO_4^-$ 57E

$Mo_2O_7^{2-}$ 69AY
$B(OH)_4^-$ 53Eb,55Ab,55Kd
$GeO(OH)_3^-$ 60AR
$H_5TeO_6^-$ 59EF,61AT
Other references: 12J,23Ma,29Kb,37Ra,
        51M,53EM,56TS,57BWa,57JS,59GJ,59Kd,
        60Ca,60CLa,63DL,63FL,63LR,64BR,65BY,
        670W

$$H_2S$$

<u>Hydrogen sulfide</u>   (<u>hydrosulfuric acid</u>)

Metal ion	Equilibrium	Log K 20°, 0.1	Log K 25°, 1.0	Log K 25°, 0	ΔH 25°, 0	ΔS 25°, 0
H^+	HL/H.L		13.8	13.9 ±0.1	-12	23
	H_2L/HL.H	6.83	6.61	7.02 ±0.04	-5.3	14
	H_2L/H_2L(g)			-0.99		
Mn^{2+}	M.L/ML(s,pink)			-10.5		
	M.L/ML(s,green)			-13.5 ±0.0	5	-45
Fe^{2+}	M.L/ML(s)			-18.1 -0.1	10	-49
Co^{2+}	M.L/ML(s,α)			-21.3		
	M.L/ML(s,β)			-25.6		
Ni^{2+}	M.L/ML(s,α)			-19.4		
	M.L/ML(s,β)			-24.9		
	M.L/ML(s,γ)			-26.6		
Cu^{2+}	M.L/ML(s)			-36.1 ±0.2	35	-48
Cu^+	M^2.L/M_2L(s)			-48.5 -0.5	52	-48
Ag^+	MHL/M.HL	13.6	13.30[j]			
	M(HL)$_2$/M.(HL)^2	17.7	17.17[j]			
	MHL/ML.H	8.3				
	M(HL)$_2$/MHL$_2$.H	9.5				
	$M_2H_2L_3 \cdot H_2L$/(M(HL)$_2$)^2		3.2[j]			
	M^2.L/M_2L(s)		-49.7[j]	-50.1 ±0.0	66	-8
CH_3Hg^+	ML/M.L	21.0				
	M_2L/ML.M	16.3				
	M_3L/M_2L.M	7				
Tl^+	MHL/M.HL		2.27			
	M_2HL/MHL.M		5.77			
	M_2OH(HL)$_3$/M^2.OH.(HL)^3		14.96			
	M_2(OH)$_2$(HL)$_2$/M^2.(OH)^2.(HL)^2		16.70			
	M^2.L/M_2L(s)		-21.1	-21.2 +0.3		
Zn^{2+}	ML/M.OH.HL		19.0			
	M.L/ML(s,α)		-24.4	-24.7 ±0.0	20	-46
	M.L/ML(s,β)			-22.5		

[j] 20°, 1.0

F. GROUP VI LIGANDS

Hydrogen sulfide (continued)

Metal ion	Equilibrium	Log K 20°, 0.1	Log K 25°, 1.0	Log K 25°, 0	ΔH 25°, 0	ΔS 25°, 0
Cd^2	MHL/M.HL		7.6			
	$M(HL)_2/M.(HL)^2$		14.6			
	$M(HL)_3/M.(HL)^3$		16.5			
	$M(HL)_4/M.(HL)^4$		18.9			
	M.L/ML(s)		-25.8	-27.0 ±0.1	25	-40
Hg^{2+}	$M(HL)_2/M.(HL)^2$		37.71[j]			
	$M(HL)_2/MHL_2.H$		6.19[j]			
	$MHL_2/ML_2.H$		8.30[j]			
	M.L/ML(s,black)		-51.0[j]	-52.7	63	-30
	M.L/ML(s,red)			-53.3		
Sn^{2+}	M.L/ML(s)			-25.9		
Pb^{2+}	M.L/ML(s)			-27.5 -0.6	31	-22
In^{3+}	$M^2.L^3/M_2L_3(s)$			-69.4		
Sb(III)	$(M(OH)_2)^2.L^3/M_2L_3(s)$			-90.8		
Bi^{3+}	$M^2.L^3/M_2L_3(s)$			-100 +1		

[j] 20°, 1.0

Bibliography:

H^+ 46K,52L,53KZ,58SG,59EGa,59Md,61LH, 61MSa,62Pc,66SW,67EM

Mn^{2+}-Cu^+,Sn^{2+},Pb^{2+},Bi^{3+} 53R,59C

Ag^+ 52GGF,53R,58SG,59C,66SW

CH_3Hg^+ 65SS

Tl^+ 53R,59KK,66GK,72GR

Zn^{2+} 53R,59C,67GSa

Cd^{2+} 53R,59C,64ST

In^{3+} 62TS

Sb(III)   52Lb,53R

Other references: OOP,OOWC,04A,06K,09BZ, 14TG,22JC,31K,32WM,33J,34ZR,38E,49KLa, 50SS,51Y,53A,56BL,56KR,57TMb,58Gb, 60ASc,60Bb,60MTF,60ZK,61EAa,63CMa, 62DGa,63Ca,64GM,64WSa,65D,66AD, 66KGS,67Gb,67GR,68HRa,71EGa,71G

$$\overset{\text{O}}{\underset{\text{HO-S-OH}}{\|}}$$

H_2O_3S		Hydrogen sulfite		(sulfurous acid)		H_2L

Metal ion H^+	Equilibrium	Log K 25°, 0.5	Log K 25°, 1.0	Log K 25°, 0	ΔH 25°, 0	ΔS 25°, 0
H^+	HL/H.L	6.79^h	6.34 6.36^e	7.18 ±0.03	$(3)^r$	43
	H_2L/HL.H	1.62^a	1.37 1.72^e	1.91 ±0.02	4.0	22
	H_2L/SO_2(g)		0.03 0.06^e	0.09	$(-6)^s$	(-20)
Ca^{2+}	M.L/ML(s)		-4.80 -5.04^l	-6.5		
Ce^{3+}	ML/M.L			8.04		
Cu^+	ML/M.L		7.85			
	$ML_2/M.L^2$		8.7			
	$ML_3/M.L^3$		9.4			
Ag^+	ML/M.L	5.4^d		5.60		
	$ML_2/M.L^2$	7.8^d		8.68		
	$ML_3/M.L^3$			9.00		
	$M^2.L/M_2L$(s)			-13.82		
CH_3Hg^+	ML/M.L	8.11^h				
Cd^{2+}	$ML_2/M.L^2$		4.2			
Hg^{2+}	$ML_2/M.L^2$		22.85^l	24.07^n		
	$ML_3/M.L^3$			25.96^n		

---

[a] 25°, 0.1; [d] 25°, 2.0; [e] 25°, 3.0; [h] 20°, 0.1; [l] 18°, 3.0; [n] 18°, 0; [r] 10-50°, 0;
[s] 25-50°, 0

Bibliography:
H^+ 12L,26SN,32RZ,34JL,37P,40Y,41TG,57CB, Cd^{2+} 57T
         58FN,58NR,63Sb,64AF,64DR,68SI,73KR Hg^{2+} 55TB

Ca^{2+} 58RB   Other references: 04D,07KB,11J,19K,30CM,
Ce^{3+} 50MS   30RH,31MM,32BR,32D,38BC,55Rb,56FP,
Cu^+ 55TSR   58Mc,59SY,60Ga,61EA,64DR,65FA,67ZO,
Ag^+ 55TSR,56CD   68LF,72BB
CH_3Hg^+ 65SS

$$HSO_4^-$$

HO_4S^-		Hydrogen sulfate ion		(sulfuric acid)		HL^-
Metal ion	Equilibrium	Log K 25°, 0.5	Log K 25°, 1.0	Log K 25°, 0	ΔH 25°, 0	ΔS 25°, 0
H^+	HL/H.L	1.32 ±0.06	1.10 ±0.08	1.99 ±0.01	5.4 ±0.3	27
		1.55[a]±0.05	1.01[d]±0.07	0.91[e]±0.02	5.6[c]	24[c]
Li^+	ML/M.L			0.64	0.0	3
Na^+	ML/M.L			0.70 ±0.05	1.1	7
K^+	ML/M.L	0.4[a]		0.85 ±0.1	3	14
NH_4^+	ML/M.L			1.11[n]		
Be^{2+}	ML/M.L	0.72[i]		1.95		
	$ML_2/M.L^2$		1.78			
	$ML_3/M.L^3$		2.08			
Mg^{2+}	ML/M.L		1.01[q]	2.23 ±0.02	1.4 ±0.1	15
Ca^{2+}	ML/M.L		1.03[q]	2.31 -0.01	1.6 ±0.1	16
	M.L/ML(s)		-2.92	-4.62 ±0.02	0.3	-20
			-3.16[e]			
Sr^{2+}	ML/M.L	1.14		2.55		
	M.L/ML(s)			-6.50 ±0.05	0.5	-28
Ba^{2+}	ML/M.L		0.66	2.7		
	$ML_2/M.L^2$		1.42			
	M.L/ML(s)			-9.96 ±0.03	5.5	-46
Ra^{2+}	M.L/ML(s)			-10.37[o]		
Sc^{3+}	ML/M.L	2.59		4.20	7.5	44
	$ML_2/M.L^2$	3.96		5.7	13	70
Y^{3+}	ML/M.L			3.47	4.8	32
		1.24[d]			4.0[d]	19[d]
	$ML_2/M.L^2$			5.3	7	48
		1.68[d]			5.5[d]	26[d]

[a] 25°, 0.1; [c] 25°, 1.0; [d] 25°, 2.0; [e] 25°, 3.0; [i] 18°, 0.5; [n] 18°, 0; [o] 20°, 0; [q] 25°, 0.7

Hydrogen sulfate ion (continued)

Metal ion	Equilibrium	Log K 25°, 0.5	Log K 25°, 1.0	Log K 25°, 0	ΔH 25°, 0	ΔS 25°, 0
La^{3+}	ML/M.L		1.43 ±0.03	3.64 ±0.02	4.4	31
		1.29[d]			3.7[d]	18[d]
	$ML_2/M.L^2$		2.46	5.29	5.7	43
Ce^{3+}	ML/M.L	1.77 ±0.02		3.59	4.6	32
		1.24[d]			4.3[d]	20[d]
	$ML_2/M.L^2$	2.90		5.2	6.2	45
Pr^{3+}	ML/M.L			3.62	4.7	32
		1.27[d]			3.9[d]	19[d]
	$ML_2/M.L^2$	1.88[d]		4.92	10.4	57
Nd^{3+}	ML/M.L			3.64	4.7	32
		1.26[d]			4.2[d]	19[d]
	$ML_2/M.L^2$	1.79[d]		5.1	8	50
Pm^{3+}	ML/M.L	1.34[d]			(4)[r]	(20)[d]
	$ML_2/M.L^2$	1.88[d]				
Sm^{3+}	ML/M.L			3.67 ±0.01	4.9	33
		1.30[d]			4.3[d]	20[d]
	$ML_2/M.L^2$	1.91[d]		5.2	7.0	47
Eu^{3+}	ML/M.L	1.87	1.54	3.67	4.9	33
		2.23[a]			3.9[d]	19[d]
		1.37[d]				
	$ML_2/M.L^2$	2.73	(2.69)	5.41	7.1	49
		1.96[d]			6.3[d]	30[d]
Gd^{3+}	ML/M.L	1.90		3.66	4.8	33
		1.33[d]			4.0[d]	20[d]
	$ML_2/M.L^2$	2.84		5.21	7.4	49
		1.75[d]			5.7[d]	27[d]
Tb^{3+}	ML/M.L			3.64	4.7	33
		1.27[d]			4.2[d]	20[d]
	$ML_2/M.L^2$			5.15	7.8	50
		1.89[d]			5.8[d]	28[d]

---

[a] 25°, 0.1; [d] 25°, 2.0; [r] 0-55°, 2.0

Hydrogen sulfate ion (continued)

Metal ion	Equilibrium	Log K 25°, 0.5	Log K 25°, 1.0	Log K 25°, 0	ΔH 25°, 0	ΔS 25°, 0
Dy^{3+}	ML/M.L			3.62	4.9	33
		1.23[d]			4.4[d]	20[d]
	$ML_2/M.L^2$			4.8	10	56
		1.72[d]			5.7[d]	27[d]
Ho^{3+}	ML/M.L			3.59	4.9	33
		1.24[d]			4.2[d]	20[d]
	$ML_2/M.L^2$			4.9	9	53
		1.76[d]			5.9[d]	28[d]
Er^{3+}	ML/M.L			3.59	4.9	33
		1.23[d]			4.2[d]	20[d]
	$ML_2/M.L^2$			5.1	8	50
		1.71[d]			5.7[d]	27[d]
Tm^{3+}	ML/M.L			3.59	4.8	33
		1.15[d]			4.2[d]	19[d]
	$ML_2/M.L^2$			5.14	7	47
		1.59[d]			5.2[d]	25[d]
Yb^{3+}	ML/M.L			3.58 ±0.02	4.7	32
		1.15[d]			4.1[d]	19[d]
	$ML_2/M.L^2$			5.2	7	47
		1.59[d]			5.3[d]	25[d]
Lu^{3+}	ML/M.L		1.29	3.52	4.7	32
		1.09[d]			4.2[d]	19[d]
	$ML_2/M.L^2$		1.9	5.3	6	44
		1.61[d]			5.8[d]	27[d]
Ac^{3+}	ML/M.L		1.20			
	$ML_2/M.L^2$		1.85			
Pu^{3+}	ML/M.L		1.26			
	$M(HL)_2/M.(HL)^2$		1.00			
Am^{3+}	ML/M.L	1.86	1.57			
		1.43[d]			(4)[s]	(20)[d]
	$ML_2/M.L^2$	2.82	2.66			
		1.85[d]				

[d] 25°, 2.0; [s] 0-55°, 2.0

Hydrogen sulfate ion (continued)

Metal ion	Equilibrium	Log K 25°, 0.5	Log K 25°, 1.0	Log K 25°, 0	ΔH 25°, 0	ΔS 25°, 0
Cm^{3+}	$ML/M.L$	1.86			(4)[s]	(20)[d]
		1.34[d]				
	$ML_2/M.L^2$	2.7				
		1.86[d]				
Cf^{3+}	$ML/M.L$	1.36[d]			(5)[s]	(20)[d]
	$ML_2/M.L^2$	2.07[d]				
Th^{4+}	$ML/M.L$	3.22[d]±0.01			5.0[d]	32[d]
	$ML_2/M.L^2$	5.53[d]±0.06			9.6[d]	58[d]
U^{4+}	$ML/M.L$	3.42[d]+0.1				
	$ML_2/M.L^2$	5.82[d]±0.07				
Np^{4+}	$ML/M.L$	3.51[d]−0.07	3.41[e]	3.53[m]		
	$ML_2/M.L^2$		5.42[e]	5.92[m]		
Pu^{4+}	$ML/M.L$		3.66			
$Pa(V)$	$MOOHL/MOOH.L$		2.21[e]			
UO_2^{2+}	$ML/M.L$	1.65[t]	1.81 ±0.0	2.95	5.0	30
					4.4[c]	24[c]
	$ML_2/M.L^2$		2.5 ±0.2	4.0	8.4[c]	41[c]
	$ML_3/M.L^3$		3.7[j]			
NpO_2^{2+}	$ML/M.L$	2.07	1.82[j]	3.27		
		2.20[a]				
	$ML_2/M.L^2$	3.4	2.62[j]			
		3.8[a]				
Mn^{2+}	$ML/M.L$			2.26 −0.02	2.1 ±0.1	17
Fe^{2+}	$ML/M.L$			2.2	1.6	15
Co^{2+}	$ML/M.L$		0.23[e]	2.36 +0.05	1.4 ±0.1	16
Ni^{2+}	$ML/M.L$	1.0[k]	0.57	2.32	1.5 −0.1	16
			0.26[e]			
	$ML_2/M.L^2$		1.42			

---

[a] 25°, 0.1; [c] 25°, 1.0; [d] 25°, 2.0; [e] 23°, 3.0; [j] 20°, 1.0; [k] 20°, 2.0; [m] 20°, 4.0 assuming HL/H.L = 0.83; [s] 0-55°, 2.0; [t] 25°, 2.7

Hydrogen sulfate ion (continued)

Metal ion	Equilibrium	Log K 25°, 0.5	Log K 25°, 1.0	Log K 25°, 0	ΔH 25°, 0	ΔS 25°, 0
Cu^{2+}	ML/M.L		0.95	2.36 ±0.03	2.1 ±0.4	18
			0.70[e]		1.7[d]	
	$M.(OH)^{1.5}.L^{0.25}/M(OH)_{1.5}L_{0.25}(s)$					
			-16.86[j]	-17.16 ±0.04	(8)[y]	(-50)
Ru^{2+}	ML/M.L	1.88[i]	1.70[j]	2.72[o]	(0)[v]	(10)
		1.30[k]				
V^{3+}	M./M.L		1.45			
Cr^{3+}	ML/M.L		2.60[x]			
Mn^{3+}	ML/M.L	1.20[t]	1.63[f]	1.90[g]		
Fe^{3+}	ML/M.L	2.24 ±0.1	2.02 ±0.1	4.04 ±0.1	(6)[w]	(30)[b]
			1.93[e]			
	$ML_2/M.L^2$		2.11[e]	5.38		
Ru^{3+}	ML/M.L	2.04[d]				
	$ML_2/M.L^2$	3.57[d]				
TiO^{2+}	ML/M.L		2.15[e]	2.5		
			2.26[f]	2.47[g]		
Zr^{4+}	ML/M.L	3.67[d]		3.68[m]		
	$ML_2/M.L^2$	6.40[d]		6.4[m]		
	$ML_3/M.L^3$	7.4[d]		7.5[m]		
Hf^{4+}	ML/M.L	3.04[d]				
	$ML_2/M.L^2$	5.44[d]				
VO^{2+}	ML/M.L			2.44 ±0.04	4.1	25
VO_2^+	ML/M.L		0.97[j]			
Ag^+	ML/M.L	0.31[d]	0.23[e]	1.3	1.5	11
	$ML_2/M.L^2$	0.19[d]	0.00[e]			
	$ML_3/M.L^3$	0.40[d]				
	$M^2.L/M_2L(s)$			-4.83 ±0.03	4.1	-8
Hg_2^{2+}	ML/M.L	1.30				
	$ML_2/M.L^2$	3.54				
	M.L/ML(s)		-4.46[e]	-6.13 ±0.04	1.3	-24

[b] 25°, 0.5; [d] 25°, 2.0; [e] 25°, 3.0; [f] 25°, 4.0; [g] 25°, 5.0; [i] 20°, 0.5; [j] 20°, 1.0;
[k] 20°, 2.0; [m] 20°, 4.0, assuming HL/H.L = 0.83; [o] 20°, 0; [t] 23°, 2.7; [v] 20-35°, 0;
[w] 1-35°, 0.5; [x] 48°, 1.0; [y] 25-75°, 0

Hydrogen sulfate ion (continued)

Metal ion	Equilibrium	Log K 25°, 0.5	Log K 25°, 1.0	Log K 25°, 0	ΔH 25°, 0	ΔS 25°, 0
Tl^+	ML/M.L			1.37 ±0.07	-0.2	6
Zn^{2+}	ML/M.L	0.93	0.89	2.38	1.5 ±0.1	16
		0.76^d	0.70^e			
	$ML_2/M.L^2$	1.9	1.2			
		1.2^d	0.7^e			
	$ML_3/M.L^3$		1.7			
		$(1.1)^d$	0.9^e			
	$ML_4/M.L^4$		1.7			
		1.4^d	0.9^e			
Cd^{2+}	ML/M.L	1.08	0.95	2.46	2.3 ±0.1	19
		0.86^d	0.71^e±0.01		1.9^d	10^d
	$ML_2/M.L^2$	2.0	1.6			
		1.3^d	0.9^e ±0.1			
	$ML_3/M.L^3$	2.7	1.8			
		1.6^d	1.0^e ±0.1			
	$ML_4/M.L^4$		(2.3)			
		1.5^d	1.0^e ±0.1			
Hg^{2+}	ML/M.L	1.34				
	$ML_2/M.L^2$	2.4				
Pb^{2+}	ML/M.L		$(0.74)^e$	2.75 -0.1		
	$ML_2/M.L^2$		1.99^e			
	M.L/ML(s)		-6.20	-7.79 ±0.02	$(3)^y$	(-26)
In^{3+}	ML/M.L	1.78^k	1.85^j			
	$ML_2/M.L^2$	1.88^k	2.60^j			
	$ML_3/M.L^3$	2.36^k	3.00^j			
Tl^{3+}	ML/M.L		2.27^e		-2.7^e	1^e
Bi^{3+}	ML/M.L		1.98^e		$(3)^z$	$(19)^e$
	$ML_2/M.L^2$		3.41^e		$(7)^z$	$(39)^e$
	$ML_3/M.L^3$		4.08^e		$(11)^z$	$(56)^e$
	$ML_4/M.L^4$		4.34^e		$(13)^z$	$(64)^e$
	$ML_5/M.L^5$		4.60^e			

---

^{d} 25°, 2.0; ^{e} 25°, 3.0; ^{j} 20°, 1.0; ^{k} 20°, 2.0; ^{y} 0-50°, 0; ^{z} 15-65°, 3.0

Hydrogen sulfate ion (continued)

Bibliography:

H^+ 10NS,37P,40S,51M,51ZA,52DJ,54S,55ER,
         56R,58NNa,59Z,61RS,62AMa,63DS,62YI,63RF,
         64F,65KYa,66CI,66MJ,66VL,67VLe,68Aba,
         69ZL,71AKa,71EM,72CG,73CD,73P,73S,74T

Li^+-K^+ 30RD,50JM,62AMa,66CL,68TH,69IE,74MV

NH_4^+ 31BR

Be^{2+} 62BK,66KW,67SSa

Mg^{2+} 27D,40MS,51DJ,58NN,66AP,68KP,69IEa,70L,
            73AM,73HP,73P

Ca^{2+} 32MD,33LHS,53BG,58NR,69KP,70GG,70L,73A,
            74MVa

Sr^{2+} 23B,56SZ,68CS

Ba^{2+} 10M,23B,33LHS,40CB,58R,60T,65L,66SSH

Ra^{2+} 33NT

$Sc^{3+}-Lu^{3+}$ 30Da,50JM,51CM,51M,54SJ,62Bc,62MM,
                    65Sa,66AM,66AMa,66AP,67CC,67FD,67KI,
                    68ALN,69IEa,72HS,73FP,74Pa

Ac^{3+} 68SMR

Pu^{3+} 67NR

$Am^{3+}-Cf^{3+}$ 65Sa,67CC,68ALN

Th^{4+} 51ZA,59Z

U^{4+} 52SH,55DW

Np^{4+} 54SH,62ST,63Ma,66AB

Pu^{4+} 51RL

Pa(V)   66Ga

UO_2^{2+} 49BM,51A,57DM,60Mb,71AKa,71BL

NpO_2^{2+} 68ABa,70AWa

$Mn^{2+}-Cu^{2+}$ 27D,38OG,47J,49NT,50F,51NL,56BD,
                    58BB,58Sb,59NN,60BB,61PF,61TO,63TS,
                    65SMb,68MM,69BG,69IEa,70L,70MM,73AM,
                    73HP,73P

Ru^{2+} 66VL,67VLc

V^{3+} 72KMN

Cr^{3+} 62FT

Mn^{3+} 73GT

Fe^{3+} 53WD,59M,60K,62DS,63W,67M,73NP

Ru^{3+} 68LK

TiO^{2+} 69VV

Zr^{4+} 49CMa,63AK,69N

Hf^{4+} 63Pa,65DK,67EME

VO^{2+} 63SW,66KW,71BL

VO_2^+ 66I

Ag^+ 30RD,43VM,52L,54TO,60SL,65HWa,
         67CC,67LH

Hg_2^{2+} 46IS,53L,57BL,70SP

Tl^+ 30BD,30RD,53BG

Zn^{2+} 27D,58NN,69IEa,70L,71FCK,73AM,
            73FC,73HP,73P

Cd^{2+} 27D,69BG,69IEa,70L,71FCK,71FCM,
            73FC,73P

Hg^{2+} 46IS,57PT

Pb^{2+} 31CL,31LP,34L,46TM,53L,55S,60RKa,
            61RS,70GNa,72BH

In^{3+} 54S,54Sc,66DR,68AL

Tl^{3+} 67MK

Bi^{3+} 71FK

Other references: 03B,07P,29B,30LM,32HJ,
    32MD,38D,41L,42Nc,50N,52F,52JM,52S,
    53CT,53Na,53NA,53SK,54BB,54D,54DP,
    54K,54NK,54Se,55LR,56HD,56Ka,57BM,
    57K, 58Aa,58Ja,58KV,58MF,59K,59Ka,
    59PL,59WL,60C,60H,60LP,60LS,60PB,
    61BT,61HA,61MC,61S,62AY,62BS,62BW,
    62ET,62JP,62NA,62RE,63AM,63BL,62Hc,
    63KV,63LM,63NP,63TU,63VRa,64BM,64BP,
    64LP,65AK,65HS,65LB,65NT,65P,64T,
    66BA,66E,66Fa,66MB,66MS,66NH,66S,
    67LN,67Ma,67VLd,67W,68M,68N,68P,
    68PS,68WSa,69BMN,69FP,69MN,69NPa,
    69PK,69RP,69SPa,69SS,69VS,69YM,70M,
    70PH,70TR,71MD,71MKA,72MC,72TSa,73Ab,
    73IVa,73PR,73PRa,73RM,74FP,74HI,74MC,
    74MS

$$H_2S_2O_3$$

| $H_2O_3S_2$ | | Hydrogen thiosulfate | (thiosulfuric acid) | | H_2L |

Metal ion	Equilibrium	Log K 25°, 0.5	Log K 25°, 1.0	Log K 25°, 0	ΔH 25°, 0	ΔS 25°, 0
H^+	HL/H.L	1.0 1.3[a]	0.8	1.6 ±0.1		
	H_2L/HL.H			0.6		
Na^+	ML/M.L	0.04		0.53 ±0.05	1.1[b]	4[b]
K^+	ML/M.L	0.28		0.96 ±0.04	2.0[b]	8[b]
Mg^{2+}	ML/M.L	0.56		1.82 ±0.03	0.4[b]	4[b]
Ca^{2+}	ML/M.L	0.70		1.98 ±0.07	0.6[b]	5[b]
Sr^{2+}	ML/M.L			2.04		
Ba^{2+}	ML/M.L			2.27 ±0.06	2.6	19
	M.L/ML(s)			-4.79		
La^{3+}	ML/M.L		0.8	2.99		
Mn^{2+}	ML/M.L	0.67		1.95	0.5[b]	5[b]
Co^{2+}	ML/M.L	0.77		2.05	0.5[b]	5[b]
Ni^{2+}	ML/M.L	0.78		2.06	0.4[b]	5[b]
Fe^{3+}	ML/M.L	1.98[a]	1.18[r]			
Cu^+	ML/M.L		10.35[s]			
	ML_2/M.L^2		12.27[s]			
	ML_3/M.L^3		13.71[s]			
Ag^+	ML/M.L			8.82[o] 7.36[f]		
	ML_2/M.L^2	12.78[d]	12.63	13.67[o] 12.72[f]	-19	-2
	ML_3/M.L^3	13.1[d]	12.8	14.2[o] 13.5[f]		
	M_2L_4/M^2.L^4			26.3[f]		
	M_3L_5/M^3.L^5			39.8[f]		
	M_6L_8/M^6.L^8			78.6[f]		
CH_3Hg^+	ML/M.L	10.75[a]			-11.7[h]	10[a]

[a] 25°, 0.1; [b] 25°, 0.5; [d] 25°, 2.0; [f] 25°, 4.0; [h] 20°, 0.1; [o] 20°, 0; [r] 18°, 1.0;
[s] 25°, 1.6

Hydrogen thiosulfate (continued)

Metal ion	Equilibrium	Log K 25°, 0.5	Log K 25°, 1.0	Log K 25°, 0	ΔH 25°, 0	ΔS 25°, 0
Tl^+	$ML/M.L$			0.86^f		
	$ML_2/M.L^2$			0.72^f		
	$ML_3/M.L^3$			0.2^f		
	$M_2L_4/M^2.L^4$			2.54^f		
	$M^2.L/M_2L(s)$			-4.54^f		
Zn^{2+}	$ML/M.L$	(1.12)	(0.62)	2.35 ± 0.06	3.1	21
			0.96^e		2.2^b	$(13)^b$
	$ML_2/M.L^2$		1.94^e			
	$ML_3/M.L^3$		3.3^e			
	$M_2L_2/M^2.L^2$		5.84^e			
Cd^{2+}	$ML/M.L$	3.2^u	2.82	3.92 ± 0.03	1.3	22
			$2.74^e \pm 0.00$		0.0^b	15^b
	$ML_2/M.L^2$		4.57	6.3 ± 0.1	-1.5^c	16^c
			$4.70^e \pm 0.05$			
	$ML_3/M.L^3$		6.4		-3.4^c	18^c
			$6.9^e \pm 0.1$			
	$ML_4/M.L^4$		7.1^e			
	$M_2L_2/M^2.L^2$		11.18^e			
Hg^{2+}	$ML_2/M.L^2$		(29.93)	29.23 ± 0.05		
	$ML_3/M.L^3$		(33.26)	30.6 ± 0.3	-38.5^c	$(23)^c$
Pb^{2+}	$ML/M.L$		2.42^e			
	$ML_2/M.L^2$		4.86^e			
	$ML_3/M.L^3$		6.2^e			
	$ML_4/M.L^4$		6.2^e			

^{b} 25°, 0.5; ^{c} 25°, 1.0; ^{e} 25°, 3.0; ^{f} 25°, 4.0; ^{u} 30°, 0.1

Bibliography:
H^+ 42YH,51DM,53Pb
Na^+-Ba^{2+} 49DW,51DM,55GM,74Aa
La^{3+} 51M,64BM
Mn^{2+}-Ni^{2+} 51DM,74Aa
Fe^{3+} 30S,57MN
Cu^+ 55TSL
Ag^+ 460A,53CP,54W,57CHP,58N,67BP
CH_3Hg^+ 65SS

Tl^+ 58Na
Zn^{2+} 51DM,55GM,70P,74Aa,74MS
Cd^{2+} 51DM,55GM,56Y,57YG,59MG,65HS,70P,74Aa,
Hg^{2+} 54T,61NS,70MSS,74Kb
Pb^{2+} 59DP

Other references: 11J,31CR,49Ba,53J,54NS,
54P,56A,56NM,57KP,57NM,57T,58D,58SW,
59Pb,59SD,64CL,68GF,68JG,72PR

$$SeSO_3^{2-}$$

O_3SSe^{2-} Selenosulfate ion L^{2-}

Metal ion	Equilibrium	Log K 25°, 1.0
Hg^{2+}	$ML_2/M.L^2$	36.8

Bibliography: 57T

$$HO-\overset{\overset{O}{\|}}{\underset{\underset{O}{\|}}{S}}-NH_2$$

H_3O_3NS Hydrogen amidosulfate   (sulfamic acid)   HL

Metal ion	Equilibrium	Log K 25°, 2.0	Log K 25°, 0	ΔH 25°, 0	ΔS 25°, 0
H^+	HL/H.L		0.988	-0.25	3.7
	H.L/HL(s)		-0.92		
Pu^{4+}	ML/M.L	0.11			

Bibliography:
H^+ 52KK,65HW,68C   Other references: 55HB,58Sc,58Sd,59S,
Pu^{4+} 68C   65HSE

$$\begin{array}{ccc} O & & O \\ \| & & \| \\ HO-S-O-O-S-OH \\ \| & & \| \\ O & & O \end{array}$$

| $H_2O_8S_2$ | | Hydrogen peroxodisulfate | (peroxodisulfuric acid) | | H_2L |

Metal ion	Equilibrium	Log K 25°, 0.1	Log K 25°, 0	ΔH 25°, 0	ΔS 25°, 0
K^+	ML/M.L	0.52	0.92	(2)[r]	(11)

_____

[r] 25-39°, 0

Bibliography: 66CL

$$H_2Se$$

H_2Se	Hydrogen selenide	(hydroselenic acid)	H_2L

Metal ion	Equilibrium	Log K 25°, 1.0	Log K 25°, 0
H^+	HL/H.L	(11.6)	15.0^o
	$H_2L/HL.H$	3.48	3.89
Mn^{2+}	ML/M.OH.HL	8.0	
	M.L/ML(s)	-12.1	(-11.5)
Fe^{2+}	M.L/ML(s)		(-26.0)
Co^{2+}	M.L/ML(s)		(-31.2)
Ni^{2+}	M.L/ML(s)		(-32.7)
Cu^{2+}	M.L/ML(s)		(-48.1)
Cu^+	$M^2.L/M_2L(s)$		(-60.8)
Ag^+	$M_2L/M^2.OH.HL$	48.5	
	$MOHL_2/M.OH.L^2$	24.1	
	$M^2.L/M_2L(s)$	-53.8	(-63.7)
Tl^+	$M^2.L/M_2L(s)$		(-33.1)
Zn^{2+}	M.L/ML(s)		(-29.4)
Cd^{2+}	M.L/ML(s)		(-35.2)
Hg^{2+}	ML/M.OH.HL	51.2	
	$ML_2/M.(OH)^2.(HL)^2$	61.0	
	$MHL_2/M.OH.(HL)^2$	52.8	
	M.L/ML(s)	-56.6	(-64.5)
Sn^{2+}	M.L/ML(s)		(-38.4)
Pb^{2+}	M.L/ML(s)		(-42.1)

---

^{o} 22°, 0

Bibliography:

H^+ 41Ha,58W,70MG

Mn^{2+} 64BU,70MGa

Fe^{2+}-Cu^+,Tl^+,Zn^{2+},Sn^{2+},Pb^{2+} 64BU

Ag^+ 64BU,70MG

Hg^{2+} 64BU,71MG

Other references: 13B,23Ha,42L,48LN,57W

$$\overset{\displaystyle O}{\underset{\displaystyle HO-Se-OH}{\|}}$$

H_2O_3Se		Hydrogen selenite		(selenous acid)		H_2L
Metal ion	Equilibrium	Log K 25°, 0.3	Log K 25°, 1.0	Log K 25°, 0	ΔH 25°, 0	ΔS 25°, 0
H^+	HL/H.L	7.94	7.78		-1.20^c	31.6^c
			8.05^e		-1.26^e	32.6^e
	$H_2L/HL.H$	2.35	2.27		1.5^c	15^c
			2.61^e		1.3^e	16^e
	$H_2L_2/(HL)^2$	1.6	0.17		1.2^c	5^c
			-0.61^e		-0.9^e	-6^e
	$H_2L_2/HL_2.H$	8.0	7.7		-0.6^c	33^c
			7.70^e		-1.1^e	32^e
	$H_3L_2/H_2L_2.H$	2.8	2.97		1.3^c	18^c
			3.53^e		1.4^e	21^e
	$H_4L_2/H_3L_2.H$	2.4	2.10		1.5^c	15^c
			1.89^e		-0.1^e	8^e
Mg^{2+}	$M.L/ML(H_2O)_6(s)$			-5.36^o		
Sr^{2+}	M.L/ML(s)			-6.10^o		
Ba^{2+}	M.L/ML(s)			-6.57		
Mn^{2+}	M.L/ML(s)			-7.27^o		
Co^{2+}	M.L/ML(s)			-7.08^o		
Cu^{2+}	$M.L/ML(H_2O)_2(s)$			-7.78^o		
Fe^{3+}	MHL/M.HL		2.81		$(6)^r$	$(30)^c$
Ag^+	ML/M.L		2.4			
	$ML_2/M.L^2$		3.76			
	$M^2.L/M_2L(s)$			-15.55	10.7	-35
Cd^{2+}	$ML_2/M.L^2$		5.1			
Hg^{2+}	$ML_2/M.L^2$		12.5			
	M.L/ML(s)		-13.82			

---

^{c} 25°, 1.0; ^{e} 25°, 3.0; ^{o} 20°, 0; ^{r} 20-40°, 1.0

Hydrogen selenite (continued)

Bibliography:

H^+ 66SC,71BSa,72AB

Mg^{2+} 66LSa

Sr^{2+} 63SL

Ba^{2+} 65LA

Mn^{2+} 66LS

Co^{2+} 64SLM

Cu^{2+} 65LS

Fe^{3+} 65HI

Ag^+ 62SLK,69MG

Cd^{2+},Hg^{2+} 57T

Other references: 09RP,20B,21RK,31BR,
        32BR,33R,39Ha,52Lb,57KC,59Me,61LPa,
        66PD,67KM,68RV,68SI

$$HSeO_4^-$$

HO_4Se^-		Hydrogen selenate ion	(selenic acid)			HL^-
Metal ion	Equilibrium	Log K 25°, 0.5	Log K 25°, 1.0	Log K 25°, 0	ΔH 25°, 0	ΔS 25°, 0
H^+	HL/H.L			1.70 ±0.05	5.6 ±0.1	27
Ca^{2+}	M.L/ML(s)			−3.09	−2.0	−21
Sr^{2+}	M.L/ML(s)			−4.40	0.1	−20
Ba^{2+}	M.L/ML(s)			−7.46	5.4	−16
Sc^{3+}	ML/M.L	1.78			(−2)[r]	(1)[b]
	$ML_2/M.L^2$	2.64				
Mn^{2+}	ML/M.L			2.43	3.5	23
Co^{2+}	ML/M.L			2.70	2.9	22
Ni^{2+}	ML/M.L			2.67	3.5	23
Ag^+	$M^2.L/M_2L(s)$			−8.91	(−10.4)	(−76)
Tl^+	$M^2.L/M_2L(s)$			−4.00	(10.3)	(16)
Zn^{2+}	ML/M.L		0.73	2.19		
	$ML_2/M.L^2$		1.35			
Cd^{2+}	ML/M.L			2.27		
Pb^{2+}	M.L/ML(s)			−6.84	3.8	−19

---

[b] 25°, 0.5;   [r] 25-45°, 0.5

Bibliography:

H^+ 64Na,65CD,70GN,74MBa
Ca^{2+} 59SS
Sr^{2+} 59SZ,63Sc
Ba^{2+} 58SSa,59SK
Mn^{2+}-Ni^{2+} 70GN
Sc^{3+} 67KI
Ag^+ 59SZF

Tl^+ 58Sa
Zn^{2+} 34Ba,74MS
Cd^{2+} 34Ba
Pb^{2+} 55Sb,59SK

Other references: 42GK,50PA,53SKa,59Ba, 68WSa

$$H_2Te$$

| H_2Te | Hydrogen telluride | (hydrotelluric acid) | H_2L |

Metal ion	Equilibrium		Log K 25°, 0
Mn^{2+}	M.L/ML(s)		(-15.9)
Fe^{2+}	M.L/ML(s)		(-30.0)
Co^{2+}	M.L/ML(s)		(-37.4)
Ni^{2+}	M.L/ML(s)		(-38.1)
Cu^+	$M^2.L/M_2L(s)$		(-62.3)
Ag^+	$M^2.L/M_2L(s)$		(-71.7)
Tl^+	$M^2.L/M_2L(s)$		(-39.2)
Zn^{2+}	M.L/ML(s)		(-33.3)
Cd^{2+}	M.L/ML(s)		(-41.5)
Hg^{2+}	M.L/ML(s)		(-69.6)
Sn^{2+}	M.L/ML(s)		(-44.7)
Pb^{2+}	M.L/ML(s)		(-46.3)

Bibliography: 64BU

Other references: 13B,23Ha,48LN,52Lb, 63Pb

$$Te(OH)_4$$

H_4O_4Te <u>Hydrogen tellurite</u>   (<u>tellurous acid</u>) H_2L

Metal ion	Equilibrium	Log K 25°, 0.5	Log K 25°, 1.0	Log K 25°, 0
H^+	HL/H.L	9.00 9.36[a]	8.60	
Ag^+	$M^2.L/M_2L(s)$			−2.43

---

[a] 25°, 0.1

Bibliography:

H^+ 73NS

Ag^+ 65GP

Other references: 20B,24K,25S,54IA,56DZb,
68NKa,68SI,71BG,71GB,74MB,74NK

HF

HF		Hydrogen fluoride	(hydrofluoric acid)			HL

Metal ion	Equilibrium	Log K 25°, 0.5	Log K 25°, 1.0	Log K 25°, 0	ΔH 25°, 0	ΔS 25°, 0
H^+	HL/H.L	2.93 ±0.02	2.96 ±0.02	3.17 ±0.01	3.20 ±0.02	25.3
		2.92[a]±0.03			2.89[c]±0.03	23.3[c]
		3.13[d]±0.02	3.30[e]±0.02			
	HL_2/HL.L	0.59 ±0.02	0.59 ±0.1	0.5 ±0.1	1.0	5
			0.86[e]±0.2		0.9[c] ±0.1	6[c]
Li^+	M.L/ML(s)			-2.77	1.1	-9
Be^{2+}	ML/M.L	4.71	4.99		(0)[s]	(20)[c]
		4.99[k]				
	ML_2/M.L^2	8.32	8.80		(-1)[s]	(40)[c]
		8.78[k]				
	ML_3/M.L^3	11.1	11.6		(-2)[s]	(40)[c]
		11.7[k]				
	ML_4/M.L^4	13.4[k]	13.1		(-2)[s]	(50)[c]
Mg^{2+}	ML/M.L	1.31 ±0.01	1.32 ±0.06	1.8	3.2[c]	17[c]
		1.46[a]				
	M.L^2/ML_2(s)			-8.18	(-2)[r]	(-40)
Ca^{2+}	ML/M.L	0.6 ±0.1	0.58 ±0.05	1.1	3.5[c]	15[c]
	M.L^2/ML_2(s)			-10.41	(4)[r]	(-30)
Sr^{2+}	ML/M.L		0.1 ±0.1		4[c]	14[c]
	M.L^2/ML_2(s)			-8.54	(1)[r]	(-40)
Ba^{2+}	ML/M.L		-0.3 ±0.1		(4)[t]	(12)[c]
	M.L^2/ML_2(s)			-5.76	(1)[r]	(-20)
Sc^{3+}	ML/M.L	6.18 ±0.01		7.1	(0)[u]	(30)[b]
	ML_2/M.L^2	11.46 ±0.02		12.9	(-1)[u]	(50)[b]
	ML_3/M.L^3	15.5 ±0.0		17.4	(-2)[u]	(60)[b]
	ML_4/M.L^4	18.4 +0.1		20.3		
Y^{3+}	ML/M.L	3.90 ±0.03	3.60	4.8	8.3[c]	44[c]
	ML_2/M.L^2	7.13 ±0.03		8.5		
	ML_2/M.L^3	10.3 ±0.0		12.1		

[a] 25°, 0.1; [b] 25°, 0.5; [c] 25°, 1.0; [d] 25°, 2.0; [e] 25°, 3.0; [k] 20°, 2.0; [r] 0-27°, 0; [s] 0-60°, 1.0; [t] 15-60°, 1.0; [u] 15-35°, 0.5

Hydrogen fluoride (continued)

Metal ion	Equilibrium	Log K 25°, 0.5	Log K 25°, 1.0	Log K 25°, 0	ΔH 25°, 0	ΔS 25°, 0
La^{3+}	ML/M.L	2.69 ±0.01	2.67	3.6	4.0[c]	26[c]
Ce^{3+}	ML/M.L	3.13	2.81	4.0	4.8[c]	29[c]
Pr^{3+}	ML/M.L		3.01		5.7[c]	33[c]
Nd^{3+}	ML/M.L		3.09		6.8[c]	37[c]
Pm^{3+}	ML/M.L		(3.16)			
Sm^{3+}	ML/M.L		3.12		9.4[c]	46[c]
Eu^{3+}	ML/M.L	3.40 -0.01	3.19		9.2[c]	45[c]
Gd^{3+}	ML/M.L	3.40 ±0.03	3.31	4.3	8.9[c]	45[c]
Tb^{3+}	ML/M.L		3.42		7.5[c]	41[c]
Dy^{3+}	ML/M.L		3.46		7.0[c]	39[c]
Ho^{3+}	ML/M.L		3.52		7.3[c]	40[c]
Er^{3+}	ML/M.L		3.54		7.4[c]	41[c]
Tm^{3+}	ML/M.L		3.56		8.7[c]	45[c]
Yb^{3+}	ML/M.L		3.58		9.6[c]	48[c]
Lu^{3+}	ML/M.L		3.61		9.5[c]	48[c]
Ac^{3+}	ML/M.L	2.72				
	$ML_2/M.L^2$	5.22				
	$ML_3/M.L^3$	7.9				
Am^{3+}	ML/M.L	3.39				
	$ML_2/M.L^2$	6.11				
	$ML_3/M.L^3$	9.0				
Cm^{3+}	ML/M.L	3.34				
	$ML_2/M.L^2$	6.18				
	$ML_3/M.L^3$	9.1				

---

[c] 25°, 1.0

Hydrogen fluoride (continued)

Metal ion	Equilibrium	Log K 25°, 0.5	Log K 25°, 1.0	Log K 25°, 0	ΔH 25°, 0	ΔS 25°, 0
Th^{4+}	$ML/M.L$	7.59 ±0.04	7.80^e	8.44 ±0.2	$(-1)^r$	(40)
				8.2^m		
	$ML_2/M.L^2$	13.44 ±0.02	13.82^e	15.08 ±0.2	$(-2)^r$	(70)
				14.7^m		
	$ML_3/M.L^3$	17.9	18.8^e	19.8 ±0.4	$(-3)^r$	(90)
	$ML_4/M.L^4$			23.2	$(-4)^r$	(100)
	$M.L^4/ML_4(s)$		-28.3^e			
Pa^{4+}	$ML/M.L$		8.03^e			
	$ML_2/M.L^2$		14.86^e			
U^{4+}	$ML/M.L$			9.0^m		
	$ML_2/M.L^2$			15.7^m		
	$ML_3/M.L^3$			21.2^m		
Np^{4+}	$ML/M.L$			8.3^m		
	$ML_2/M.L^2$			14.5^m		
	$ML_3/M.L^3$			20.3^m		
	$ML_4/M.L^4$			25.1^m		
Pu^{4+}	$ML/M.L$		6.77			
$Pa(V)$	$MOL/M(OH)_3.HL$		3.56^e			
	$MOL/_2MOL.L$		7.39^e			
	$MOL_3/MOL_2.L$		6.56^e			
UO_2^{2+}	$ML/M.L$	4.3	4.54 −0.1		0.4^c	22^c
		4.55^d		5.0^f		
	$ML_2/M.L^2$		7.97 −0.08		0.5^c	38^c
	$ML_3/M.L^3$		10.55 −0.09		0.6^c	51^c
	$ML_4/M.L^4$		12.0 −0.2		0.1^c	55^c
NpO_2^{2+}	$ML/M.L$	4.04	3.85^j	4.6		
		4.12^a				
	$ML_2/M.L^2$	7.00	6.97^j			
		7.01^a				
Mn^{2+}	$ML/M.L$		0.7 ±0.1			
Fe^{2+}	$ML/M.L$		0.8			

[a] 25°, 0.1; [c] 25°, 1.0; [d] 25°, 2.0; [e] 25°, 3.0; [f] 25°, 4,0, assuming HL/H.L = 3.5; [j] 20°, 1.0; [m] 20°, 4.0 H^+, assuming HL/H.L = 3.5; [r] 5-45°, 0

Hydrogen fluoride (continued)

Metal ion	Equilibrium	Log K 25°, 0.5	Log K 25°, 1.0	Log K 25°, 0	ΔH 25°, 0	ΔS 25°, 0
Co^{2+}	ML/M.L		0.4			
Ni^{2+}	ML/M.L		0.5 ±0.1			
Cu^{2+}	ML/M.L	(0.7)	0.9 ±0.1	(1.2)	(1)[t]	(7)[b]
Cr^{3+}	ML/M.L	4.36		5.2		
	$ML_2/M.L^2$	7.70				
	$ML_3/M.L^3$	10.2				
Mn^{3+}	ML/M.L	5.65[d]				
Fe^{3+}	ML/M.L	5.18 ±0.04	5.18	6.0	2.3[b]	31[b]
	$ML_2/M.L^2$	9.13 ±0.04	9.07		3.5[b]	54[b]
	$ML_3/M.L^3$	11.9 ±0.1	12.1		4.4[b]	69[b]
VO^{2+}	ML/M.L		3.37 -0.09		1.9[c]	22[c]
	$ML_2/M.L^2$		5.74 -0.3		3.5[c]	38[c]
	$ML_3/M.L^3$		7.29 -0.2		4.9[c]	50[c]
	$ML_4/M.L^4$		8.1 -0.7		6.4[c]	59[c]
Zr^{4+}	ML/M.L			9.8		
		8.94[d]		9.4[m]		
	$ML_2/M.L^2$	16.4[d]		17.2[m]		
	$ML_3/M.L^3$	22.4[d]		23.7[m]		
	$ML_4/M.L^4$			29.5[m]		
	$ML_5/M.L^5$			23.5[m]		
	$ML_6/M.L^6$			28.3[m]		
Hf^{4+}	ML/M.L			9.0[m]		
	$ML_2/M.L^2$			16.5[m]		
	$ML_3/M.L^3$			23.1[m]		
	$ML_4/M.L^4$			28.8[m]		
	$ML_5/M.L^5$			34.0[m]		
	$ML_6/M.L^6$			38.0[m]		
VO_2^+	ML/M.L			3.04[j]		
	$ML_2/M.L^2$			5.60[j]		
	$ML_3/M.L^3$			6.9[j]		
	$ML_4/M.L^4$			7.0[j]		

[b] 25°, 0.5; [c] 25°, 1.0; [d] 25°, 2.0; [j] 20°, 1.0; [m] 20°, 4.0 H^+, assuming HL/H.L = 3.5; [t] 15-35°, 0.5

Hydrogen fluoride (continued)

Metal ion	Equilibrium	Log K 25°, 0.5	Log K 25°, 1.0	Log K 25°, 0	ΔH 25°, 0	ΔS 25°, 0
$Nb(OH)_2^{3+}$	$ML_5/ML_4 \cdot L$		2.51[e]			
Ag^+	$ML/M \cdot L$	-0.17	-0.32	0.4	(-3)[s]	(-10)[b]
CH_3Hg^+	$ML/M \cdot L$	1.50[h]				
Tl^+	$ML/M \cdot L$			0.10		
$(CH_3)_3Sn^+$	$ML/M \cdot L$		2.28			
	$ML_2/M \cdot L^2$		2.89			
$(CH_3)_3Pb^+$	$ML/M \cdot L$		0.81			
$(C_2H_5)_3Pb^+$	$ML/M \cdot L$		0.54			
Zn^{2+}	$ML/M \cdot L$	0.73	0.78 ±0.03	1.15	3.8	18
			0.8[e]		2.0[c]	10[c]
					1.8[e]	10[e]
Cd^{2+}	$ML/M \cdot L$		0.46		1.2[c]	6[c]
			0.57[e]		1.0[e]	6[e]
	$ML_2/M \cdot L^2$		0.53		0.5[c]	4[c]
			0.85[e]			
Hg^{2+}	$ML/M \cdot L$	1.03		1.6	(1)[s]	(8)[b]
Sn^{2+}	$ML/M \cdot L$		4.08			
	$ML_2/M \cdot L^2$		6.68			
	$ML_3/M \cdot L^3$		9.5			
$(CH_3)_2Sn^{2+}$	$ML/M \cdot L$		3.70			
	$ML_2/M \cdot L^2$		6.57			
	$ML_3/M \cdot L^3$		8.0			
Pb^{2+}	$ML/M \cdot L$	1.26[d]	1.44 ±0.04			
	$ML_2/M \cdot L^2$	2.55[d]	2.54			
	$M \cdot L^2/ML_2(s)$	-6.60[d]	-6.26	-7.44	(5)[r]	(-20)[d]
$(CH_3)_2Pb^{2+}$	$ML/M \cdot L$		1.73			
$(C_2H_5)_2Pb^{2+}$	$ML/M \cdot L$		1.54			
$(C_3H_7)_2Pb^{2+}$	$ML/M \cdot L$		1.61			

[b] 25°, 0.5; [c] 25°, 1.0; [d] 25°, 2.0; [e] 25°, 3.0; [h] 20°, 0.1; [r] 9-27°, 0; [s] 15-35°, 0.5

Hydrogen fluoride (continued)

Metal ion	Equilibrium	Log K 25°, 0.5	Log K 25°, 1.0	Log K 25°, 0	ΔH 25°, 0	ΔS 25°, 0
B(III)	$M(OH)_3L/M(OH)_3 \cdot L$		-0.30			
	$M(OH)_2L_2 \cdot OH/M(OH)_3 \cdot L^2$		-6.27			
	$MOHL_3 \cdot (OH)^2/M(OH)_3 \cdot L^3$		-14.2			
	$ML_4 \cdot (OH)^3/M(OH)_3 \cdot L^4$		-21.6			
Al^{3+}	$ML/M \cdot L$	6.11 ± 0.03	6.09	7.0	0.7^c	30^c
		$6.43^a \pm 0.03$			1.1^a	33^a
	$ML_2/M \cdot L^2$	11.12 ± 0.1		12.6		
		$11.63^a \pm 0.04$			2.0^a	60^a
	$ML_3/M \cdot L^3$	15.0 ± 0.3		16.7		
		$15.5^a \pm 0.0$			2.2^a	78^a
	$ML_4/M \cdot L^4$	18.0 ± 0.8		19.1		
		$18.3^a \pm 0.4$			2.2^a	91^a
	$ML_5/M \cdot L^5$	19.4			1.8^a	97^b
	$ML_6/M \cdot L^6$	19.8				
Ga^{3+}	$ML/M \cdot L$	4.49	4.38	5.9	2.5^c	28^c
	$ML_2/M \cdot L^2$	8.00				
	$ML_3/M \cdot L^3$	10.5				
In^{3+}	$ML/M \cdot L$	3.75	3.70 ± 0.03	4.6	2.6	30
		3.74^d			$2.2^c \pm 0.0$	24^c
	$ML_2/M \cdot L^2$	6.5 ± 0.1	6.4 ± 0.1	8.1	5	50
		6.6^d			$3^c \pm 1$	40^c
	$ML_3/M \cdot L^3$	8.6	8.6 ± 0.0	10.3	7	70
		9.0^d			$5^c \pm 2$	60^c
	$ML_4/M \cdot L^4$	9.9	$9.8 +0.1$	11.5	9	80
		10.3^d			8^c	70^c
CH_3Sn^{3+}	$ML/M \cdot L$	5.10				
	$ML_2/M \cdot L^2$	9.85				
	$ML_3/M \cdot L^3$	14.0				
	$ML_4/M \cdot L^4$	17.1				
	$ML_5/M \cdot L^5$	19.3				

^{a} 25°, 0.1; ^{b} 25°, 0.5; ^{c} 25°, 1.0; ^{d} 25°, 2.0

Hydrogen fluoride (continued)

Metal ion	Equilibrium	Log K 25°, 0.5	Log K 25°, 1.0	Log K 25°, 0	ΔH 25°, 0	ΔS 25°, 0
Sb^{3+}	$ML/M.L$	3.0^v				
	$ML_2/M.L^2$	5.7^v				
	$ML_3/M.L^3$	8.3^v				
	$ML_4/M.L^4$	10.9^v				
Bi^{3+}	$ML/M.L$	1.42^v				
$Ge(IV)$	$ML_4/M(OH)_4 \cdot (HL)^4$	7.30				
	$MHL_5/M(OH)_4 \cdot (HL)^5$	8.94				

---

^{v} 30°, 2.0

Bibliography:

H^+ 42B,43BW,47BD,53HJ,53HW,54AL,54CT,56AL,
58AN,59KP,63C,63Ea,63EMK,64FR,65BC,65CG,
65SS,67WC,69B,70HW,70PM,70W,71AK,71Nc,
72B,73KB,73VK

Li^+ 64SH

Be^{2+} 61HG,65BG

$Mg^{2+}-Ba^{2+}$ 23B,54CT,55P,68TW,69AL,69GS,70E,
71BH,71CV

Sc^{3+} 55P,59KP,69ALa

Y^{3+} 59KP,59Se,61PG,67WC,69AL,69ALa

$La^{3+}-Lu^{3+}$ 55P,57KH,67WC,69AL,70AL

Ac^{3+} 70AL

Am^{3+},Cm^{3+} 69ALa

Th^{4+} 49DR,50DS,51ZA,69Na,70Bc,71KMF

$Pa^{4+},Pa(V)$ 66Ga

U^{4+} 69GV,69Na

Np^{4+} 66AB

Pu^{4+} 49Mb,55P

UO_2^{2+} 54ALa,54DP,56AL,69GV,71AK

NpO_2^{2+} 68AB,70AWa

Mn^{2+} 65CG,72BHa

Fe^{2+},Co^{2+} 72BHa

Ni^{2+} 56AR,72BHa

Cu^{2+} 56AR,58CP,72BHa

Cr^{3+} 52WT,55P

Mn^{3+} 48T

Fe^{3+} 42BG,49DR,53HW,53Sc,55P,55Y,56CH,
59Se,61Y,67WC

VO^{2+} 58AN,71AK

Zr^{4+} 49CMa,55P,63AK,67Na,69N,72B

Hf^{4+} 67N

VO_2^+ 69IV

$Nb(OH)_2^{3+}$ 71Nc

Ag^+ 52LM,61CP

CH_3Hg^+ 65SS

Tl^+ 53BG

$(CH_3)_3Sn^+,(CH_3)_2Sn^{2+},CH_3Sn^{3+}$ 66CP

$(CH_3)_3Pb^+,(C_2H_5)_3Pb^+,(CH_3)_2Pb^{2+},$
$(C_2H_5)_2Pb^{2+},(C_3H_7)_2Pb^{2+}$ 70PM

Zn^{2+} 56AR,58CP,63MHa,69G,71CD

Cd^{2+} 43L,66G

Hg^{2+} 55P

Sn^{2+} 61CP,70BT

Pb^{2+} 23B,61SR,64MH,65BC,71B,72H

$B(III)$ 71GH

Al^{3+} 43BO,53LJ,55P,59KG,69B,71AM,71WT

Ga^{3+} 55Y,55P,71WT

In^{3+} 54HK,54Sa,55P,68AL,69R,71WT,74VK

Hydrogen fluroide (continued)

Sb^{3+} 70Ba

Bi^{3+} 69Ba

Ge(IV)   63BP

Other references: 12P,24DH,25RH,27A,31KP,
        36RB,37J,37RP,39L,39R,41ID,46R,46Ra,
        48RS,48W,49L,49TD,50K,50MKa,50TK,51Da,
        51EU,51MS,51W,52JK,52K,52WT,53HJ,54BBa,
        54Fa,54JK,54SD,54SE,55DW,55M,55RU,56Kg.
        56TK,57Sd,57TV,58PL,59KGb,59Ta,59TL,

59WP,60CC,60Ka,60KG,60KV,60SB,60TV,61D,
61Kd,61MF,61TD,62Ba,62CM,62LN,62NL,62SE,
62VF,63MHa,63TN,63VH,63VR,64FC,64RK,64VH,
65DK,65HS,65SG,64SK,65VW,66BF,66LN,66N,
66PP,67Aa,67BN,67HR,67I,67KR,67LD,67LN,
67MF,67PM,67VK,67VS,68FH,68HSb,68IZ,68K,
68Ka,68KK,68KKa,68KKb,68PM,68SR,68V,69B,
69Bb,69DK,69KK,69KKP,69SG,69VSa,70B,70BH,
70BO,71B,71NL,71PMP,72B,72H,72LO,73J,73MSa,
74MG

Cl^-

Cl^- <u>Chloride ion</u> L^-

Metal ion	Equilibrium	Log K 25°, 0.5	Log K 25°, 1.0	Log K 25°, 0	ΔH 25°, 0	ΔS 25°, 0
K^+	$ML/M.L$			(-0.7)		
Rb^+	$ML/M.L$			(-0.55) ±0.2		
Cs^+	$ML/M.L$			(-0.39) ±0.1		
$(CH_3)_4N^+$	$ML/M.L$			(0.04)		
$(C_4H_9)_4N^+$	$ML/M.L$			(0.40)		
Be^{2+}	$ML/M.L$	-0.3^q		$(-0.8)^f$		
	$ML_2/M.L^2$			-0.7^f		
Mg^{2+}	$ML/M.L$		-1.0^e			
Ba^{2+}	$ML/M.L$			$(-0.13)^n$		
Sc^{3+}	$ML/M.L$	0.04^q		-0.12^f		
	$ML_2/M.L^2$	-0.1^q		-0.8^f		
Y^{3+}	$ML/M.L$		-0.1			
La^{3+}	$ML/M.L$		-0.1 ±0.0	-0.2^f	1.2	4^c
	$ML_2/M.L^2$			-0.6^f		
Ce^{3+}	$ML/M.L$		-0.1 ±0.0			
	$ML_2/M.L^2$		-0.5			
Pr^{3+}	$ML/M.L$		-0.1^j			
Eu^{3+}	$ML/M.L$		-0.1 ±0.0	-0.15^f	$(0)^s$	$(0)^c$
	$ML_2/M.L^2$		-0.7 ±0.2	-0.7^f		
Tm^{3+}	$ML/M.L$		-0.1^j			
Yb^{3+}	$ML/M.L$		-0.2^j			
Lu^{3+}	$ML/M.L$		-0.4^j	-0.35^f		
	$ML_2/M.L^2$			-0.6^f		
Ac^{3+}	$ML/M.L$		-0.1	-0.04^f		
	$ML_2/M.L^2$		-0.6	$(-1.0)^f$		
Pu^{3+}	$ML/M.L$		-0.1			

---

[c] 25°. 1.0; [e] 25°, 3.0; [f] 25°, 4.0; [j] 20°, 1.0; [q] 20°, 0.7; [n] 18°, 0; [s] 0-48°, 1.0

Chloride ion (continued)

Metal ion	Equilibrium	Log K 25°, 0.5	Log K 25°, 1.0	Log K 25°, 0	ΔH 25°, 0	ΔS 25°, 0
Am^{3+}	ML/M.L		$-0.1^j \pm 0.1$	-0.15^f		
	$ML_2/M.L^2$			-0.7^f		
Bk^{3+}	ML/M.L		-0.02			
Es^{3+}	ML/M.L		-0.02			
Th^{4+}	ML/M.L	0.30 ± 0.05	0.18	1.38		
		0.08^d		$0.17^f \pm 0.06$		
	$ML_2/M.L^2$	-1.0^d		$-0.9^f \pm 0.0$		
Pa^{4+}	$ML_2/M.L^2$		0.0^e			
U^{4+}	ML/M.L	0.26^d	0.30^j		$(-5)^t$	$(-16)^d$
Np^{4+}	ML/M.L	0.15^i	-0.04^j			
		0.04^k				
	$ML_2/M.L^2$	-0.15^k	-0.24^j			
	$ML_3/M.L^3$		-0.5^j			
Pu^{4+}	ML/M.L		0.14	0.15^m		
	$ML_2/M.L^2$		-0.17	0.08^m		
	$ML_3/M.L^3$			-1.0^m		
UO_2^{2+}	ML/M.L	-0.06^d	-0.10^j	0.21	$(4)^t$	$(13)^d$
NpO_2^{2+}	ML/M.L	-0.3	-0.09^u			
	$ML_2/M.L^2$		-0.8^u			
PuO_2^{2+}	ML/M.L	$0.10^d-0.01$			$(2)^t$	$(7)^d$
	$ML_2/M.L^2$	$-0.35^d-0.07$			$(4)^t$	$(12)^d$
Mn^{2+}	ML/M.L		0.04			
Co^{2+}	ML/M.L		-0.05			
		-0.14^d	$-0.26^e \pm 0.02$		0.5^d	1^d
Ni^{2+}	ML/M.L		0.00			
		$-0.21^d \pm 0.04$	-0.57^e		0.5^d	1^d
Cu^{2+}	ML/M.L	0.09^d	-0.06^e	0.40	1.6^d	6^d
	$M.(OH)^{1.5}.L^{0.5}/M(OH)_{1.5}L_{0.5}(s)$	-17.16	-17.3 ± 0.1		$(9)^r$	(-50)

---

^{d} 25°, 2.0; ^{e} 25°, 3.0; ^{f} 25°, 4.0; ^{i} 20°, 0.5; ^{j} 20°, 1.0; ^{k} 20°, 2.0; ^{m} 20°, 4.0; ^{r} 25-75°, 0; ^{t} 10-40°, 2,0; ^{u} 10°, 3.0

Chloride ion (continued)

Metal ion	Equilibrium	Log K 25°, 0.5	Log K 25°, 1.0	Log K 25°, 0	ΔH 25°, 0	ΔS 25°, 0
Re^{2+}	ML/M.L	2.0				
Ti^{3+}	ML/M.L	0.34[v]				
Cr^{3+}	ML/M.L	-0.4[d]	-0.5	0.05[f]	6.6[g]	22[f]
Mn^{3+}	ML/M.L	0.9[d]	1.12[e]			
Fe^{3+}	ML/M.L	0.64	0.63 ±0.03	1.48 ±0.00	5.6	26
		0.7[d]	0.8[e]	1.0[f]	4.3[c]	17[c]
	$ML_2/M.L^2$		0.75 ±0.05	2.13		
	$ML_3/M.L^3$		-0.7			
Co^{3+}	ML/M.L		1.42[e]			
VO^{2+}	ML/M.L		0.04[j]			
Zr^{4+}	ML/M.L	0.30[d]		0.92[w]		
	$ML_2/M.L^2$			1.32[w]		
	$ML_3/M.L^3$			1.51[w]		
	$ML_4/M.L^4$			1.2[w]		
Hf^{4+}	ML/M.L	0.38[d]	0.34[l]			
	$ML_2/M.L^2$	0.07[d]	-0.02[l]			
	$ML_3/M.L^3$	-0.6[d]				
Cu^{+}	ML/M.L			2.70[g]		
	$ML_2/M.L^2$	5.19[h]		5.5[o]		
				6.00[g]		
	$ML_3/M.L^3$			5.7[o]		
				6.0[g]		
	$M_2L_4/M^2.L^4$			13.1[g]		
	M.L/ML(s)			-6.73		
				-7.38[g]		

---

[c] 25°, 1.0; [d] 25°, 2.0; [e] 25°, 3.0; [f] 25°, 4.0; [g] 25°, 5.0; [h] 20°, 0.1; [j] 20°, 1.0; [l] 20°, 3.0 $HClO_4$; [o] 20°, 0; [v] 40°, 0.5; [w] 20°, 6.5 H^+

Chloride ion (continued)

Metal ion	Equilibrium	Log K 25°, 0.5	Log K 25°, 1.0	Log K 25°, 0	ΔH 25°, 0	ΔS 25°, 0
Ag^+	ML/M.L	3.08^a	3.36	3.31 ±0.00		
			3.45^m	3.70^g		
	$ML_2/M.L^2$	5.08^a	5.20	5.25 ±0.01		
			5.67^m	5.62^g		
	$ML_3/M.L^3$		6.0^m	6.4^g		
	$ML_4/M.L^4$		6.0^m	6.1^g		
	M.L/ML(s)	-9.62	-9.74	-9.74 ±0.00	15.7	8
			-10.05^e	-10.40^m		
Hg_2^{2+}	$M.L^2/ML_2$(s)	-16.88		-17.91 ±0.03	22^s ±3	-8^b
CH_3Hg^+	ML/M.L	5.18^a	5.32		-6.0^h	4^a
Tl^+	ML/M.L	0.11 ±0.00	0.04 ±0.04	0.49 ±0.03	-1.5 ±0.1	-3
		-0.10^d±0.02	-0.1^e ±0.1	0.0^f ±0.1		
	$ML_2/M.L^2$		-0.1 ±0.3	0.0 ±0.2		
		-0.6^d ±0.5	-1.0^e ±0.1	-0.8^f ±0.5		
	M.L/ML(s)			-3.74 ±0.02	$(10)^r$	(15)
				-3.04^f		
$(CH_3)_3Pb^+$	ML/M.L		0.32			
$(C_2H_5)_3Pb^+$	ML/M.L		0.57			
Pd^{2+}	ML/M.L		4.47 -0.5	6.1 -0.1	-3.0^c	10^c
	$ML_2/M.L^2$		7.74 -0.3	10.7 -0.1	-5.6^c	17^c
	$ML_3/M.L^3$		10.2 -0.4	13.1 ±0.0	-8.2^c	19^c
	$ML_4/M.L^4$		11.5 ±0.6	15.4 ±0.3	-11.6^c	14^c
Pt^{2+}	$ML_2/ML.L$	4.0				
	ML_2(cis)/ML_2(trans)	0.08				
	$ML_3/ML_2.L$	2.96				
	$ML_4/ML_3.L$	1.90			$(-4)^s$	$(-5)^b$

---

^{a} 25°, 0.1; ^{b} 25°, 0.5; ^{c} 25°, 1.0; ^{d} 25°, 2.0; ^{e} 25°, 3.0; ^{f} 25°, 4.0; ^{g} 25°, 5.0; ^{h} 20°, 0.1; ^{m} 20°, 4.0; ^{r} 0-50°, 0; ^{s} 7-40°, 0.5

Chloride ion (continued)

Metal ion	Equilibrium	Log K 25°, 0.5	Log K 25°, 1.0	Log K 25°, 0	ΔH 25°, 0	ΔS 25°, 0
Zn^{2+}	ML/M.L		0.11	0.43		
		-0.49^d	-0.19^e	0.30^f	1.3^e	4^e
				0.61		
	$ML_2/M.L^2$	0.02^d	-0.6^e	0.0^f	9^e	27^e
	$ML_3/M.L^3$			0.5		
		-0.1^d	0.1^e	1.0^f	0^e	1^e
	$ML_4/M.L^4$			0.2		
				-1^f		
	$M.(OH)^{1.5}.L^{0.5}/M(OH)_{1.5}L_{0.5}(s)$			-13.4		
Cd^{2+}	ML/M.L	1.35 ± 0.02	1.35 ± 0.02	1.98 ± 0.03	0.3^b	7^b
		$1.44^d\pm0.02$	$1.54^e\pm0.05$	$1.66^f\pm0.1$	-0.1^e	7^e
	$ML_2/M.L^2$	1.7 ± 0.1	1.7 ± 0.1	2.6 ± 0.1	0.9^b	11^b
		$1.9^d \pm0.1$	$2.2^e \pm0.1$	$2.4^f \pm0.1$	0.0^e	10^e
	$ML_3/M.L^3$		1.5 ± 0.2	2.4 ± 0.1	2.4^c	15^c
		$1.9^d \pm0.1$	$2.3^e \pm0.1$	$2.8^f \pm0.3$	1.9^e	17^e
	$ML_4/M.L^4$			1.7		
			1.6^e	$2.2^f +0.3$	6.1^f	31^f
Hg^{2+}	ML/M.L	$6.74\ -0.1$	6.72		-4.8	
			7.07^e		$-5.7^b \pm0.2$	12^b
					-5.8^e	13^e
	$ML_2/M.L^2$	$13.22\ -0.2$	13.23		$-12.8^b \pm0.0$	18^b
			13.98^e		-12.3^e	23^e
	$ML_3/M.L^3$	14.1 ± 0.2	14.2		-15.0^b	14^b
			$14.7^e \pm0.0$		-13.3^e	23^e
	$ML_4/M.L^4$	15.1 ± 0.1	15.3		-14.9^b	19^b
			$16.2^e \pm0.1$		-14.8^e	25^e
	ML/MOHL.H		3.1			
	$MOHL.L/ML_2.OH$		4.23	4.09	-1.2	15
	$M(OH)_2.L/MOHL.OH$		3.8	3.77	-1.2	13

---

[b] 25°, 0.5; [c] 25°, 1.0; [d] 25°, 2.0; [e] 25°, 3.0; [f] 25°, 4.0

Chloride ion (continued)

Metal ion	Equilibrium	Log K 25°, 0.5	Log K 25°, 1.0	Log K 25°, 0	ΔH 25°, 0	ΔS 25°, 0
Sn^{2+}	$ML/M.L$			1.51		
		1.08^d	$1.17^e \pm 0.02$	1.45^f	$(3)^x$	$(15)^e$
	$ML_2/M.L^2$			2.25		
		1.72^d	$1.72^e \pm 0.02$	2.35^f	$(3)^x$	$(20)^e$
	$ML_3/M.L^3$			2.0		
		1.5^d	$1.7^e \pm 0.0$	2.5^f	$(6)^x$	$(30)^e$
	$ML_4/M.L^4$			1.5		
				2.3^f		
$(CH_3)_2Sn^{2+}$	$ML/M.L$		0.38^e			
	$ML_2/M.L^2$		-0.1^e			
Pb^{2+}	$ML/M.L$	0.9 −0.07	0.90 ± 0.06	1.59 ± 0.02	4.4	22
		$1.02^d \pm 0.02$	$1.17^e \pm 0.03$	$1.29^f \pm 0.06$		
	$ML_2/M.L^2$	1.3 −0.1	1.3 ± 0.1	1.8		
		$1.4^d \pm 0.0$	$1.7^e \pm 0.1$	$2.0^f \pm 0.2$		
	$ML_3/M.L^3$		1.4 ± 0.2	1.7		
		$1.5^d \pm 0.1$	$1.9^e \pm 0.1$	$2.3^f \pm 0.2$		
	$ML_4/M.L^4$			1.4		
		$0.7^d \pm 0.2$	$1.2^e \pm 0.2$	$1.7^f \pm 0.2$		
	$M.L^2/ML_2(s)$		-5.0^e	-4.78 ± 0.02		
$(CH_3)_2Pb^{2+}$	$ML/M.L$		0.76			
	$ML_2/M.L^2$		1.31			
$(C_2H_5)_2Pb^{2+}$	$ML/M.L$		0.96			
	$ML_2/M.L^2$		1.74			
$(C_3H_7)_2Pb^{2+}$	$ML/M.L$		0.99			
	$ML_2/M.L^2$		1.84			
$Au(III)$	$M(OH)_3L/M(OH)_4.H.L$			8.51^o		
	$M(OH)_2L_2/M(OH)_4.H^2.L^2$			16.57^o		
	$MOHL_3/M(OH)_4.H^3.L^3$			23.6^o		
	$ML_4/M(OH)_4.H^4.L^4$			29.6^o		
Ga^{3+}	$ML/M.L$	0.01^q				

---

^{d} 25°, 2.0; ^{e} 25°, 3.0; ^{f} 25°, 4.0; ^{o} 20°, 0; ^{q} 20°, 0.7; ^{x} 0-45°, 3.0

Chloride ion (continued)

Metal ion	Equilibrium	Log K 25°, 0.5	Log K 25°, 1.0	Log K 25°, 0	ΔH 25°, 0	ΔS 25°, 0
In^{3+}	ML/M.L	$2.32^q \pm 0.05$	2.20^j			
		2.45^d	2.58^e		1.2^d	15^d
	$ML_2/M.L^2$	$3.62^q \pm 0.05$	3.56^j			
		3.4^d	3.84^e		0.8^d	18^d
	$ML_3/M.L^3$	$4.0^q \pm 0.2$				
		3.7^d	4.2^e		8^d	44^d
	MOHL/ML.OH		10.3^e			
	$M_2OHL/MOHL.M$		1.6^e			
Tl^{3+}	ML/M.L	6.72		7.72	-5.5^e	14^e
			$7.10^e \pm 0.06$	7.46^f	-6.0^f	14^f
	$ML_2/M.L^2$	11.76 ± 0.07		13.48	-9.9^e	24^e
			$12.46^e \pm 0.1$	13.25^f	-10.1^f	27^f
	$ML_3/M.L^3$	14.4 ± 0.0		16.5	-11.0^e	35^e
			$15.8^e \pm 0.3$	16.7^f	-11.2^f	39^f
	$ML_4/M.L^4$	16.3 ± 0.1		18.3	-11.3^e	45^e
			$18.0^e \pm 0.3$	19.4^f	-11.3^f	51^f
	ML/MOHL.H		1.87^e			
As(III)	$M(OH)_2L/M(OH)_3.H.L$			-1.07		
	$MOHL_2/M(OH)_3.H^2.L^2$			-4.54		
	$ML_3/M(OH)_3.H^3.L^3$			-8.7		
Sb^{3+}	ML/M.L			2.3^f		
	$ML_2/M.L^2$			3.5^f		
	$ML_3/M.L^3$			4.2^f		
	$ML_4/M.L^4$			4.7^f		
	$ML_5/M.L^5$			4.7^f		
	$ML_6/M.L^6$			4.1^f		

---

^{d} 25°, 2.0; ^{e} 25°, 3.0; ^{f} 25°, 4.0; ^{j} 20°, 1.0; ^{q} 20°, 0.7

Chloride ion (continued)

Metal ion	Equilibrium	Log K 25°, 0.5	Log K 25°, 1.0	Log K 25°, 0	ΔH 25°, 0	ΔS 25°, 0
Bi^{3+}	ML/M.L	$(2.36)^k$	2.2^e		4.0	
					0.5^f	12^e
	$ML_2/M.L^2$	3.5^k	3.5^e		$(4)^x$	$(30)^e$
	$ML_3/M.L^3$	5.4^k	5.8^e		$(5)^x$	$(40)^e$
	$ML_4/M.L^4$	6.1^k	6.8^e		$(4)^x$	$(50)^e$
	$ML_5/M.L^5$	6.7^k	7.3^e			
	$ML_6/M.L^6$	6.6^k	7.4^e			
	$MOL(s).H^2/M.L$		-6.47	-7.80	$(1)^r$	(-30)
		$-6.59^d \pm 0.05$	-6.75^e		$(4)^t$	$(-20)^d$

^{d} 25°, 2.0; ^{e} 25°, 3.0; ^{f} 25°, 4.0; ^{k} 20°, 2.0; ^{r} 15-25°, 0; ^{t} 15-25°, 2.0; ^{x} 25-65°, 3.0

Bibliography:

K^+	71PJ
Rb^+	12NF,66MB,72DJ
Cs^+	12NF,27D,54GMa,71PJ
$(CH_3)_4N^+,(C_4H_9)_4N^+$	68F
Be^{2+}	65MJ,71SK
Mg^{3+}	73HHa
Ba^{2+}	35MD
Sc^{3+}	64RSM,66SH
$Y^{3+}-Lu^{3+}$	51M,62PM,63CU,64La,65S,67SS,71KN
Ac^{3+}	68SMR,69SS
Pu^{3+}	53CM
Am^{3+}	62G,62PM,65S,71KN
Bk^{3+},Es^{3+}	72HP
Th^{4+}	50DS,51ZA,52WS
$Pa^{4+},Pa(V)$	66Ga
U^{4+}	54AL,55DW
Np^{4+}	66SN
Pu^{4+}	58RA,60GN
UO_2^{2+}	51Aa,54DP,57DM
NpO_2^{2+}	55CS,70AW
PuO_2^{2+}	57NB,61RMa
Mn^{2+}	74BL
Co^{2+}	60LR,66KL,67MS,70MM,74BL

Ni^{2+}	57KL,60LR,66KL,70MM,74BL
Cu^{2+}	49NT,51Ma,51NL,58BB,60BB,66KL, 68MM,69MM
Re^{2+}	65PY
Ti^{3+}	54DQ
Cr^{3+}	58SK,64BK,66AS,67HK
Mn^{3+}	48T,74RN
Fe^{3+}	39L,42RS,55LR,61SRD,61WK,62WG,63HC, 67VLb,69F,71MH
Co^{3+}	66CN
VO^+	58AN
Zr^{4+}	49CMa,57S
Hf^{4+}	63PA,67HP
Cu^+	38L,61H,68ST,70AR,70GZ
Ag^+	38OB,52GM,52JMa,54GM,54KT,55DJ, 55Fb,57CH,57L,61K,64AJ,64WG,71AB
Hg_2^{2+}	29B,46L,47JQ,55DG,63HI
CH_3Hg^+	65SS,73BI
Tl^+	26BH,27D,27O,28RV,30BD,34CL,37RD, 53BG,55Aa,57N,57NN,58H,61KMF,61NR, 62Se,63KM,67Ka,67KPa,69BP,71BS, 71FR,72FI
$(C_6H_5)_3Sn^+,(C_6H_5)_3Pb^+$	65SM

Chloride ion (continued)

$(CH_3)_3Pb^+, (C_2H_5)_3Pb^+$ 71PM

Pd^{2+} 57DBF,63GKG,64BS,64W,65FK,66SB,68L, 68LMV,72E,72R

Pt^{2+} 66EL,67DE,70Ea

Zn^{2+} 44SL,57KL,58AS,64MK,68SM,69G,71FCK,74BL

Cd^{2+} 30RD,36HF,41L,49K,53E,53VD,57KL,58TF, 59Ma,62BD,63MKN,65M,66G,66M,67MF,68GJ, 68P,69SP,71FCK,72FKM,73HH,74BL,74EM, 74FRP

Hg^{2+} 47LJ,57M,58E,60GK,61MP,63EM,63HI,64CI, 65A,65PI,66VS,68CG

Sn^{2+} 28P,50DC,52VR,61RM,61TH,62HZ

$(CH_3)_2Sn^{2+}$ 65FMT

Pb^{2+} 30RD,44NG,55BPP,55K,55Na,57KL,57PC, 61M,63MKc,64AP,64MK,64MKb,64SM,65MKF, 66VSe,70FS,71V,72RSL,73V

Au(III) 48B

Ga^{3+} 67MA

In^{3+} 54CI,54S,54Sb,54Sc,59BK,69R,72F,72Fa

Tl^{3+} 60B,63AG,64LR,64WG,64KMb,71BS

As(III) 57Ad

Sb^{3+} 59PD

Bi^{3+} 57AG,63KMa,67VL,68VG

Other references: 03S,04BE,10M,23B,23P, 26LD,30W,31F,31FL,32N,33HJ,33NS,34BH, 34CC,35BM,35DH,36R,37M,38G,38PS,39G, 42Ba,42GN,42Nc,43O,44O,45B,56Na,48C, 49B,49DJ,49GGa,49Hb,49Jb,49RP,50B, 50KN,50MD,50N,51MS,51NK,51RL,51SSW, 52La,53BGa,53BL,53CT,53G,53GT,53N, 53P,53Pa,53WS,54SE,55GE,55M,55RC, 55WW,56G,56P,56PV,55Tb,57KS,57SL, 57TS,58Da,58GK,58Jb,58MW,58SPS,58SW, 58ZB,59CN,59Kc,59Mb,59Mc,59MC,59TC, 60BT,60FSS,60GG,60GS,60L,60Ma,60TZ, 61BT,61Ha,61MA,61MS,61S,61SM,62AP, 62BS,62DC,62FSD,62FT,62M,62MR,62MS, 62MSa,62P,62PPL,62Sb,62ST,63FD,63GK, 63KB,63MF,63MMa,63ND,64Ba,64BMa,64BP, 64ID,64NU,64SAb,64SB,64SM,64VR,65AB, 65BW,65GS,64HA,65HE,64HP,65HS,65JL, 65MRS,65NPG,66CP,66DO,66F,66L,66MSY, 66SG,66SHa,66WD,67BP,67EME,67ES,67Ga, 67IW,67KR,67L,67NK,68CF,68EPa,68L, 68LM,68MH,68NK,68NM,69BM,69Ca,69CPK, 69J,69KS,69MA,69MNM,69MP,69NPS,69PB, 69SB,69SM,69ST,70AW,70DS,70HV,70KBM, 70PH,70RG,71BHa,71BN,71D,71DC,71KB, 71KM,71MKA,71MM,71PB,71PJ,71PS,72BA, 72BBM,72DJ,72J,73SB,72TS,73GS,73L, 73Pa,74AC,74BC,74FKa,74GW,74MS,74SP

$$ClO_3^-$$

O_3Cl^- Chlorate ion L^-

Metal ion	Equilibrium	Log K 25°, 0.5	Log K 25°, 1.0	Log K 25°, 0	ΔH 25°, 0	ΔS 25°, 0
Li^+	ML/M.L			(-1.2)		
Na^+	ML/M.L			(-0.49) ±0.1		
K^+	ML/M.L			(-0.15) +0.1		
Rb^+	ML/M.L			(-0.10)		
Ba^{2+}	ML/M.L			(0.7)		
Sc^3	ML/M.L		-0.02		(0)[r]	(0)[c]
La^{3+}	ML/M.L		-0.2			
Eu^{3+}	ML/M.L	0.02[a]			(-5)[s]	(-17)[a]
Tb^{3+}	ML/M.L	-0.05[a]			(-4)[s]	(-14)[a]
Th^{4+}	ML/M.L	0.26				
Cu^{2+}	$M.(OH)^{1.5}.L^{0.5}/M(OH)_{1.5}L_{0.5}(s)$	-15.69		-15.89		
Fe^{3+}	ML/M.L		-0.40			
Ag^+	ML/M.L			(0.22)		
Tl^+	ML/M.L			0.47		
Cd^{2+}	ML/M.L		-0.26	-0.30[e]		
Pb^{2+}	ML/M.L		-0.32			
	$ML_2/M.L^2$		-0.6			

---

[a] 25°, 0.1; [c] 25°, 1.0; [e] 25°, 3.0; [r] 15-30°, 1.0; [s] 2-40°, 0.1

Bibliography:

Li^+	67ADT	Cu^{2+}	63LL
Na^+-Rb^+	31BR,66MB,72DD	Fe^{3+}	71MH
Ba^{2+}	35MD	Ag^+	48M
Sc^{3+}	72MH	Tl^+	30BD
La^{3+}	51M	Cd^{2+}	43L,56Kb
Eu^{3+},Tb^{3+}	72RC	Pb^{2+}	56Kb
Th^{4+}	50DS		

Other references: 65PY,73HH

$$ClO_4^-$$

O_4Cl^- Perchlorate ion L^-

Metal ion	Equilibrium	Log K 20°, 0.15	Log K 25°, 1.0	Log K 25°, 0	ΔH 25°, 0	ΔS 25°, 0
K^+	ML/M.L			(-0.01) ±0.00		
Rb^+	ML/M.L			(0.13)		
$(CH_3)_4N^+$	ML/M.L			0.27	0.2	2
$(C_2H_5)_4N^+$	ML/M.L			-0.08	0.0	0
$(C_3H_7)_4N^+$	ML/M.L			0.05	2.5	9
Ce^{3+}	ML/M.L		0.15[s]	-0.21[t]	(-17)[u]	(-60)[s]
Am^{3+}	ML/M.L	-0.07[d]				
Cu^{2+}	$M.(OH)^{1.7}.L^{0.3}/M(OH)_{1.7}L_{0.3}$(s)			-17.2		
Fe^{3+}	ML/M.L	0.4 ±0.1		1.15		
Tl^+	ML/M.L			0.0 ±0.2	(2)[v]	(7)

[d] 25°, 2.0; [s] 25°, 1.1; [t] 25°, 5.1; [u] 18-40°, 1.1; [v] 23-80°, 0

Bibliography:

K^+ 45J,71Da Fe^{3+} 54Se,58Ha,59Sa,60RS,69F
Rb^+ 71Da Tl^+ 37RD,66MB,67ZB
$(CH_3)_4N^+$, $(C_2H_5)_4N^+$, $(C_3H_7)_4N^+$ 69IE   Other references: 48M,52Se,54HR,55HB,
Ce^{3+} 56SW   60HR,61H,63FP,63Hc,64S,65HD,65JB,
Am^{3+} 72BC   66LV,66R,68D,68OA,70KS,73J,74J
Cu^{2+} 49NT

Br^-

Br^- Bromide ion L^-

Metal ion	Equilibrium	Log K 25°, 0.5	Log K 25°, 1.0	Log K 25°, 0	ΔH 25°, 0	ΔS 25°, 0
Cs^+	ML/M.L			(0.03)		
$(CH_3)_4N^+$	ML/M.L			(0.16) ±0.07		
$(C_2H_5)_4N^+$	ML/M.L			(0.38)		
$(C_3H_7)_4N^+$	ML/M.L			(0.49)		
Be^{2+}	ML/M.L	-0.4^q		-0.7^f		
	$ML_2/M.L^2$			-0.8^f		
Mg^{2+}	ML/M.L		-1.4^e			
Sc^{3+}	ML/M.L	-0.07^q				
	$ML_2/M.L^2$	-0.3^q				
Y^{3+}	ML/M.L		-0.15			
Ce^{3+}	ML/M.L		-0.2			
Pr^{3+}	ML/M.L		-0.2^e			
Sm^{3+}	ML/M.L		-0.2^e			
Eu^{3+}	ML/M.L		-0.2			
	$ML_2/M.L^2$		-0.4			
Ho^{3+}	ML/M.L		$(-0.6)^e$			
Er^{3+}	ML/M.L		-0.5^e			
Ac^{3+}	ML/M.L		-0.2			
	$ML_2/M.L^2$		-0.5			
U^{4+}	ML/M.L		0.18^j			
UO_2^{2+}	ML/M.L		-0.3^j	-0.2		
Co^{2+}	ML/M.L	-0.13^q				
		$(-0.11)^d$	-0.7^e		0.1^d	0^d
	$ML_2/M.L^2$	-0.4^q				
Ni^{2+}	ML/M.L	$(-0.12)^d$	-0.8^e		0.1^d	0^d
Cu^{2+}	ML/M.L	-0.07^d	-0.5^e ±0.1	-0.03	0.9^d	3^d
	$M.(OH)^{1.5}.L^{0.5}/M(OH)_{1.5}L_{0.5}(s)$	-16.70		-17.15^o		

^{d} 25°, 2.0; ^{e} 25°, 3.0; ^{f} 25°, 4.0; ^{j} 20°, 1.0; ^{o} 20°, 0; ^{q} 20°, 0.7

Bromide ion (continued)

Metal ion	Equilibrium	Log K 25°, 0.5	Log K 25°, 1.0	Log K 25°, 0	ΔH 25°, 0	ΔS 25°, 0
Fe^{3+}	ML/M.L		-0.2 ±0.1	0.6 ±0.0	(6)[r]	(20)
	$ML_2/M.L^2$		-0.5 ±0.1			
Hf^{4+}	ML/M.L		-0.1[l]			
Cu^+	$ML_2/M.L^2$			5.9		
	M.L/ML(s)			-8.3		
Ag^+	ML/M.L	4.30[a]		4.68[n]		
	$ML_2/M.L^2$	6.64[a]	7.23[g]	(7.7)[n]		
	$ML_3/M.L^3$	8.1[a]	8.3	8.7 ±0.2		
			9.2[g]			
	$ML_4/M.L^4$	8.9[a]	9.5[g]	9.0		
	M.L/ML(s)	-12.10[a]	-11.92	-12.30 ±0.02	20.2	11
			-12.62[g]			
Hg_2^{2+}	$M.L^2/ML_2(s)$	-21.29		-22.25 ±0.01	31[s] ±1	2[b]
CH_3Hg^+	ML/M.L	6.49[a]			-9.9[h]	-4[a]
Tl^+	ML/M.L	0.48	0.41	0.91 ±0.03	-3.0 ±0.5	-6
		0.38[d]	0.34[e]	0.33[f]±0.01	-5[f]	-15[f]
	$ML_2/M.L^2$			0.17[f]±0.04		
	$ML_3/M.L^3$			-0.1[f] ±0.2		
	$ML_4/M.L^4$			-0.6[f]		
	M.L/ML(s)			-5.44 ±0.03	(14)[t]	(22)
				-4.81[f]-0.01		
$(C_6H_5)_3Sn^+$	ML/M.L	3.3[p]				
$(C_6H_5)_3Pb^+$	ML/M.L	5.7[p]				
Pd^{2+}	ML/M.L		5.17		-5.1[c]	7[c]
	$ML_2/M.L^2$		9.42			
	$ML_3/M.L^3$		12.7			
	$ML_4/M.L^4$		14.9 ±1		-13.1[a]	24[c]
Zn^{2+}	ML/M.L		-0.59[e]±0.02		0.4[e]	-1[e]

[a] 25°, 0.1; [b] 25°, 0.5; [c] 25°, 1.0; [d] 25°, 2.0; [e] 25°, 3.0; [f] 25°, 4.0; [g] 25°, 5.0; [h] 20°, 0.1; [l] 20°, 3.0; [n] 18°, 0; [p] 30°, 0.1; [r] 30-40°, 0; [s] 7-40°, 0.5; [t] 5-45°, 0

Bromide ion (continued)

Metal ion	Equilibrium	Log K 25°, 0.5	Log K 25°, 1.0	Log K 25°, 0	ΔH 25°, 0	ΔS 25°, 0
Cd^{2+}	ML/M.L	1.55 ±0.05	1.57 ±0.01	2.14 ±0.02	-0.8[b]	5[b]
		1.63[d]±0.05	1.74[e]±0.09		-1.0[e] ±0.0	5[e]
	$ML_2/M.L^2$		2.1 ±0.1	3.0 ±0.1	-0.8[b]	7[b]
		2.2[d] ±0.1	2.4[e] ±0.1		-1.6[e] -1	6[e]
	$ML_3/M.L^3$		2.6 ±0.1	3.0 ±0.1	0.0[b]	12[b]
		2.8[d] ±0.2	3.3[e] ±0.2		0.2[e] -1	16[e]
	$ML_4/M.L^4$		2.6 ±0.2	2.9 ±0.2		
		3.2[d] ±0.2	3.8[e] ±0.2		0.5[e] -1	19[e]
Hg^{2+}	ML/M.L	9.00 ±0.06	9.40[e]		-10.6	
					-10.2[b]+0.1	7[b]
					-9.6[e]	11[e]
	$ML_2/M.L^2$	17.1 ±0.2	17.98[e]		-20.9	
					-21.0[b] ±0.2	8[b]
					-19.2[e]	18[e]
	$ML_3/M.L^3$	19.4 ±0.2	20.7[e]		-23.8[b]	9[b]
					-21.8[e]	22[e]
	$ML_4/M.L^4$	21.0 ±0.2	22.23[e]		-25.9	
					-27.8[b] ±0.1	3[b]
					-25.2[e]	17[e]
	$M.L^2/ML_2(s)$	-18.9			25	-3[b]
Sn^{2+}	ML/M.L		0.74	1.16 ±0.05		
		0.50[d] ±0.07	0.75[e]±0.03	0.88[f]±0.03	(1)[u]	(7)[e]
	$ML_2/M.L^2$		0.9	1.7 ±0.1		
		(1.19)[d]	1.15[e]±0.02	1.43[f]	(3)[u]	(20)[e]
	$ML_3/M.L^3$		1.2[e] ±0.1	1.5[f]	(2)[u]	(20)[e]
	$ML_4/M.L^4$		0.4[e]	1.0[f]		
$(CH_3)_2Sn^{2+}$	ML/M.L		-1.0[e]			

---

[b] 25°, 0.5; [d] 25°, 2.0; [e] 25°, 3.0; [f] 25°, 4.0; [u] 0-45°, 3.0

Bromide ion (continued)

Metal ion	Equilibrium	Log K 25°, 0.5	Log K 25°, 1.0	Log K 25°, 0	ΔH 25°, 0	ΔS 25°, 0
Pb^{2+}	$ML/M.L$	1.06	1.10 ±0.06	1.77 ±0.1		
		1.20^d	1.29^e±0.01	1.48^f±0.03	$(-1)^v$	$(3)^e$
	$ML_2/M.L^2$	1.8	1.8	2.6		
		2.0^d	2.2^e +0.1	2.5^f	$(-1)^v$	$(7)^e$
	$ML_3/M.L^3$	2.0	2.2 ±0.0	3.0		
		2.5^d ±0.1	2.9^e ±0.1	3.5^f	$(-1)^v$	$(10)^e$
	$ML_4/M.L^4$		2.0	2.3		
		2.6^d	3.1^e	3.5^f	$(-4)^v$	$(1)^e$
	$ML_5/M.L^5$	1.6^d	2.4^e	2.7^f		
	$M.L^2/ML_2(s)$			-5.68^f		
Ga^{3+}	$ML/M.L$	-0.10^q				
In^{3+}	$ML/M.L$	2.04^q	1.93			
		1.99^d		2.08^f	0.5^d	11^d
	$ML_2/M.L^2$	3.1^q	2.6			
		2.6^d		3.4^f	1.4^d	17^d
	$ML_3/M.L^3$	3.4^q		4.0^f		
	$ML_4/M.L^4$			4.8^f		
Tl^{3+}	$ML/M.L$	8.3^i	8.9	9.7^n		
			9.28^e	9.51^f	-9.0^f	13^f
	$ML_2/M.L^2$	14.6^i	16.4	16.6^n		
			16.70^e	16.88^f	-15.1^f	27^f
	$ML_3/M.L^3$	19.2^i		21.2^n		
			22.1^e	22.3^f	-19.6^f	36^f
	$ML_4/M.L^4$	22.3^i		23.9^n		
			25.7^e	26.4^f	-21.8^f	48^f

---

^{d} 25°, 2.0; ^{e} 25°, 3.0; ^{f} 25°, 4.0; ^{i} 20°, 0.4; ^{n} 18°, 0; ^{q} 20°, 0.7; ^{v} 5-65°, 3.0

Bromide ion (continued)

Metal ion	Equilibrium	Log K 25°, 0.5	Log K 25°, 1.0	Log K 25°, 0	ΔH 25°, 0	ΔS 25°, 0
Bi^{3+}	$ML/M.L$	2.37	2.22	3.06 +0.2	3.3	25
		$2.32^d \pm 0.04$	2.63^e	3.12^f	0.0^f	14^f
					0.7^w	
	$ML_2/M.L^2$	4.2	(4.4)	5.6		
		$4.4^d \pm 0.1$	5.0^e	5.7^f		
	$ML_3/M.L^3$	5.9	6.2	7.4		
		$6.3^d \pm 0.1$	6.7^e	8.2^f		
	$ML_4/M.L^4$	7.3	(7.2)	8.6		
		$7.8^d \pm 0.1$	8.1^e	10.0^f		
	$ML_5/M.L^5$	8.2	8.7	(9.2)		
		$9.2^d \pm 0.1$	$(9.0)^e$	11.9^f		
	$ML_6/M.L^6$	8.3	8.8	(8.7)		
		$9.5^d \pm 0.2$	9.8^e	11.8^f		
	$M.L/H^2.MOL(s)$	-6.24^d				
		-6.52^k				

---

^{d} 25°, 2.0; ^{e} 25°, 3.0; ^{f} 25°, 4.0; ^{k} 20°, 2.0; ^{w} 50°, 4.0

Bibliography:

Cs^+ 68HF

$(CH_3)_4N^+, (C_2H_5)_4N^+, (C_3H_7)_4N^+$ 65Lb,67Wa

Be^{2+} 65MJ,71SK

Mg^{2+} 73HHa

Sc^{3+} 64MR

Y^{3+}, Ce^{3+}, Eu^{3+} 63CU

$Pr^{3+}, Sm^{3+}, Er^{3+}$ 73KP

Ho^{3+} 66MSY,73KP

Ac^{3+} 68SMR

U^{4+} 54AL

UO_2^{2+} 51Aa,57DM

Co^{2+} 61LW,65FM,66KL

Ni^{2+} 61LW,66KL

Cu^{2+} 50Na,51NL,60LR,66KL,68MM,70MM

Fe^{3+} 39L,42RS,55LR,57YT,67M,71MH

Hf^{4+} 67HP

Cu^+ 38L

Ag^+ 38OB,53BLa,53GM,54GM,54KT,54LP,54PV, 57L,67BP

Hg_2^{2+} 29B,48BJ,63HI

CH_3Hg^+ 65SS

Tl^+ 23B,33IT,55Aa,56Ca,57N,57NN,58Ma, 60KMa,62SD,69CP,74FRI

$(C_6H_5)_3Sn^+, (C_6H_5)_3Pb^+$ 65SM

Pd^{2+} 63GKG,64SB,64FK,66BSA,66SB,67IW, 72E,72R

Zn^{2+} 44SL,69G

Cd^{2+} 39B,41L,53E,53F,57KE,62BD,65HS,66G, 67SG,73HH,74EM,74FK

Hg^{2+} 39G,49BJ,57M,58E,58ST,60GK,61MP, 63BS,64CI,65A

Sn^{2+} 28P,51DP,52V,62Ha,69FB

$(CH_3)_2Sn^{2+}$ 65FMT

Bromide ion (continued)

Pb^{2+} 55BPR,55PP,56K,61KMT,63MKb,68FS,70FS,   57SL,58Da,60EK,60FSS,60GS,61Ha,61Mc,
         72FSL,73HH   61SM,62FSD,62P,63EM,63ND,64MKa,64PB,
Ga^{3+} 67MA   64SLI,65MRI,65SMa,66DO,66LB,67KP,67MF,
In^{3+} 54CI,54S,54Sc,57BH,69R   67NP,67TG,68KTa,68SRR,69MA,69MM,69SGM,
Tl^{3+} 49B,56PV,60BT,63AG,64LR,67YK   70DS,70Eb,71BHa,71D,71EG,71KSa,71MO,
Bi^{3+} 53BGa,57AG,65JL,67LD,67VL,71FKS   71PJ,71TS,72BH,72CP,72V,72Va,73GS,73SP,
Other references: 03S,31FL,51MS,52Fa,53G,   73V
         54CV,54SE,54W,55M,55Na,56C,56SL,

$$BrO_3^-$$

O_3Br^- Bromate ion L^-

Metal ion	Equilibrium	Log K 25°, 0.5	Log K 25°, 1.0	Log K 25°, 0	ΔH 25°, 0	ΔS 25°, 0
Li^+	ML/M.L	-0.77^r	-0.82^s	-0.5		
Na^+	ML/M.L			$(-0.4) \pm 0.1$		
K^+	ML/M.L			$(-0.3)^o$		
Ba^{2+}	ML/M.L			$(0.86)^o$		
	$M.L^2/ML_2(H_2O)(s)$ -5.11					
Sc^{3+}	ML/M.L		0.65		$(-8)^t$	$(-24)^c$
	$ML_2/M.L^2$		0.75			
Eu^{3+}	ML/M.L	0.58^a			$(-3)^v$	$(-8)^a$
Tb^{3+}	ML/M.L	0.49^a			$(-4)^v$	$(-11)^a$
Th^{4+}	ML/M.L	0.81				
	$ML_2/M.L^2$	0.91				
Cu^{2+}	$M.(OH)^{1.5}.L^{0.5}/M(OH)_{1.5}L_{0.5}(s)$ -16.13			-16.53		
Fe^{3+}	ML/M.L		0.36		$(4)^t$	$(15)^c$
Ag^+	M.L/ML(s)			-4.26 ± 0.02	19.3	45
Tl^+	ML/M.L			0.3^u		
	M.L/ML(s)			$-3.78^u -0.07$	12	23
Cd^{2+}	ML/M.L		0.06^e			
Pb^{2+}	ML/M.L			1.85		
	$M.L^2/ML_2(s)$			-5.10		

---

^{a} 25°, 0.1; ^{c} 25°, 1.0; ^{e} 25°, 3.0; ^{o} 18°, 0; ^{r} 25°, 0.15; ^{s} 25°, 0.20; ^{t} 15-35, 1.0; ^{u} 30°, 0; ^{v} 2-40°, 0.1

Bibliography:

Li^+ 63RSa

Na^+ 31BR,57FK

K^+ Ba^{2+} 31BR

Sc^{3+} 72MH

Eu^{3+},Tb^{3+} 72RC

Cu^{2+} 63LLa

Th^{4+} 50DS

Fe^{3+} 71MH

Ag^+ 23B,49TL,51Mb,63RD,67SV

Tl^+ 23B,68K,69KM

Cd^{2+} 43L

Pb^{2+} 36MH

Other references: 03B,48M,74GF

I^-

Iodide ion L^-

Metal ion	Equilibrium	Log K 25°, 0.5	Log K 25°, 1.0	Log K 25°, 0	ΔH 25°, 0 (1)[r]	ΔS 25°, 0 (2)
K^+	ML/M.L			(−0.19)		
Rb^+	ML/M.L			(0.04)		
Cs^+	ML/M.L			(−0.03)		
$(CH_3)_4N^+$	ML/M.L			(0.31)		
$(C_2H_5)_4N^+$	ML/M.L			(0.46)		
$(C_3H_7)_4N^+$	ML/M.L			(0.66)		
$(C_4H_9)_4N^+$	ML/M.L			(0.78)		
Eu^{3+}	ML/M.L		−0.4			
Hf^{4+}	ML/M.L		−0.5[l]			
Cu^+	$ML_2/M.L^2$			8.9		
	$ML_3/M.L^3$			9.4[g]		
	$ML_4/M.L^4$			9.7[g]		
	M.L/ML(s)			−12.0		
Ag^+	ML/M.L		(8.1)[f]	6.58[n]		
	$ML_2/M.L^2$		11.0[f]	(11.7)[n]		
	$ML_3/M.L^3$			(13.1)[n]	−29	(−37)[n]
		13.6[d]	13.8[f]	14.1[t]		
	$ML_4/M.L^4$	14.2[d]	14.3[f]	14.4[t]		
	$M_2L_6/M^2.L^6$		29.7[f]			
	$M_3L_8/M^3.L^8$		46.4[f]			
	M.L/ML(s)		−16.35[f]	−16.08	26.5	15
Hg_2^{2+}	$M.L^2/ML_2(s)$	−27.47		−28.33 ±0.02	39[s] ±4	5[b]
CH_3Hg^+	ML/M.L	8.60[h]				
	$ML_2/M.L^2$	8.86[h]				
	M.L/ML(s)	−11.46[h]				

[b] 25°, 0.5; [d] 25°, 2.0; [f] 25°, 4.0; [g] 25°, 5.0; [h] 20°, 0.1; [l] 20°, 3.0; [n] 18°, 0; [r] 5-55°, 0; [s] 7-40°, 0.5; [t] 25°, 7.0

G. GROUP VII LIGANDS

Iodide ion (continued)

Metal ion	Equilibrium	Log K 25°, 0.5	Log K 25°, 1.0	Log K 25°, 0	ΔH 25°, 0	ΔS 25°, 0
$C_2H_5Hg^+$	$ML_2/ML.L$		-0.67			
	$ML_3/ML_2.L$		0.75			
	$M.L/ML(s)$		-4.11			
Tl^+	$ML/M.L$		$0.74^f \pm 0.02$			
	$ML_2/M.L^2$		$0.90^f \pm 0.00$			
	$ML_3/M.L^3$		$1.06^f \pm 0.02$			
	$M.L/ML(s)$		-6.73^f	-7.23 ± 0.04	18^r	27
$(C_6H_5)_3Sn^+$	$ML/M.L$	3.7^p				
$(C_6H_5)_3Pb^+$	$ML/M.L$	7.3^p				
Pd^{2+}	$ML_4/M.L^4$		24.5 ± 0.5			
Zn^{2+}	$ML/M.L$		-1.5^e			
Cd^{2+}	$ML/M.L$	1.86 ± 0.04	1.89 ± 0.02	2.28 ± 0.1	-2.3 -0.1	3
		$1.99^d \pm 0.02$	$2.13^e \pm 0.07$		$-2.5^c \pm 0.0$	0^c
					$-2.2^e \pm 0.1$	2^e
	$ML_2/M.L^2$	3.2 ± 0.1	3.2 ± 0.1	3.92 ± 0.1	-3.0^c	5^c
		$3.4^d \pm 0.1$	$3.6^e \pm 0.1$		-2.5^e	8^e
	$ML_3/M.L^3$	4.4 ± 0.1	4.5 ± 0.1	5.0 ± 0.1	-4.4^c	6^c
		$4.8^d \pm 0.1$	$5.1^e \pm 0.1$		-3.2^e	13^e
	$ML_4/M.L^4$	5.5 ± 0.1	5.6 ± 0.1	6.0 ± 0.1	-8.4^c	-3^c
		$6.1^d \pm 0.1$	$6.6^e \pm 0.1$		-7.0^e	7^e
Hg^{2+}	$ML/M.L$	12.87			-17.1 ± 0.5	
					-18.0^b	-2^b
	$ML_2/M.L^2$	23.82			-34.2^b	-6^b
	$ML_3/M.L^3$	27.6 -0.1				
	$ML_4/M.L^4$	29.8 ± 0.1			-43.3 ± 0.3	
					-44.3^b	-12^b
	$ML/MOHL.H$	4.0				
	$M.L^2/ML_2(s)$	-27.95			41.1^b	10^b

^{b} 25°, 0.5; ^{c} 25°, 1.0; ^{d} 25°, 2.0; ^{e} 25°, 3.0; ^{f} 25°, 4.0; ^{p} 30°, 0.1; ^{r} 10-26°, 0

Iodide ion (continued)

Metal ion	Equilibrium	Log K 25°, 0.5	Log K 25°, 1.0	Log K 25°, 0	ΔH 25°, 0	ΔS 25°, 0
Sn^{2+}	ML/M.L		0.70^f			
	$ML_2/M.L^2$		1.13^f			
	$ML_3/M.L^3$		2.1^f			
	$ML_4/M.L^4$		2.3^f			
	$ML_6/M.L^6$		2.6^f			
	$ML_8/M.L^8$		2.1^f			
	$M.L^2/ML_2(s)$		-5.08^f			
Pb^{2+}	ML/M.L	1.30^d	1.26	1.92 ±0.1		
			1.69^e			
	$ML_2/M.L^2$	2.4^d	2.8	3.2		
	$ML_3/M.L^3$	3.1^d	3.4	3.9		
	$ML_4/M.L^4$	4.4^d	3.9	4.5		
			5.3^g			
	$M.L^2/ML_2(s)$	-7.61^d		-8.10 ±0.09	$(15)^r$	(10)
Ga^{3+}	ML/M.L	-0.2^q				
In^{3+}	ML/M.L	1.64^q				
		0.99^d			-0.7^d	2^d
	$ML_2/M.L^2$	2.56^q				
		2.26^d			0.8^d	14^d
Tl^{3+}	$ML_4/M.L^4$		35.7^f			
Bi^{3+}	ML/M.L	3.63				
	$ML_4/M.L^4$	15.0^k				
	$ML_5/M.L^5$	16.8^k				
	$ML_6/M.L^6$	18.8^k				
	$M.L^3/ML_3(s)$	-18.09^k				

^{d} 25°, 2.0; ^{e} 25°, 3.0; ^{f} 25°, 4.0; ^{g} 25°, 5.0; ^{k} 20°, 2.0; ^{q} 20°, 0.7; ^{r} 0-60°, 0

Iodide ion (continued)

Bibliography:

K^+ 68AT

Rb^+ 64FF

Cs^+ 68HF

$(CH_3)_4N^+, (C_2H_5)_4N^+, (C_3H_7)_4N^+, (C_4H_9)_4N^+$ 68F

Eu^{3+} 63CU

Hf^{4+} 67HP

Cu^+ 59FS

Ag^+ 38OB,54KT,54W,56La,56LP,57L,62FSV

Hg_2^{2+} 29B,38L,49QS,63HI

CH_3Hg^+ 63Sb,65SS

$C_2H_5Hg^+$ 65BB

Tl^+ 21JS,23B,37DR,57N,60KMa

$(C_6H_5)_3Sn^+, (C_6H_5)_3Pb^+$ 65SM

Pd^{2+} 63GKG,65FK

Zn^{2+} 69G

Cd^{2+} 38BV,41L,56QP,60AM,64VG,66G,67SG,67VM,
             68G,68GJ,69FD,69VP,70DS,74EM,74FK

Hg^{2+} 39G,49QS,52YA,54W,57M,57MV,58E,
             60GK,61MP,63EM,63HI,64CI,73Aa

Sn^{2+} 68HJ

Pb^{2+} 31F,44N,45N,55BPR,55PP,56KE,60FSS,
             60HT,60NM,61T,70FS

Ga^{3+} 67MA

In^{3+} 54CI,54S,69R

Tl^{3+} 66J

Bi^{3+} 57AG,57FH

Other references: 03S,23B,31FL,33HJ,36HB,
             49SBa,51MS,53G,54CV,54SE,54YS,55M,
             56SL,57KM,57TS,60CL,60GG,60L,60TM,
             64BL,64EH,65HS,65NP,65SL,67BP,67CP,
             67EH,67LD,67MF,67MFR,67PI,68GY,
             69EP,71BHa,71K,71PJ,72FKS,72FSa,
             72FSb

$$\overset{\text{O}}{\underset{\text{HO-I=O}}{\|}}$$

HO$_3$I			Hydrogen iodate	(iodic acid)		HL
Metal ion	Equilibrium	Log K 25°, 0.5	Log K 25°, 1.0	Log K 25°, 0	ΔH 25°, 0	ΔS 25°, 0
H$^+$	HL/H.L			0.77 ±0.03	2.4	12
Na$^+$	ML/M.L			(−0.48)		
K$^+$	ML/M.L			(−0.26)±0.04		
Mg^{2+}	ML/M.L			0.72 ±0.00		
Ca^{2+}	ML/M.L			0.89		
	M.L^2/ML$_2$(s)	−5.07	−4.89	−6.15 ±0.02	21	43
		−4.70[d]	−4.84[e]	−5.06[f]		
Sr^{2+}	ML/M.L			1.00		
	M.L^2/ML$_2$(s)	−5.40	−5.29	−6.48		
		−5.29[d]	−5.30[e]	−5.37[f]		
Ba^{2+}	ML/M.L			1.10		
	M.L^2/ML$_2$(s)	−7.76	−7.60	−8.81 ±0.01		
		−7.43[d]	−7.35[e]	−7.39[f]		
Y^{3+}	M.L^3/ML$_3$(s)			−10.15 +0.2	2.2[r]	−39
La^{3+}	M.L^3/ML$_3$(s)			−10.99 +0.07	6.9[r]	−27
Ce^{3+}	ML/M.L	1.22[a]		1.90		
	M.L^3/ML$_3$(s)			−10.86	6.8[r]	−27
Pr^{3+}	ML/M.L	1.18[a]				
	M.L^3/ML$_3$(s)			−10.89 +0.2	6.7[r]	−27
Nd^{3+}	M.L^3/ML$_3$(s)			−11.02 +0.1	6.4[r]	−29
Pm^{3+}	ML/M.L	1.12[a]		1.81		
Sm^{3+}	M.L^3/ML$_3$(s)			−11.30 +0.1	5.8[r]	−32
Eu^{3+}	ML/M.L	1.15[a]−0.2		1.83	(3)[s]	(16)[a]
	M.L^3/ML$_3$(s)			−11.41 +0.1	5.1[r]	−35
Gd^{3+}	M.L^3/ML$_3$(s)			−11.37 +0.2	4.2[r]	−38

---

[a] 25°, 0.1; [d] 25°, 2.0; [e] 25°, 3.0; [f] 25°, 4.0; [r] 25°, 0.2; [s] 0−40°, 0.1

Hydrogen iodate (continued)

Metal ion	Equilibrium	Log K 25°, 0.5	Log K 25°, 1.0	Log K 25°, 0	ΔH 25°, 0	ΔS 25°, 0
Tb^{3+}	ML/M.L	1.20^a-0.3				
	$M.L^3/ML_3(s)$			-11.19 +0.08	4.1^r	-38
Dy^{3+}	$M.L^3/ML_3(s)$			-11.04 +0.1	3.6^r	-39
Ho^{3+}	$M.L^3/ML_3(s)$			-10.87 +0.2	3.2^r	-39
Er^{3+}	ML/M.L	1.26^a		1.96		
	$M.L^3/ML_3(s)$			-10.71 +0.3	3.1^r	-39
Tm^{3+}	ML/M.L	1.33^a		2.02		
	$M.L^3/ML_3(s)$			-10.55 +0.2	2.6^r	-40
Yb^{3+}	ML/M.L	1.18^a		1.88		
	$M.L^3/ML_3(s)$			-10.40 +0.2	2.3^r	-40
Lu^{3+}	$M.L^3/ML_3(s)$			-10.25	2.0^r	-40
Th^{4+}	ML/M.L	2.88				
	$ML_2/M.L^2$	4.80				
	$ML_3/M.L^3$	(7.17)				
	$M.L^4/ML_4(s)$	-14.62				
UO_2^{2+}	$ML_2/M.L^2$	2.73^r				
	$ML_3/M.L^3$	3.67^r				
	$M.L^2/ML_2(s)$	-7.01^r				
Cu^{2+}	$M.L^2/ML_2(s)$			-7.13 ±0.01	6.8	-10
	$M.(OH)^{1.5}.L^{0.5}/M(OH)_{1.5}L_{0.5}(s)$			-17.56		
Cr^{3+}	$ML_2/M.L^2$	2.11				
	$M.L^3/ML_3(s)$	-5.3				
Ag^+	ML/M.L		0.19	0.63	5.1	20
	$ML_2/M.L^2$			1.90	-5.2	-9
	M.L/ML(s)		-7.08	-7.51 ±0.01	12	-6
Hg_2^{2+}	$M.L^2/ML_2(s)$			-17.89		
Tl^+	ML/M.L		0.15			
	M.L/ML(s)			-5.51 ±0.00	13	19
Zn^{2+}	$M.L^2/ML_2(s)$			-5.41		

---

[a] 25°, 0.1; [r] 25°, 0.2

Hydrogen iodate (continued)

Metal ion	Equilibrium	Log K 25°, 0.5	Log K 25°, 1.0	Log K 25°, 0	ΔH 25°, 0	ΔS 25°, 0
Cd^{2+}	$ML/M.L$		0.51			
	$ML_2/M.L^2$		1.52			
	$M.L^2/ML_2(s)$			-7.64		
Pb^{2+}	$M.L^2/ML_2(s)$	-11.48[t]		-12.61		

---

[t] 35°, 0.3

Bibliography:

H^+ 27O,34AR,34K,39NR,41LLa,44HB,67PP

Na^+ 31BR

K^+ 31BR,48M,59S

Mg^{2+} 30D,38WD

Ca^{2+} 34K,38WD,49DW,53BG,74FRa

Sr^{2+} 52CM,74FRa

Ba^{2+} 35MD,38D,39NR,43DV,49DW,74FRa

$Y^{3+}-Yb^{3+}$ 63LM,66FP,69BC,72RC,73CB

Th^{4+} 50DS,61SF

UO_2^{2+} 59KSN

Cu^{2+} 48K,51LW,51Ma,62LL,63RB

Cr^{3+} 69MH

Ag^+ 23B,38KL,41DS,41LL,51Mb,56RM,

Hg_2^{2+} 29B

Tl^+ 29LG,53BG,72BH

Zn^{2+} 50S

Cd^{2+} 50S,72BH

Pb^{2+} 23B,64SM

Other references: 02NK,03RD,05S,09HS,12S,
                   43T,52Sd,53NA,59B,59HJ,59R,62ML,
                   65DB,65K,67KR,72BBa,74GF

$$H_5IO_6$$

H_5O_6I Hydrogen periodate   (periodic acid) H_5L

Metal ion	Equilibrium	Log K 25°, 0	ΔH 25°, 0	ΔS 25°, 0
H^+	$[H_4L + IO_4]/H_3L.H$	8.29 ±0.04	(2)[r]	(45)
	$H_5L/[H_4L + IO_4]$	1.58 ±0.03		
	IO_4/H_4L	1.45 ±0.02	(11)[s]	(44)
K^+	$MIO_4/M.IO_4$	(0.24)		
	$M.IO_4/MIO_4(s)$	-3.43	15.1	35
Cs^+	$M.IO_4/MIO_4(s)$	-2.65	13.1	32
Cu^{2+}	$M^2.(OH)^3.H_4L/M_2HL(s)$	-42.6		
	$M^5.(OH)^8.(H_4L)^2/M_5L_2(s)$	-110.1		
Cd^{2+}	$M^2.(OH)^3.H_4L/M_2HL(s)$	-42.0		
	$M^5.(OH)^8.(H_4L)^2/M_5L_2(s)$	-109.5		

[r] 25-45°, 0; [s] 5-45°, 0

Bibliography:

H^+ 54N,65BL,66SV,68KD

K^+ 48M,51CH

Cs^+ 68KD

Cu^{2+} 54N

Cd^{2+} 55Rc

Other references: 03RD,48IN,53SH,60LY,61L,
                   64LW,69HSc,68MF,69BWa

III. PROTONATION VALUES FOR OTHER LIGANDS

Ligand	Equilibrium	Log K 25°, 0	ΔH 25°, 0	ΔS 25°, 0	Bibliography
Hydrogen niobate[*]	HL/H.L	13.8[e]			64Nb,
$(H_8Nb_6O_{19})$, H_8L	H_2L/HL.H	10.88[e]			other references:
					56LPa,60LSV
Hydrogen trithiocarbonate	HL/H.L	8.22[o]	(-3)[r]	(30)	63GKa
(H_2CS_3), H_2L	H_2L/HL.H	2.68[o]			
Hydrogen perthiocarbonate	HL/H.L	7.24	(-10)[r]	(0)	66GW
(H_2CS_4), H_2L					
Hydrogen triselenocarbonate	HL/H.L	7.13	(-10)[r]	(-1)	67GD
(H_2CSe_3), H_2L	H_2L/HL.H	1.16[s]			
Hydrogen germanate	HL/H.L	12.6			26M,29P,31SH,32LM,
$(Ge(OH)_4)$, H_2L		11.7[c]			63Ia,63IS,64HK,
		12.4[e]			74MB, other
	H_2L/HL.H	9.3			references: 26RS,
		9.02[b]±0.00			32GM,48C,55La,57A,
		9.02[c]			58KT,60A,62NF,64GZ,
	$H_{13}L_8/(H_2L)^8.(OH)^3$				66AN
		27.8			
		29.3[b] ±0.2			
		30.4[c]			
	H_2L/GeO_2 (s,hexagonal)				
		-1.37 +0.01			
	H_2L/GeO_2 (s,tetragonal)				
		-4.37			
Hydrogen peroxophosphate	HL/H.L	12.8[a]			60FB,65BE
(H_3PO_5), H_3L	H_2L/HL.H	5.5[a]			
	$H_3L/H_2L.H$	1.1[a]			

[a] 25°, 0.1; [b] 25°, 0.5; [c] 25°, 1.0; [e] 25°, 3.0; [o] 20°, 0; [r] 0-25°, 0; [s] 0°, 0; [*] metal constants were also reported but are not included in the compilation of stability constants.

III. Protonation Values (continued)

Ligand	Equilibrium	Log K 25°, 0	ΔH 25°, 0	ΔS 25°, 0	Bibliography
Hydrogen thiophosphate	HL/H.L	9.99[i]			69MKb,69PN
(H_3PO_3S), H_3L	H_2L/HL.H	5.83			other reference:
		5.38[a]			65NS
		5.23[b]			
		5.25[i]			
		5.04[c]			
	H_3L/H_2L.H	1.52[i]			
Hydrogen tetrathiophosphate	HL/H.L	6.5[i]			69PN
(H_3PS_4), H_3L	H_2L/HL.H	3.4[i]			
	H_3L/H_2L.H	1.7[i]			
Hydrogen amidophosphate	HL/H.L	8.63	(−5)[t]	(−20)	61ICa,68LW,69PN
$(H_2NPO_3H_2)$, H_2L		8.02[i]			
		8.28[c,u]			
	H_2L/HL.H	3.08	(0)[t]	(−10)	
		2.59[i]			
		(3.3)[c,u]			
Hydrogen diamidophosphate	HL/H.L	4.85[i]			69PN
$((NH_2)_2PO_2H)$, HL	H_2L/HL.H	1.03[i]			
Hydrogen diamidothiophosphate	HL/H.L	4.2[i]			69PN
$((NH_2)_2PSOH)$, HL	H_2L/HL.H	1.9[i]			
Hydrogen fluorophosphate[*]	HL/H.L	5.12			61RT
(H_2PO_3F), HL		4.72[v]	(2)[w]	(30)	other reference:
		4.47[c]			29L
Hydrogen arsenite	HL/H.L	9.29	−6.6	20	50JW,59AR,61AT,
(arsenous acid)		9.13[a]			64SSW, other
$(As(OH)_3)$, HL		9.09[b]			references: 13WS,
		9.11[x]			28H,40GH,40IA
	HL/$(As_4O_6)^{0.25}$(s)				
		−0.69			

[a] 25°, 0.1; [b] 25°, 0.5; [c] 25°, 1.0; [i] 20°, 0.5; [t] 0-40°, 0.5; [u] $(CH_3)_4NBr$ used as background electrolyte; [v] 25°, 0.25; [w] 0-65°, 0.25; [x] 25°, 1.5; [*] metal constants were also reported but are not included in the compilation of stability constants.

III. Protonation Values (continued)

Ligand	Equilibrium	Log K 25°, 0	ΔH 25°, 0	ΔS 25°, 0	Bibliography
Hydrogen arsenate[*]	HL/H.L	11.50	-4.4	38	53AA,58Mc,59FM,
(arsenic acid)	$H_2L/HL.H$	6.96 ±0.02	-0.8	29	64SSL,64SSW, other
(H_3AsO_4), H_3L		6.39[c]			references: 13WS,
	$H_3L/H_2L.H$	2.24 ±0.06	1.7	16	28H,42TL,56C,56Ca,
					56CS
Hydrogen antimonate	HL/H.L	2.72			63LMa,74MB
(antimonic acid)		2.47[b,u]			
$(Sb(OH)_5)$, HL	$Sb_{12}(OH)_{64}.H^4/(HL)^{12}$				
		20.34			
		23.06[b,u]			
	$Sb_{12}(OH)_{64}/Sb_{12}(OH)_{65}.H$				
		3.62			
	$Sb_{12}(OH)_{65}/Sb_{12}(OH)_{66}.H$				
		4.83			
	$Sb_{12}(OH)_{66}/Sb_{12}(OH)_{67}.H$				
		5.82			
Hydrogen tetrasulfide	HL/H.L	6.3[h]			60SF
(H_2S_4), H_2L	$H_2L/HL.H$	3.8[h]			
Hydrogen pentasulfide	HL/H.L	5.7[h]			60SF
(H_2S_5), H_2L	$H_2L/HL.H$	3.5[h]			
Hydrogen hydroxylamidosulfate	HL/H.L	1.48[y]			65CW
(hydroxylamine-O-sulfonic acid)					
(H_2NOSO_3H), HL					
Hydrogen peroxosulfate	HL/H.L	9.86[n]			65Ka,
(H_2SO_5), H_2L					other reference:
					63GR

---

[b] 25°, 0.5; [c] 25°, 1.0; [h] 20°, 0.1; [n] 19°, 0; [u] $(CH_3)_4NClO_4$ used as background electrolyte; [y] 45°, 1.0; [*] metal constants were also reported but are not included in the compilation of stability constants.

III. Protonation Values (continued)

Ligand	Equilibrium	Log K 25°, 0	ΔH 25°, 0	ΔS 25°, 0	Bibliography
Hydrogen tellurate[*]	HL/H.L	11.00 ±0.05	(-9)[r]	(20)	57Aa,59EF,60AT,
(telluric acid)		10.46[c]			62EE,66Bb,71KBa,
(Te(OH)$_6$), H$_2$L	H$_2$L/HL.H	7.66 ±0.05	(-7)[r]	(10)	72KMB,73KM,
		7.30[c]			other references:
	H$_3$L$_2$.H/(H$_2$L)^2	-6.84			20B,32BR,48F,53La,
		-6.31[c]			53ST,61L,62LYa,
	H$_3$L$_2$/H$_2$L$_2$.H	7.14[c]			66KC,71BG
	H$_2$L$_2$/HL$_2$.H	9.48[c]			
Hydrogen hypochlorite	HL/H.L	7.53 ±0.02	-3.3	23	37P,40H,46AM,57C,
(hypochlorous acid)					66Ma, other
(HOCl), HL					references: 04S,
					22NW,24Sa,33BD,
					33D,33G,33IM,33RA,
					37SB,38SG,40Ha,
					42Hb,42S,52Lc,
					57MF,58Fa,58FM
Hydrogen chlorite[*]	HL/H.L	1.95 ±0.01	4.1	23	37P,65LP,68HR
(chlorous acid)		1.66[b]			other references:
(HOClO), HL		1.61[c]			44T,56H,52Lc,54Da,
					64GK
Hydrogen hypobromite	HL/H.L	8.63 ±0.03	(-7)[r]	(15)	56KT,58AT,64FM,
(hypobromous acid)					other references:
(HOBr), HL					38SG,39KH,57MF
Hydrogen hypoiodite	HL/H.L	10.64			51BG,58Ca,
(hypoiodous acid)	I/HL.H	1.54			other references:
(HOI), HL					22F,25M,42Sa,

---

[b] 25°, 0.5; [c] 25°, 1.0; [r] 10-60°, 0; [*] metal constants were also reported but are not included in the compilation of stability constants.

IV. LIGANDS CONSIDERED BUT NOT INCLUDED

Ligand	Bibliography
Hydrogen tantalate ($Ta(OH)_5$)	60LSV,63BLN
Hydrogen octacyanowolframate (IV) ($H_4W(CN)_8$)	71SKa
Hydrogen octacyanowolframate (V) ($H_3W(CN)_8$)	71SKa
Hydrogen manganate (VI) (H_2MnO_4)	24SS,60LYa
Manganate (VII) ion (permanganate ion) (MnO_4^-)	43L,56C,60BC
Hydrogen technetate (VII) (pertechnetic acid) ($HTcO_4$)	53CS,63RH,63SK
Hydrogen pentacarbonylmanganate (-I) ($HMn(CO)_5$)	58HW
Hydrogen tetracarbonylferrate (-II) ($H_2Fe(CO)_4$)	49KS,52HHa
Hexachloroiridate(III) ion ($IrCl_6^{3-}$)	73KT
Hexabromoiridate(III) ion ($IrBr_6^{3-}$)	70KT
Cyanamide ($C(NH)_2$)	54SS
Nitrogen oxide (NO)	24MH,61Tc
Nitramide (NO_2NH_2)	27BKa
Hydrogen α-oxyhyponitrite ($H_2N_2O_3$)	63SRM
Hydrogen pentaphosphate ($H_7P_5O_{16}$)	50VC
Hydrogen hexaphosphate ($H_8P_6O_{19}$)	60IC,61I,61ICa
Hydrogen tetradecaphosphate ($H_{16}P_{14}O_{43}$)	60IC,61I
Hydrogen hexacontaphosphate ($H_{62}P_{60}O_{121}$)	60IC,61I,61ICa
Hydrogen μ-disulfidohexaoxodiphosphate ($H_2O_3PS_2PO_3H_2$)	65NS
Hydrogen fluorotriphosphate ($H_4P_3O_9F$)	65Fb
Hydrogen dithionite ($H_2S_2O_4$)	11J,64LRC,67BC
Hydrogen imidobis(fluorosulfate) ($NH(SO_2F)_2$)	65Ra

V. BIBLIOGRAPHY

Russian translations have the page of the original in parentheses.

00A   S. Arrhenius, Z. Phys. Chem., 1893, 11, 805

00B   M. Berthelot, Ann. Chim. Phys., 1877, 10, 433

00Ba   R. Bach, Z. Phys. Chem., 1892, 9, 241

00Bb   G. Bredig, Z. Phys. Chem., 1894, 13, 289

00Bc   G. Bodlander, Z. Phys. Chem., 1900, 35, 23

00BD   D. Berthelot and M. Delepine, Compt. Rend. Soc. Chim. France, 1899, 129, 326

00D   F. Dolezalek, Z. Elektrochem., 1899, 5, 533

00H   A. Hantzsch, Chem. Ber., 1899, 32, 3066

00KH   F. Kohlrausch and A. Heydweiller, Ann. Phys., 1894, 53, 209; Z. Phys. Chem., 1894, 14, 317

00L   H. Ley, Z. Phys. Chem., 1899, 30, 193

00M   J. L. R. Morgan, Z. Phys. Chem., 1895, 17, 513

00N   W. Nernst, Z. Phys. Chem., 1894, 14, 155

00P   T. Paul, Chem. Ztg., 1899, 23, 535

00S   M. Schumann, Chem. Ber., 1900, 33, 527

00T   J. Thomsen, Thermochemische Untersuchungen I, Leipzig, 1882, 162, 403

00W   C. A. West, J. Chem. Soc., 1900, 77, 705

00WC   J. Walker and W. Cormack, J. Chem. Soc., 1900, 77, 5

01B   G. Bodlander, Festschrift fur R. Dedekind, Braunschweig, 1901, p. 153

01E   C. L. von Ende, Z. Anorg. Allg. Chem., 1901, 26, 129

01L   M. G. Levi, Gazz. Chim. Ital., 1901, 31, No. 2, 1

01W   K. Winkelblech, Z. Phys. Chem., 1901, 36, 546

02B   A. A. Blanchard, Z. Phys. Chem., 1902, 41, 681

02BF   G. Bodlander and R. Fittig, Z. Phys. Chem., 1902, 39, 597

02NK   A. A. Noyes and D. A. Kohr, Z. Phys. Chem., 1902, 42, 336

02S   K. Schick, Z. Phys. Chem., 1902, 42, 155

03AC   R. Abegg and A. J. Cox, Z. Phys. Chem., 1903, 46, 1

03B   W. Bottger, Z. Phys. Chem., 1903, 46, 521

03E   H. Euler, Chem. Ber., 1903, 36, 1854, 3400

03RD   V. Rothmund and C. Drucker, Z. Phys. Chem., 1903, 46, 827

03S   M. S. Sherrill, Z. Phys. Chem., 1903, 43, 705; 1904, 47, 103

04A   F. Auerbach, Z. Phys. Chem., 1904, 49, 217

04BE   G. Bodlander and W. Eberlein, Z. Anorg. Allg. Chem., 1904, 39, 197

04D   K. Drucker, Z. Phys. Chem., 1904, 49, 563

04S   J. Sand, Z. Phys. Chem., 1904, 48, 610; ibid, 1905, 50, 465

05AS   R. Abegg and H. Schafer, Z. Anorg. Allg. Chem., 1905, 45, 293

05G   H. Grossmann, Z. Anorg. Allg. Chem., 1905, 43, 356

05S   V. Sammet, Z. Phys. Chem., 1905, 53, 641

05SA   J. F. Spencer and R. Abegg, Z. Anorg. Allg. Chem., 1905, 44, 379 (see also 64SM)

06B   N. Bjerrum, Kgl. Danske Vid. Selsk., Nat. Math. Afd., 1906, 4, 1; Z. Phys. Chem.,
            1907, 59, 336

06Ba   E. Bauer, Z. Phys. Chem., 1906, 56, 215

06GE   H. Goldschmidt and M. Eckhardt, Z. Phys. Chem., 1906, 56, 385

06K   J. Knox, Z. Elektrochem., 1906, 12, 477

07K   C. W. Kanolt, J. Amer. Chem. Soc., 1907, 29, 1402

07KB   W. Kerp and E. Bauer, Arb. Kaiser Gesundh., 1907, 26, 297

07L   H. Lunden, J. Chim. Phys., 1907, 5, 574

07P   M. Pleissner, Arb. Kaiser Gesundh., 1907, 26, 384

07S   M. S. Sherrill, J. Amer. Chem. Soc., 1907, 29, 1641

08B   N. Bjerrum, Diss., Copenhagen, 1908

08D   H. G. Denham, J. Chem. Soc., 1908, 93, 41

V. BIBLIOGRAPHY

08M   E. Muller, Z. Elektrochem., 1908, 14, 76

09A   A. J. Allmand, J. Chem. Soc., 1909, 95, 2151

09AB   G. A. Abbott and W. C. Bray, J. Amer. Chem. Soc., 1909, 31, 729

09BZ   L. Bruner and J. Zawadzki, Z. Anorg. Allg. Chem., 1909, 65, 136; ibid, 1910, 67, 454

09HS   A. E. Hill and J. P. Simmons, Z. Phys. Chem., 1909, 67, 594

09L   J. Lundberg, Z. Phys. Chem., 1909, 69, 442

09LB   R. Lorenz and A. Bohi, Z. Phys. Chem., 1909, 66, 733

09RP   A. Rosenheim and M. Pritze, Z. Anorg. Allg. Chem., 1909, 63, 275

09SF   O. Sackur and E. Fritzmann, Z. Elektrochem., 1909, 15, 842

09SL   J. M. Spencer and M. LePla, Z. Anorg. Allg. Chem., 1909, 65, 10

10A   A. J. Allmand, Z. Elektrochem., 1910, 16, 254

10B   N. Bjerrum, Z. Phys. Chem., 1910, 73, 724

10BS   K. Beck and P. Stegmuller, Arb. Kaiser Besundh., 1910, 34, 446

10M   A. C. Melcher, J. Amer. Chem. Soc., 1910, 32, 50

10NK   A. A. Noyes, Y. Kato, and R. B. Sosman, Z. Phys. Chem., 1910, 73, 1

10NS   A. A. Noyes and M. A. Stewart, J. Amer. Chem. Soc., 1910, 32, 1133

10W   J. K. Wood, J. Chem. Soc., 1910, 97, 878

11AV   F. Ageno and E. Valla, Atti. Accad. Lincei, Rend. Classe Sci. Fis. Mat. Nat., 1911, 20, 706

11J   K. Jellinek, Z. Phys. Chem., 1911, 76, 257

12J   R. A. Joyner, Z. Anorg. Allg. Chem., 1912, 77, 103

12K   A. Kirschner, Z. Phys. Chem., 1912, 79, 245

12L   J. Lindner, Monat. Chem., 1912, 33, 613

12NF   A. A. Noyes and K. G. Falk, J. Amer. Chem. Soc., 1912, 34, 454

12P   H. Pick, Nernst-Festschrift, Halle, 1912, p. 360

12S   J. F. Spencer, Z. Phys. Chem., 1912, 80, 701

13AP   F. Auerbach and H. Pick, Arb. Kaiser Gesundh., 1913, 45, 113

13B   L. Bruner, Z. Elektrochem., 1913, 19, 861

13K   C. Kullgren, Z. Phys. Chem., 1913, 85, 466

14MG   L. Michaelis and T. Garmendia, Biochem. Z., 1914, 67, 431

14MR   L. Michaelis and P. Rona, Biochem. Z., 1914, 67, 182

14TG   A. Thiel and H. Gessner, Z. Anorg. Allg. Chem., 1914, 86, 1

14W   G. Weissenberger, Z. Phys. Chem., 1914, 88, 257

15J   J. Johnston, J. Amer. Chem. Soc., 1915, 37, 2001

16O   E. Oliveri-Mandala, Gazz. Chim. Ital., 1916, 46, 298

16V   K. A. Vesterberg, Z. Anorg. Allg. Chem., 1916, 94, 371

17B   H. Bassett, Jr., J. Chem. Soc., 1917, 111, 620

17K   I. M. Kolthoff, Chem. Weekblad, 1917, 14, 1016

17LB   G. N. Lewis, T. B. Brighton, and R. L. Sebastian, J. Amer. Chem. Soc., 1917, 39,
         2245

17SL   C. A. Seyler and P. V. Lloyd, J. Chem. Soc., 1917, 111, 138, 994

19K   I. M. Kolthoff, Chem. Weekblad , 1919, 16, 1154

19L   N. Lofman, Z. Anorg. Allg. Chem., 1919, 107, 241

20B   E. Blanc, J. Chim. Phys., 1920, 18, 28

20F   G. Fuseya, J. Amer. Chem. Soc., 1920, 42, 368

20K   I. M. Kolthoff, Pharm. Weekblad, 1920, 57, 514

20LL   A. B. Lamb and A. T. Larson, J. Amer. Chem. Soc., 1920, 42, 2024

20M   A. D. Mitchell, J. Chem. Soc., 1920, 117, 957

21G   S. Glasstone, J. Chem. Soc., 1921, 119, 1689, 1914

21JS   G. Jones and W. C. Schumb, Proc. Amer. Acad. Sci., 1921, 56, 199 (see also 64SM)

21LF   A. B. Lamb and G. R. Fonda, J. Amer. Chem. Soc., 1921, 43, 1154

21RK   A. Rosenheim and L. Krause, Z. Anorg. Allg. Chem., 1921, 118, 177

22AR   M. P. Applebey and R. D. Reid, J. Chem. Soc., 1922, 121, 2129

22F   A. Furth, Z. Elektrochem., 1922, 28, 57

22JC   K. Jellinek and J. Czerwinski, Z. Phys. Chem., 1922, 102, 438

22M   H. Menzel, Z. Phys. Chem., 1922, 100, 276

22NW   A. A. Noyes and T. A. Wilson, J. Amer. Chem. Soc., 1922, 44, 1630

22O   Y. Osaka, Mem. Coll. Sci. Kyoto, 1922, 5, 131

23B   W. Bottger, Landolt-Bornstein Tabellen, 1923, Hw II, 1185

23H   J. Heyrovsky, Trans. Faraday Soc., 1923, 19, 692

23Ha   M. de Hlasko, J. Chim. Phys., 1923, 20, 167

V. BIBLIOGRAPHY

23K   I. M. Kolthoff, Rec. Trav. Chim., 1923, 42, 973

23LR   G. N. Lewis and M. Randall, Thermodynamics, McGraw-Hill, New York, 1923, p. 487

23M   A. E. Mitchell, J. Chem. Soc., 1923, 123, 1887

23Ma   H. Menzel, Z. Phys. Chem., 1923, 105, 402

23P   H. C. Parker, J. Amer. Chem. Soc., 1923, 45, 2017

24B   H. T. S. Britton, J. Chem. Soc., 1924, 125, 1572

24DH   C. W. Davies and L. J. Hudleston, J. Chem. Soc., 1924, 125, 260

24JG   K. Jellinek and H. Gordon, Z. Phys. Chem., 1924, 112, 207

24JJ   F. Jirsa and H. Jelinek, Z. Elektrochem., 1924, 30, 286, 534

24K   J. Kasarnowsky, Z. Phys. Chem., 1924, 109, 287

24Ka   E. Klarmann, Z. Anorg. Allg. Chem., 1924, 132, 289

24MH   W. Manchot and H. Haunschild, Z. Anorg. Allg. Chem., 1924, 140, 22

24PW   E. B. R. Prideaux and A. T. Ward, J. Chem. Soc., 1924, 125, 69

24PWa   E. B. R. Prideaux and A. T. Ward, J. Chem. Soc., 1924, 125, 423

24RK   H. Remy and A. Kuhlmann, Z. Anal. Chem., 1924, 65, 161

24S   R. Schuhmann, J. Amer. Chem. Soc., 1924, 46, 52

24Sa   F. G. Soper, J. Chem. Soc., 1924, 125, 2227

24SS   H. I. Schlesinger and H. B. Siems, J. Amer. Chem. Soc., 1924, 46, 1965

25B   J. Breszina, Rec. Trav. Chim., 1925, 44, 520

25Ba   H. T. S. Britton, J. Chem. Soc., 1925, 127, 2110, 2148, 2796, 2956

25DS   M. K. Domonovitsch and O. V. Sarubina, Biochem. Z., 1925, 163, 464

25G   J. K. Gjaldbaek, Z. Anal. Chem. 1925, 144, 269

25H   J. Heyrovsky, Chem. Listy, 1925, 19, 168

25HL   L. E. Holt, Jr., V. K. LaMer, and H. B. Chown, J. Biol. Chem., 1925, 64, 509

25HW   R. W. Harman and F. P. Worley, Trans. Faraday Soc., 1925, 20, 502

25M   H. D. Murray, J. Chem. Soc., 1925, 127, 882

25MM   L. Michaelis and M. Mizutani, Z. Phys. Chem., 1925, 116, 135

25S   R. Schuhmann, J. Amer. Chem. Soc., 1925, 47, 356

25Sa   S. J. Smrz, Rec. Trav. Chim., 1925, 44, 580

25W   H. J. deWijs, Rev. Trav. Chim., 1925, 44, 663

25WR   W. G. Whitman, R. P. Russell, and G. H. B. Davis, J. Amer. Chem. Soc., 1925, 47, 70

26B   N. Bjerrum, Ergebn. Exakt. Naturwiss., 1926, 5, 125

26Ba   W. Bottger, Landolt-Bornstein Tabellen, 1926, Eg I, 662

26BH   J. A. V. Butler and E. S. Hiscocks, J. Chem. Soc., 1926, 2554

26H   G. Hagg, Z. Anorg. Allg. Chem., 1926, 155, 21

26HB   F. L. Hahn and K. Brunngasser, Z. Anorg. Allg. Chem., 1926, 153, 88

26LD   E. Lange and F. Durr, Z. Phys. Chem., 1926, 121, 361

26M   J. H. Muller, Proc. Amer. Phil. Soc., 1926, 65, 183

26RH   A. G. Rees and L. J. Hudleston, J. Chem. Soc., 1926, 1334

26RS   W. A. Roth and O. Schwartz, Chem. Ber., 1926, 59, 338

26SN   M. S. Sherrill and A. A. Noyes, J. Amer. Chem. Soc., 1926, 48, 1861

27A   M. Aumeras, Comp. Rend. Acad. Sci. Paris, 1927, 184, 1650

27B   H. T. S. Britton, J. Chem. Soc., 1927, 614

27BK   J. N. Bronsted and C. V. King, Z. Phys. Chem., 1927, 130, 699

27BKa   J. N. Bronsted and C. V. King, J. Amer. Chem. Soc., 1927, 49, 193

27C   E. J. Cohn, J. Amer. Chem. Soc., 1927, 49, 173

27D   C. W. Davies, Trans. Faraday Soc., 1927, 23, 351

27Da   H. M. Dawson, J. Chem. Soc., 1927, 1290

27DJ   H. G. Dietrich and J. Johnston, J. Amer. Chem. Soc., 1927, 49, 1419

27H   R. W. Harman, J. Phys. Chem., 1927, 31, 616

27K   I. M. Kolthoff, Rec. Trav. Chim., 1927, 46, 350

27KB   I. M. Kolthoff and W. Bosch, Rec. Trav. Chim., 1927, 46, 180

27L   E. Lane, Z. Anorg. Allg. Chem., 1927, 165, 325

27O   L. Onsager, Phys. Z., 1927, 28, 277

27Oa   A. Olander, Z. Phys. Chem., 1927, 129, 1

27S   E. Sadolin, Z. Anorg. Allg. Chem., 1927, 160, 133

27SH   J. Sendroy, Jr. and A. B. Hastings, J. Biol. Chem., 1927, 71, 783

27WB   A. C. Walker, U. B. Bray, and J. Johnston, J. Amer. Chem. Soc., 1927, 49, 1235

28BV   J. N. Bronsted and K. Volqvartz, Z. Phys. Chem., 1928, 134, 97

28FM   R. Fricke and K. Meyring, Z. Anorg. Allg. Chem., 1928, 176, 325

V. BIBLIOGRAPHY

28H   W. S. Hughes, J. Chem. Soc., 1928, 491

28HE   H. von Halban and J. Eisenbrand, Z. Phys. Chem., 1928, 132, 401

28J   P. Job, Ann. Chim. (France), 1928, 9, 113

28Ja   P. Job, Compt. Rend. Acad. Sci. Paris, 1928, 186, 1546

28KB   I. M. Kolthoff and W. Bosch, Rec. Trav. Chim., 1928, 47, 826

28M   C. Morton, J. Chem. Soc., 1928, 1401

28P   M. Prytz, Z. Anorg. Allg. Chem., 1928, 172, 147

28Pa   J. Piater, Z. Anorg. Allg. Chem., 1928, 174, 321

28Pb   M. Prytz, Z. Anorg. Allg. Chem., 1928, 174, 355

28RL   A. J. Rabinowitsch and E. Laskin, Z. Phys. Chem., 1928, 134, 387

28RS   M. Randall and H. M. Spencer, J. Amer. Chem. Soc., 1928, 50, 1572

28RV   M. Randall and W. V. A. Vietti, J. Amer. Chem.Soc., 1928, 50, 1526

28S   R. N. J. Saal, Rec. Trav. Chim., 1928, 47, 264

28SH   W. C. Stadie and E. R. Hawes, J. Biol. Chem., 1928, 77, 241

29B   A. E. Brodsky, Z. Elektrochem., 1929, 35, 833

29BU   N. Bjerrum and A. Unmack, Kgl. Danske Vid. Selsk., Math. Fys. Medd., 1929, 9, No. 1

29FJ   G. L. Frear and J. Johnston, J. Amer. Chem. Soc., 1929, 51, 2082

29G   E. C. Gilbert, J. Phys. Chem., 1929, 33, 1235

29JM   M. Jowett and H. Millet, J. Amer. Chem. Soc., 1929, 51, 1004

29K   W. D. Kline, J. Amer. Chem. Soc., 1929, 51, 2093

29Ka   I. N. Kugelmass, Biochem. J., 1929, 23, 587

29Kb   V. A. Kargin, Z. Anorg. Allg. Chem., 1929, 183, 77

29KH   A. Klemenc and E. Hayek, Monat. Chem., 1929, 53/54, 407

29L   W. Lange, Chem. Ber., 1929, 62, 793

29LG   V. K. LaMer and F. H. Goldman, J. Amer. Chem. Soc., 1929, 35, 833

29MJ   H. Millet and M. Jowett, J. Amer. Chem. Soc., 1929, 51, 997

29O   A. Olander, Z. Phys. Chem., 1929, 144, 49

29P   W. Pugh, J. Chem. Soc., 1929, 1994

29S   B. Schrager, Coll. Czech. Chem. Comm., 1929, 1, 275; Chem. News, 1929, 138, 354

29T   H. Topelmann, J. Prakt. Chem., 1929, 121, 320

30BD   H. E. Blayden and C. W. Davies, J. Chem. Soc., 1930, 949

30BH   L. Birckenbach and K. Hutter. Z. Anorg. Allg. Chem., 1930, 190, 1

30CM   W. B. Campbell and O. Maass, Canad. J. Res., 1930, 2, 42

30D   C. W. Davies, J. Chem. Soc., 1930, 2410

30Da   C. W. Davies, J. Chem. Soc., 1930, 2421

30E   L. W. Elder, Jr., Trans. Electrochem. Soc., 1930, 57, 383

30HK   F. L. Hahn and R. Klockman, Z. Phys. Chem., 1930, A146, 373

30HKa   F. L. Hahn and R. Klockmann, Z. Phys. Chem., 1930, A151, 80

30HO   H. S. Harned and B. B. Owen, J. Amer. Chem. Soc., 1930, 52, 5091

30K   F. K. V. Koch, J. Chem. Soc., 1930, 2053

30LM   E. Lange and J. Monheim, Z. Phys. Chem., 1930, A150, 349

30M   K. Masaki, Bull. Chem. Soc. Japan, 1930, 5, 345

30Ma   C. Morton, Quart. J. Pharm. Pharmacol., 1930, 3, 438

30N   P. Nylen, Diss., Uppsala Univ., 1930

30NB   R. F. Newton and M. G. Bolinger, J. Amer. Chem. Soc., 1930, 52, 921

30R   E. J. Roberts, J. Amer. Chem. Soc., 1930, 52, 3877

30RD   E. C. Righellato and C. W. Davies, Trans. Faraday Soc., 1930, 26, 592

30RH   M. Randall and J. O. Halford, J. Amer. Chem. Soc., 1930, 52, 178

30S   H. Schmid, Z. Phys. Chem., 1930, A148, 321

30W   W. F. K. Wynne-Jones, J. Chem. Soc., 1930, 1064

31B   J. Bjerrum, Kgl. Danske Vid. Selsk.,Math. Fys. Medd., 1931, 11, No. 5

31BD   N. Bjerrum and C. R. Dahm. Z. Phys. Chem., 1931, Bodenstein-Festband, 627

31BDa   H. T. S. Britton and E. N. Dodd, J. Chem. Soc., 1931, 2332

31BR   W. H. Banks, E. C. Righellato, and C. W. Davies, Trans. Faraday Soc., 1931, 27, 621

31CL   I. A. Cowperthwaite and V. K. LaMer, J. Amer. Chem. Soc., 1931, 53, 4333

31CR   E. Carriere and Raulet, Comp. Rend. Acad. Sci. Paris, 1931, 192, 423

31F   H. Fromherz, Z. Phys. Chem., 1931, A153, 376

31FA   E. H. Fawcett and S. F. Acree, J. Res. Nat. Bur. Stand., 1931, 6, 757

31FL   H. Fromherz and K. H. Lih, Z. Phys. Chem., 1931, A153, 321, 376

31K   I. M. Kolthoff, J. Phys. Chem., 1931, 35, 2711

31KE   I. M. Kolthoff and R. Elmquist, J. Amer. Chem. Soc., 1931, 53, 1217

31KK   I. M. Kolthoff and T. Kameda, J. Amer. Chem. Soc., 1931, 53, 832

31KP   P. Kubelka and V. Pristoupil, Z. Anorg. Allg. Chem., 1931, 197, 391

31L   J. W. H. Lugg, Trans. Faraday Soc., 1931, 27, 297

31LP   V. K. LaMer and G. Parks, J. Amer. Chem. Soc., 1931, 53, 2040

31N   A. V. Novoselova, Zh. Obsh. Khim., 1931, 1, 668

31MM   O. M. Morgan and O. Maass, Canad. J. Res., 1931, 5, 162

31P   M. Prytz, Z. Anorg. Allg. Chem., 1931, 197, 103; 200, 133

31SH   R. Schwarz and E. Huf, Z. Anorg. Allg. Chem., 1931, 203, 188

32B   J. Bjerrum, Kgl. Danske Vid. Selsk., Math. Fys. Medd., 1932, 11, No. 10

32BR   H. T. S. Britton and R. A. Robinson, Trans. Faraday Soc., 1932, 28, 531

32D   D. S. Davis, Chem. Met. Eng., 1932, 39, 615

32E   G. Endres, Z. Anorg. Allg. Chem., 1932, 205, 321

32HJ   K. Hass and K. Jellinek, Z. Phys. Chem., 1932, A162, 153

32IS   F. Ishikawa and E. Shibata, Sci. Reports, Tohoku Univ., 1932, 21, 499

32JP   M. Jowett and H. I. Price, Trans. Faraday Soc., 1932, 28, 668

32LM   A. W. Laubengayer and D. S. Morton, J. Amer. Chem. Soc., 1932, 54, 2303

32M   K. Murata, Kogyo Kagaku Zasshi, 1932, 35, Suppl., 523B

32Ma   J. Muus, Z. Phys. Chem., 1932, A159, 268

32MD   R. W. Money and C. W. Davies, Trans. Faraday Soc., 1932, 28, 609

32N   E. W. Newman, J. Amer. Chem. Soc., 1932, 54, 2195

32RF   M. Randall and M. Frandsen, J. Amer. Chem. Soc., 1932, 54, 40

32RZ   W. A. Roth and H. Zeumer, Z. Elektrochem., 1932, 38, 164

32WM   R. H. Wright and O. Maass, Canad. J. Res., 1932, 6, 94, 588

33AT   M. Aumeras and A. Tamisier, Bull. Soc. Chim. France, 1933, 53, 97

33BD   H. T. S. Britton and E. N. Dodd, Trans. Faraday Soc., 1933, 29, 537

33BW   H. T. S. Britton and B. M. Wilson, J. Chem. Soc., 1933, 1050

33D   G. F. Davidson, J. Textile Inst., 1933, 24, 185

33FM   R. Fricke and K. Meyring, Z. Anorg. Allg. Chem., 1933, 214, 269

33G   J. M. Gallart, An. Real. Soc. Espana Fis. Quim., 1933, 31, 422

33HH   H. S. Harned and W. J. Hamer, J. Amer. Chem. Soc., 1933, 55, 2194, 4496

33HJ   K. Hass and K. Jellinek, Z. Phys. Chem., 1933, A162, 153

33IM   J. W. Ingham and J. Morrison, J. Chem. Soc., 1933, 1200

33IT   F. Ishikawa and Y. Terui, Bull. Inst. Phys. Chem. Res. Tokyo, 1933, 12, 755

33J   K. Jellinek, Lehrbuch der physikalischen Chemie, 3nd ed., Enke, Stuttgart, Vol. IV,
        1933, p. 62

33KA   P. A. Kriukov and G. P. Awsejewitsch, Z. Elektrochem., 1933, 39, 884

33LH   M. LeBlanc and O. Harnapp, Z. Phys. Chem., 1933, A166, 321

33LHS   W. M. Latimer, J. F. G. Hicks, Jr., and P. W. Schufz, J. Chem. Phys., 1933, 1, 620

33N   L. F. Nims, J. Amer. Chem. Soc., 1933, 55, 1946

33NS   S. von Naray-Szabo and Z. Szabo, Z. Phys. Chem., 1933, A166, 228

33NT   B. Nikitin and P. Tolmatscheff, Z. Phys. Chem., 1933, A167, 260

33R   P. Rumpf, Compt. Rend. Acad. Sci. Paris, 1933, 197, 686

33RA   A. Rius and V. Arnal, An. Real. Soc. Espana Fis. Quim., 1933, 31, 497

33T   A. Tamisier, Bull. Soc. Chim. France. 1933, 53, 157

34AR   E. Abel, O. Redlich, and P. Hersch, Z. Phys. Chem., 1934, A170, 112

34B   J. Bjerrum, Kgl. Danske Vid. Selsk., Math. Frs. Medd., 1934, 12, No. 15

34Ba   W. H. Banks, J. Chem. Soc., 1934, 1010

34Bb   W. H. Bennett, J. Phys. Chem., 1934, 38, 573

34BG   H. T. S. Britton and W. L. German, J. Chem. Soc., 1934, 1156

34BH   W. C. Bray and A. V. Hershey, J. Amer. Chem. Soc., 1934, 56, 1889

34BY   G. E. K. Branch, D. L. Yabroff, and B. Bettman, J. Amer. Chem. Soc., 1934, 56, 937

34CC   K. S. Chang and Y. T. Cha, J. Chinese Chem. Soc., 1934, 2, 298

34CL   I. A. Cowperthwaite, V. K. LaMer, and J. Barksdale, J. Amer. Chem. Soc., 1934, 56,
        544

34FW   E. P. Flint and L. S. Wells, J. Res. Nat. Bur. Stand., 1934, 12, 751

34GM   R. O. Griffith and A. McKeown, Trans. Faraday Soc., 1934, 30, 530

34GS   E. A. Guggenheim and T. D. Schindler, J. Phys. Chem., 1934, 38, 533

34JL   H. F. Johnstone and P. W. Leppla, J. Amer. Chem. Soc., 1934, 56, 2233

34K   G. Kilde, Z. Anorg. Allg. Chem., 1934, 218, 113

34L   A. Lebettre, J. Chim. Phys., 1934, 31, 348

34La   R. W. Lawrence, J. Amer. Chem. Soc., 1934, 56, 776

34LS   G. N. Lewis and P. W. Schutz, J. Amer. Chem. Soc., 1934, 56, 1913

34M   A. Maffei, Gazz. Chim. Ital., 1934, 64, 149

34N   L. F. Nims, J. Amer. Chem. Soc., 1934, 56, 1110

34NR   J. D. Neuss and W. Rieman, III, J. Amer. Chem. Soc., 1934, 56, 2238

34O   B. B. Owen, J. Amer. Chem. Soc., 1934, 56, 1695

34Oa   B. B. Owen, J. Amer. Chem. Soc., 1934, 56, 2785

34R   A. Ringbom, Acta Acad. Aboensis, Math. Phys., 1934, 8, No. 5

34ZR   H. Zeumer and W. A. Roth, Z. Elektrochem., 1934, 40, 777

35BM   A. S. Brown and D. A. MacInnes, J. Amer. Chem. Soc., 1935, 57, 459

35BW   H. T. S. Britton and W. C. Williams, J. Chem. Soc., 1935, 796

35CM   J. Y. Cann and G. B. Mueller, J. Amer. Chem. Soc., 1935, 57, 2525

35D   C. W. Davies, J. Chem. Soc., 1935, 910

35DH   T. DeVries and E. M. Hattox, Proc. Indiana Acad. Sci., 1935, 44, 138

35HR   F. Halla and F. Ritter, Z. Phys. Chem., 1935, A175, 63, 396

35KA   K. K. Kelley and C. T. Anderson, Bur. Mines Bull., No. 384, 1935

35KT   I. M. Kolthoff and W. J. Tomsicek, J. Phys. Chem., 1935, 39, 955

35MB   D. A. MacInnes and D. Belcher, J. Amer. Chem. Soc., 1935, 57, 1683

35MD   G. Macdougall and C. W. Davies, J. Chem. Soc., 1935, 1416

35O   B. B. Owen, J. Amer. Chem. Soc., 1935, 57, 1526

35SM   T. Shedlovsky and D. A. MacInnes, J. Amer. Chem., Soc., 1935, 57, 1705

35W   O. Weider, Chem. Ber., 1935, 68, 1423

36C   M. Chatelet, J. Chim. Phys., 1936, 33, 313

36E   E. Endredy, Math. Natur. Anz. Ungar. Akad. Wiss., 1936, 54, 459

36H   A. Holmquist, Svenk Kem. Tidskr., 1936, 48, 106

36HB   A. V. Hershey and W. C. Bray, J. Amer. Chem. Soc., 1936, 58, 1760

36HD   E. M. Hattox and T. DeVries, J. Amer. Chem. Soc., 1936, 58, 2126

36HF   H. S. Harned and M. E. Fitzgerald, J. Amer. Chem. Soc., 1936, 58, 2624

36MH   F. H. MacDougall and E. J. Hoffman, J. Phys. Chem., 1936, 40, 317

36MJ   L. A. McDowell and H. L. Johnston, J. Amer. Chem. Soc., 1936, 58, 2009

36R   R. A. Robinson, Trans. Faraday Soc., 1936, 32, 743

36RB   I. G. Ryss and N. P. Bakina, Dokl. Akad. Nauk. SSSR, 1936, 11, 107

36RR   O. Redlich and P. Rosenfeld, Monat. Chem., 1936, 67, 223

36S   G. Schwarzenbach, Helv. Chim. Acta, 1936, 19, 178

36SE   G. Schwarzenbach, A. Epprecht, and H. Erlenmeyer, Helv. Chim. Acta, 1936, 19, 1292

36SH   M. S. Sherrill and A. J. Haas, Jr., J. Amer. Chem. Soc., 1936, 58, 952

36W   W. F. K. Wynne-Jones, Trans. Faraday Soc., 1936, 32, 1397

36WS   G. C. Ware, J. B. Spulnik, and E. C. Gilbert, J. Amer. Chem. Soc., 1936, 58, 1605

37BH   A. A. Browman and A. B. Hastings, J. Biol. Chem., 1937, 119, 241

37C   L. H. N. Cooper, Proc. Roy Soc. (London), 1937, B124, 299

37CB   J. A. Cranston and H. F. Brown, J. Roy. Tech. Coll.(Glasgow), 1937, 4, 54

37D   C. W. Davies, J. Amer. Chem. Soc., 1937, 59, 1760

37DR   C. W. Davies and R. A. Robinson, Trans. Faraday Soc., 1937, 33, 633

37J   A. T. Jensen, Z. Phys. Chem., 1937, A180, 93

37M   M. Moller, J. Phys. Chem., 1937, 41, 1123

37N   P. Nylen, Z. Anorg. Allg. Chem., 1937, 230, 385

37P   K. S. Pitzer, J. Amer. Chem. Soc., 1937, 59, 2365

37PS   K. S. Pitzer and W. V. Smith, J. Amer. Chem. Soc., 1937, 59, 2633

37Q   M. Quintin, Compt. Rend. Acad. Sci. Paris, 1937, 204, 968

37R   R. A. Robinson, J. Amer. Chem. Soc., 1937, 59, 84

37Ra   M. E. Rumpf, Ann. Chim. (France), 1937, 8, 456

37RD   R. A. Robinson and C. W. Davies, J. Chem. Soc., 1937, 574

37RP   W. A. Roth, H. Pahlke, A. Bertram, and E. Borger, Z. Elektrochem., 1937, 43, 350

37SB   A. Skrabal and A. Berger, Monat. Chem., 1937, 70, 168

37SBP   W. V. Smith, O. L. I. Brown, and K. S. Pitzer, J. Amer. Chem. Soc., 1937, 59, 1213

37SM   H. Schmid, R. Marchgraber, and F. Dunkl, Z. Elektrochem., 1937, 43, 337.

38BC   W. B. Beazley, W. B. Campbell, and O. Maass, Dominion Forest Service Bull., No. 93, 1938

38BV   R. G. Bates and W. C. Vosburgh, J. Amer. Chem. Soc., 1938, 60, 137

38CF   B. N. Cacciapuoti and F. Ferla, Atti Accad. Naz. Lincei, Rend. Classe Sci. Fis. Mat Nat., 1938, 28, 385

V. BIBLIOGRAPHY

38CH   J. Curry and C. L. Hazelton, J. Amer. Chem. Soc., 1938, 60, 2771, 2773

38D   C. W. Davies, J. Chem. Soc., 1938, 2093

38E   A. G. Epprecht, Helv. Chim. Acta, 1938, 21, 205

38EW   D. H. Everett and W. F. K. Wynne-Jones, Proc. Roy. Soc. (London), 1938, A169, 190

38G   E. Guntelberg, Thesis, Copenhagen, 1938

38GH   A. B. Garrett and A. E. Hirschler, J. Amer. Chim. Soc., 1938, 60, 299

38JL   H. L. Johnston and H. L. Leland, J. Amer. Chem. Soc., 1938, 60, 1439

38KL   I. M. Kolthoff and J. J. Lingane, J. Phys. Chem., 1938, 42, 133

38L   W. M. Latimer, Oxidation Potentials, Prentice-Hall, New York, 1938, 163, 171

38LJ   A. B. Lamb and A. G. Jacques, J. Amer. Chem. Soc., 1938, 60, 1215

38O   Y. Oka. Nippon Kagaku Zasshi, 1938, 59, 971

38OB   B. B. Owen and S. R. Brinkley, Jr., J. Amer. Chem. Soc., 1938, 60, 2233

38OG   B. B. Owen and R. W. Gurry, J. Amer. Chem. Soc., 1938, 60, 3074

38PO   J. N. Pearce and L. D. Ough, J. Amer. Chem. Soc., 1938, 60, 80

38PS   A. Pinkus and R. Schepmans, Bull. Soc. Chim. Belges, 1938, 47, 337

38R   O. Redlich, Z. Phys. Chem., 1938, A182, 42

38SG   E. A. Shilov and J. N. Gladtchikova, J. Amer. Chem. Soc., 1938, 60, 490

38T   J. E. Thygesen, Z. Anorg. Allg. Chem., 1938, 237, 101

38TN   A. C. Taylor and L. F. Nims, J. Amer. Chem. Soc., 1938, 60, 262 (see 64SM)

38WD   W. C. Wise and C. W. Davies, J. Chem. Soc., 1938, 273

39B   R. G. Bates, J. Amer. Chem. Soc., 1939, 61, 308

39G   A. B. Garrett, J. Amer. Chem. Soc., 1939, 61, 2744

39Ga   M. Gorman, J. Amer. Chem. Soc., 1939, 61, 3342

39GH   A. B. Garrett and W. W. Howell, J. Amer. Chem. Soc., 1939, 61, 1730

39GV   A. B. Garrett, S. Vellenga, and C. M. Fontana, J. Amer. Chem. Soc., 1939, 61, 367

39H   H. Hagisawa, Bull. Inst. Phys. Chem. Res. Tokyo, 1939, 18, 260, 368

39Ha   H. Hagisawa, Bull. Inst. Phys. Chem. Res. Tokyo, 1939, 18, 648

39HJ   E. Hogge and H. L. Johnston, J. Amer. Chem. Soc., 1939, 61, 2154

39KH   M. Kiese and A. B. Hastings, J. Amer. Chem. Soc., 1939, 61, 1291

39L   F. Lindstrand, Diss., Lund, 1939 (see 64SM)

39LZ   W. M. Latimer and H. W. Zimmermann, J. Amer. Chem. Soc., 1939, 61, 1550

39NR   S. Naidich and J. E. Ricci, J. Amer. Chem. Soc., 1939, 61, 3268

39R   W. A. Roth, Ann. Chem., 1939, 542, 35

40BC   H. F. Brown and J. A. Cranston, J. Chem. Soc., 1940, 578

40CB   E. Cohen and J. J. A. Blekkingh Jr., Z. Phys. Chem., 1940, A186, 257

40GH   A. B. Garrett, O. Holmes, and A. Laube, J. Amer. Chem. Soc., 1940, 62, 2024

40GM   R. O. Griffith and A. McKeown, Trans. Faraday Soc., 1940, 36, 766

40GR   I. Greenwald, J. Redish, and C. A. Kibrick, J. Biol. Chem., 1940, 135, 65

40H   H. Hagisawa, Bull. Inst. Phys. Chem. Res. Tokyo, 1940, 19, 1220

40Ha   G. Holst, Svensk Kem. Tidskr., 1940, 52, 258

40IA   F. Ishikawa and I. Aoki, Bull. Inst. Phys. Chem. Res. Tokyo, 1940, 19, 136

40KH   M. Kiese and A. B. Hastings, J. Biol. Chem., 1940, 132, 267

40MS   W. A. Mason and W. J. Shutt, Proc. Roy. Soc., (London), 1940, A175, 257

40Q   M. Quintin, Compt. Rend Acad. Sci. Paris, 1940, 210, 625

40RE   P. S. Roller, Jr., and G. Ervin, Jr., J. Amer. Chem. Soc., 1940, 62, 461

40S   C. R. Singleterry, Thesis, Univ. Chicago, 1940 (see 62YI)

40SF   M. von Stackelberg and H. von Freyhold, Z. Elektrochem., 1940, 46, 120

40Y   N. Yui, Bull. Inst. Phys. Chem. Tokyo, 1940, 19, 1229

41B   J. Bjerrum, Metal Amine Formation in Aqueous Solution, P. Haase and Son,
       Copenhagen, 1941

41Ba   C. Brosset, Svensk Kem. Tidskr., 1941, 53, 434

41DS   P. F. Derr, R. M. Stockdale, and W. C. Vosburgh, J. Amer. Chem. Soc., 1941, 63,
       2670

41FS   R. K. Fox, D. F. Swinehart, and A. B. Garrett, J. Amer. Chem. Soc., 1941, 63, 1779

41G   I. Greenwald, J. Biol. Chem., 1941, 141, 789

41GH   A. B. Garrett and R. E. Heiks, J. Amer. Chem. Soc., 1941, 63, 562

41H   H. Hagisawa, Bull. Inst. Phys. Chem. Res. Tokyo, 1941, 20, 251

41Ha   H. Hagisawa, Bull. Inst. Phys. Chem. Res. Tokyo, 1941, 20, 384

41HS   H. S. Harned and S. R. Scholes, Jr., J. Amer. Chem. Soc., 1941, 63, 1706

41ID   R. W. Ivett and T. DeVries, J. Amer. Chem. Soc., 1941, 63, 2821

41K   J. Kasper, Thesis, John Hopkins Univ., Baltimore, Md., 1941

41L   I. Leden, Z. Phys. Chem., 1941, A188, 160

41LK   O. E. Lanford and S. J. Kiehl, J. Phys. Chem., 1941, 45, 300

41LL   N. C. C. Li and Y. T. Lo, J. Amer. Chem. Soc., 1941, 63, 394

41LLa   N. C. C. Li and Y. T. Lo, J. Amer. Chem. Soc., 1941, 63, 397

41M   I. G. Murgulescu, Bull. Sci. Ecole Polytech. Timisoara, 1941, 10, 379

41Ma   T. Moeller, J. Amer. Chem. Soc., 1941, 63, 1206, 2625

41NZ   B. V. Nekrasov and G. V. Zotov, Zh. Prik. Khim., 1941, 14, 264

41SW   R. H. Stokes, J. M. Wilson, and R. A. Robinson, Trans. Faraday Soc., 1941, 37, 566

41TG   H. V. Tartar and H. H. Garretson, J. Amer. Chem. Soc., 1941, 63, 808

41TY   K. Takahashi and N. Yui, Bull. Inst. Phys. Chem. Res. Tokyo, 1941, 20, 521

41Y   N. Yui, Bull. Inst. Phys. Chem. Res. Tokyo, 1941, 20, 256

41Ya   N. Yui, Bull. Inst. Phys. Chem. Res. Tokyo, 1941, 20, 390

42B   C. Brosset, Svensk Kem. Tidskr., 1942, 54, 155

42Ba   C. Brosset, Diss., Stockholm, 1942

42BG   C. Brosset and B. Gustaver, Svensk Kem. Tidskr., 1942, 54, 185

42D   T. W. Davis, Ind. Eng. Chem. Anal., 1942, 14, 709

42DM   L. S. Darken and H. F. Meier, J. Amer. Chem. Soc., 1942, 64 621

42GK   R. W. Gelbach and G. B. King, J. Amer. Chem. Soc., 1942, 64, 1054

42GL   M. Gorman and P. A. Leighton, J. Amer. Chem. Soc., 1942, 64, 719

42GN   A. B. Garrett, M. V. Noble, and S. Miller, J. Chem. Educ., 1942, 19, 485

42H   L. J. Heidt, J. Phys. Chem., 1942, 46, 624

42Ha   J. Harbo, Farm. Tidende, 1942, No. 30

42Hb   J. Hoye, Kgl. Norske Vid. Sel. For., 1942, 14, 1

42KP   I. M. Kolthoff, R. W. Perlich, and D. Weiblen, J. Phys. Chem., 1942, 46, 561

42LK   O. E. Lanford and S. J. Kiehl, J. Amer. Chem. Soc., 1942, 64, 291

42MR   T. Moeller and P. W. Rhymer, J. Phys. Chem., 1942, 46, 477

42N   R. Nasanen, Z. Phys. Chem., 1942, A190, 183

42Na   R. Nasanen, Z. Phys. Chem., 1942, A191, 54

42Nb   R. Nasanen, Ann. Acad. Sci. Fenn., 1942, A59, No. 2

42Nc   R. Nasanen, Ann. Acad. Sci. Fenn., 1942, A59, No. 7

42RS   E. Rabinowitch and W. H. Stockmayer, J. Amer. Chem. Soc., 1942, 64, 335

42S   A. Skrabal, Z. Elektrochem., 1942, 48, 314

42Sa   A. Skrabal, Chem. Ber., 1942, 75, 1570

42SH   T. Suzuki and H. Hagisawa, Bull. Inst. Phys. Chem. Res. Tokyo, 1942, 21, 601

42TL   N. A. Tanandev and R. A. Lovi, Zh. Prik. Khim., 1942, 15, 214

42W   R. C. Wells, J. Washington Acad. Sci., 1942, 32, 321

42YH   N. Yui and H. Hagisawa, Bull. Inst. Phys. Chem. Res. Tokyo, 1942, 21, 597

43B   D. Bezier, Bull Soc. Chim. France, 1943, 10, 329

43BA   R. G. Bates and S. F. Acree, J. Res. Nat. Bur. Stand., 1943, 30, 129

43BO   C. Brosset and J. Orring, Svensk. Kem. Tidskr., 1943, 55, 101

43BR   G. L. Beyer and W. Rieman, III, J. Amer. Chem. Soc., 1943, 65, 971

43BW   C. Brosset and U. Wahlberg, Svensk Kem. Tidskr., 1943, 55, 335

43DV   P. F. Derr and W. C. Vosburgh, J. Amer. Chem. Soc., 1943, 65, 2408

43HD   H. S. Harned and R. Davis, Jr., J. Amer. Chem. Soc., 1943, 65, 2030

43L   I. Leden, Diss., Lund, 1943 (see 64SM)

43N   R. Nasanen, Suomen Kem., 1943, B16, 1

43O   H. Olerup, Svensk Kem. Tidskr., 1943, 55, 324

43OK   B. B. Owen and E. J. King, J. Amer. Chem. Soc., 1943, 65, 1612

43OKa   Y. Oka, K. Kawagaki, and R. Kadoya, Nippon Kagaku Zasshi, 1943, 64, 718

43P   K. J. Pedersen, Kgl. Danske Vid. Selsk., Mat. Fys. Medd., 1943, 20, No. 7

43RB   O. Redlich and J. Bigeleisen, J. Amer. Chem. Soc., 1943, 65, 1883

43SK   M. S. Sherrill, C. B. King, and R. C. Spooner, J. Amer. Chem. Soc., 1943, 65, 170

43T   I. Takacs, Magyar Kem. Foly., 1943, 49, 33, 100

43TH   H. Tabor and A. B. Hastings, J. Biol. Chem., 1943, 148, 627

43VM   W. C. Vosburgh and R. S. McClure, J. Amer. Chem. Soc., 1943, 65, 1060

44A   M. B. Alpert, Thesis, Yale Univ., 1944

44B   J. Bjerrum, Kgl. Danske Vid. Selsk., Math. Fys. Medd., 1944, 21, No. 4

44C   C. S. Chu, J. Chinese Chem. Soc., 1944, 11, 113

44F   W. Feitknecht, Helv. Chim. Acta, 1944, 27, 771

44HB   H. v. Halban and J. Brull, Helv. Chim. Acta, 1944, 27, 1719

44KN   S. Kilpi and R. Nasanen, Suomen Kem., 1944, B17, 9

44L   I. Leden, Svensk. Kem. Tidskr., 1944, 56, 31

44La   F. Lindstrand, Svensk. Kem. Tidskr., 1944, 56, 251, 282

44MD   G. G. Manov, N. J. DeLollis, and S. F. Acree, J. Res. Nat. Bur. Stand., 1944, 33,
        287

44MK   T. Moeller and H. E. Kremers, J. Phys. Chem., 1944, 48, 395

44N   R. Nasanen, Suomen Kem., 1944, B17, 11

44Na   R. Nasanen, Suomen Kem., 1944, B17, 31

44NG   M. V. Noble and A. B. Garret, J. Amer. Chem. Soc., 1944, 66, 231

44O   H. Olerup, Diss., Lund, 1944

44SL   L. G. Sillen and B. Liljequist, Svensk. Kem. Tidskr., 1944, 56, 85

44T   K. Tachiki, Nippon Kagaku Zasshi, 1944, 65, 346

45B   J. Bjerrum, Kem. Maanedsblad, 1945, 26, 24

45D   C. W. Davies, J. Chem. Soc., 1945, 460

45G   I. Greenwald, J. Biol. Chem., 1945, 161, 697

45H   G. Holst, Svensk Papperstidn, 1945, 48, 23

45J   J. H. Jones, J. Amer. Chem. Soc., 1945, 67, 855

45M   J. Mereadie, Comp. Rend. Acad. Sci. Paris, 1945, 221, 581

45N   R. Nasanen, Suomen Kem., 1945, B18, 45

45Na   R. Nasanen, Ann. Acad. Sci. Fenn., 1945, AII, No. 13

45P   K. J. Pedersen, Kgl. Danske Vid. Selsk., Mat. Fys. Medd., 1945, 22, No. 10

46AM   R. W. Asmussen and L. T. Muus, Trans. Danske Acad. Tech. Sci., 1946, 3

46H   Y. Hentola, Diss., Helsinki-Helsingfors, 1946

46IS   G. Infeldt and L. G. Sillen, Svensk. Kem. Tidskr., 1946, 58, 104

46K   H. Kubli, Helv. Chim. Acta, 1946, 29, 1962

46KD   A. F. Kapustinsky and I. P. Dezideryeva, Trans. Faraday Soc., 1946, 42, 69

46KS   A. Kossiakoff and D. V. Sickman, J. Amer. Chem. Soc., 1946, 68, 442

46L   J. T. Law, Thesis, Univ. New Zealand, 1946, p. 35

46M   T. Moeller, J. Phys. Chem., 1946, 50, 242

46N   R. Nasanen, Suomen Kem., 1946, B19, 90

46Na   R. Nasanen, Ann. Acad. Sci. Fenn., 1946, AII, No. 17

460A   A. Olander and O. Adelsohn, Svensk. Kem. Tidskr., 1946, 58, 33

46R   I. G. Ryss, Dokl. Akad. Nauk SSSR, 1946, 52, 417

46Ra   I. G. Ryss, Zh. Obsh. Khim., 1946, 16, 331

46SC   P. Souchay and G. Carpeni, Bull. Soc. Chim. France, 1946, 160

46TM   I. V. Tananaev and I. B. Mizetskaya, Zh. Anal. Khim., 1946, 1, 6

47BD   H. H. Broene and T. DeVries, J. Amer. Chem. Soc., 1947, 69, 1644

47GD   M. Geloso and P. Deschamps, Compt. Rend. Acad. Sci. Paris, 1947, 225, 742

47HK   W. E. Harris and I. M. Kolthoff, J. Amer. Chem. Soc., 1947, 69, 446

47J   J. C. James, Thesis, Univ. London, 1947

47JQ   A. Jonsson, I. Qvarfort, and L. G. Sillen, Acta Chem. Scand., 1947, 1, 461

47LJ   B. Lindgren, A. Jonsson, and L. G. Sillen, Acta Chem. Scand., 1947, 1, 479

47ML   D. A. MacInnes and L. G. Longsworth, MDDC-911, 1947

47N   R. Nasanen, Acta Chem. Scand., 1947, 1, 763

47R   F. Rivenq, Bull. Soc. Chim. France, 1947, 971

47SF   P. Souchay and J. Faucherre, Bull. Soc. Chim. France, 1947, 529

48B   N. Bjerrum, Bull. Soc. Chim. Belges, 1948, 57, 432

48BJ   P. O. Bethge, I. Jonevall-westoo, and L. G. Sillen, Acta Chem. Scand., 1948, 2, 828

48C   O. A. Chaltykyan, Zh. Obsh. Khim., 1948, 18, 1626

48DJ   C. W. Davies and J. C. James, Proc. Roy. Soc. (London), 1948, A195, 116

48F   J. Faucherre, Comp. Rend. Acad. Sci. Paris, 1948, 227, 1367

48GF   M. Geloso and J. Faucherre, Compt. Rend. Acad. Sci. Paris, 1948, 227, 200, 430

48HS   L. J. Heidt and M. E. Smith, J. Amer. Chem. Soc., 1948, 70, 2476

48IN   M. F. Ivanova and M. B. Neiman, Dokl. Akad. Nauk. SSSR, 1948, 60, 1005

48K   R. M. Keefer, J. Amer. Chem. Soc., 1948, 70, 476

48KA   G. J. R. Krige and R. Arnold, J. S. African Chem. Inst., 1948, 1, 61

48LN   J. J. Lingane and L. W. Niedrach, J. Amer. Chem. Soc., 1948, 70, 4115

48M   C. B. Monk, J. Amer. Chem. Soc., 1948, 70, 3281

48RS   I. G. Ryss, M. M. Slufskaya, and S. D. Palevskaya, Zh. Fiz. Khim., 1948, 22, 1322

48SD   D. I. Stock and C. W. Davies, Trans. Faraday Soc., 1948, 44, 856

48SP   P. Souchay and D. Peschanski, Bull. Soc. Chim. France, 1948, 439

48T   H. Taube, J. Amer. Chem. Soc., 1948, 70, 3928

48W   C. A. Wamser, J. Amer. Chem. Soc., 1948, 70, 1209

49A   S. Ahrland, Acta Chem. Scand., 1949, 3, 374

49Aa   S. Ahrland, Acta Chem. Scand., 1949, 3, 1067

49B   R. Benoit, Bull. Soc. Chim. France, 1949, 518

49Ba   T. E. Brehmer, Finska Kim. Medd., 1949, 58, 79

49Bb   N. Bjerrum in Niels Bjerrum Selected Papers, Munksgaard, Copenhagen, 1949, p. 344
          (see 64SM)

49BM   R. H. Betts and R. K. Michels, J. Chem. Soc., 1949, S286

49BP   R. G. Bates and G. D. Pinching, J. Res. Nat. Bur. Stand., 1949, 42, 419

49BPa   R. P. Bell and J. E. Prue, J. Chem. Soc., 1949, 362

49CM   R. E. Connick and W. H. McVey, J. Amer. Chem. Soc., 1949, 71, 1534

49CMa   R. E. Connick and W. H. McVey, J. Amer. Chem. Soc., 1949, 71, 3182

49DC   T. DeVries and D. Cohen, J. Amer. Chem. Soc., 1949, 71, 1114 (see 64SM)

49DJ   C. W. Davies and A. L. Jones, Disc. Faraday Soc., 1949, 5, 103

49DM   C. W. Davies and C. B. Monk, J. Chem. Soc., 1949, 413

49DR   H. W. Dodgen and G. K. Rollefson, J. Amer. Chem. Soc., 1949, 71, 2600

49DW   C. W. Davies and P. A. H. Wyatt, Trans. Faraday Soc., 1949, 45, 770

49E   E. Eriksson, Kgl. Lantbruks-Hogskol Ann., 1949, 16, 39, 72

49EG   M. G. Evans, P. George, and N. Uri, Trans. Faraday Soc., 1949, 45, 230

49EP   U. R. Evans and M. J. Pryor, J. Chem. Soc., 1949, S157

49EU   M. G. Evans and N. Uri, Trans. Faraday Soc., 1949, 45, 224

49GG   K. H. Gayer and A. B. Garrett, J. Amer. Chem. Soc., 1949, 71, 2973

49GGa   R. M. Garrels and F. T. Gucker, Jr., Chem. Rev., 1949, 44, 117

49H   J. C. Hindman, Nat. Nucl. Ener. Ser., 1949, IV-14B, 370

49Ha   J. C. Hindman, Nat. Nucl. Ener. Ser., 1949, IV-14B, 388

49Hb   J. C. Hindman, Nat. Nucl. Ener. Ser., 1949, IV-14B, 405

49J   W. Jaenicke, Z. Naturforsch, 1949, 4a, 353

49Ja   J. C. James, Trans. Faraday Soc., 1949, 45, 855

49Jb   J. C. James, J. Amer. Chem. Soc., 1949, 71, 3243

49JM   H. W. Jones, C. B. Monk, and C. W. Davies, J. Chem. Soc., 1949, 2693

49K   E. L. King, J. Amer. Chem. Soc., 1949, 71, 319

49Ka   E. L. King, Nat. Nucl. Ener. Ser., 1949, IV-14B, 638

49KD   K. A. Kraus and J. R. Dam, Nat. Nucl. Ener. Ser., 1949, IV-14B, 466

49KDa   K. A. Kraus and J. R. Dam, Nat. Nucl. Ener. Ser., 1949, IV-14B, 478, 528

49KH   W. D. Kingery and D. N. Hume, J. Amer. Chem. Soc., 1949, 71, 2393

49KK   I. A. Korshunov and E. F. Khrulkova, J. Gen. Chem. USSR, 1949, 19,   (2045)

49KL   N. Konopik and O. Leberl, Monat. Chem., 1949, 80, 655

49KLa   N. Konopik and O. Leberl, Monat. Chem., 1949, 80, 781

49KN   K. A. Kraus, F. Nelson, and G. L. Johnson, J. Amer. Chem. Soc., 1949, 71, 2510

49KS   P. Krumholz and H. M. A. Stettiner, J. Amer. Chem. Soc., 1949, 71, 3035

49L   S. Lacroix, Ann. Chim. (France), 1949, 4, 5

49M   C. B. Monk, J. Chem. Soc., 1949, 423

49Ma   C. B. Monk, J. Chem. Soc., 1949, 427

49Mb   C. K. McLane, Nat. Nucl. Ener. Ser., 1949, IV-14B, 414

49NT   R. Nasanen and V. Tamminen, J. Amer. Chem. Soc., 1949, 71, 1994

49OS   A. R. Olson and T. R. Simonson, J. Chem. Phys., 1949, 17, 1322

49QS   I. Qvarfort and L. G. Sillen, Acta Chem. Scand., 1949, 3, 505

49RP   L. Riccoboni, P. Papoff, and G. Arich, Gazz. Chim. Ital., 1949, 79, 547

49RR   L. B. Rogers and C. A. Reynolds, J. Amer. Chem. Soc., 1949, 71, 2081

49S   J. Sutton, J. Chem. Soc., 1949, S275

49SB   A. G. Stromberg and I. E. Bykov, J. Gen. Chem. USSR, 1949, 19,   (245)

49SBa   A. G. Stromberg and I. E. Bykov, J. Gen. Chem. USSR, 1949, 19,   (1816)

49T   B. Topley, Quart. Rev., 1949, 3, 345

49TD   I. V. Tananaev and E. N. Deichman, Izvest. Akad. Nauk. SSSR, Otdel. Khim. Nauk.,
            1949, 144

49TL   I. V. Tananaev, Yu. L. Lelchuk, and B. Kh. Petrovitskay, J. Gen. Chem. USSR,
            1949, 19,   (1207)

49YP   K. B. Yatsimirskii and L. L. Pankova, J. Gen. Chem. USSR, 1949, 19, 569 (617)

49Z   J. Zurc., Diss., Univ. Zurich, 1949

49ZN   E. L. Zebroski and F. K. Neumann, KAPL-184, 1949

50A   T. V. Arden, J. Chem. Soc., 1950, 882

V. BIBLIOGRAPHY

50AF   N. V. Akselrud and Ya. A. Fialkov, Ukr. Khim. Zh., 1950, 16, 283

50B   J. Badoz-Lambling, Bull. Soc. Chim. France, 1950, 552

50BL   J. Bjerrum and C. G. Lamm, Acta Chim. Scand., 1950, 4, 997

50BP   R. G. Bates and G. D. Pinching, J. Amer. Chem. Soc., 1950, 72, 1393

50BQ   P. Bernheim and M. Quintin, Compt. Rend. Acad. Sci. Paris, 1950, 230, 388

50BW   R. P. Bell and G. M. Waind, J. Chem. Soc., 1950, 1979

50CJ   C. V. Cole and M. L. Jackson, Proc. Soil Sci. Soc. Amer., 1950, 15, 84

50DC   F. R. Duke and W. G. Courtney, Iowa State Coll. J. Sci., 1950, 24, 397

50DS   R. A. Day, Jr., and R. W. Stoughton, J. Amer. Chem. Soc., 1950, 72, 5662

50ET   D. A. Everest and H. Terry, J. Chem. Soc., 1950, 2282

50F   S. Fronaeus, Acta Chem. Scand., 1950, 4, 72

50Fa   T. D. Farr, TVA Chem. Eng. Rept., No. 8, 1950, p. 26, 27, 52

50FG   S. C. Furman and C. S. Garner, J. Amer. Chem. Soc., 1950, 72, 1785

50GG   K. H. Gayer and A. B. Garrett, J. Amer. Chem. Soc., 1950, 72, 3921

50H   T. Hirata, Reports Res. Sci. Dept. Kyushu Univ., 1950, 1, 199

50Ha   B. C. Haldar, Curr. Sci., 1950, 19, 244, 283

50JM   I. L. Jenkins and C. B. Monk, J. Amer. Chem. Soc., 1950, 72, 2695

50JMa   H. W. Jones and C. B. Monk, J. Chem. Soc., 1950, 3475

50JMb   J. C. James and C. B. Monk, Trans. Faraday Soc., 1950, 46, 1041

50JW   E. Jozefowicz, S. Witekowa, and W. Zubranska, Rocz. Chem., 1950, 24, 64

50K   K. E. Kleiner, J. Gen. Chem. USSR, 1950, 20,   (1747)

50KN   K. A. Kraus and F. Nelson, J. Amer. Chem. Soc., 1950, 72, 3901

50LO   H. A. Laitinen and E. I. Onstott, J. Amer. Chem. Soc., 1950, 72, 4729

50M   L. Meites, J. Amer. Chem. Soc., 1950, 72, 184

50MD   H. McConnell and N. Davidson, J. Amer. Chem. Soc., 1950, 72, 3164, 3168

50MK   T. Moeller and G. L. King, J. Phys. Chem., 1950, 54, 999

50MKa   G. E. Moore and K. A. Kraus, ORNL-795, 1950

50MS   S. W. Mayer and S. D. Schwartz, J. Amer. Chem. Soc., 1950, 72, 5106

50N   R. Nasanen, Acta Chem. Scand., 1950, 4, 140

50Na   R. Nasanen, Acta Chem. Scand., 1950, 4, 816

50PA   A. V. Pamfilov and A. L. Agafonova, Zh. Fiz. Khim., 1950, 24, 1147

50S   F. Saegusa, Nippon Kagaku Zasshi, 1950, 71, 223

50SS   P. Souchay and R. Schaal, Bull. Soc. Chim. France, 1950, 819

50SSa P. Souchay and R. Schaal, Bull. Soc. Chim. France, 1950, 824

50SZ   G. Schwarzenbach and J. Zurc, Monat Chem., 1950, 81, 202

50TK   Sh. T. Talipov and V. A. Khadeev, J. Gen. Chem. USSR, 1950, 20,   (774),
         (783)

50VC   J. R. Van Wazer and D. A. Campanella, J. Amer. Chem. Soc., 1950, 72, 655

50VK   M. G. Vladimirova and I. A. Kakovskii, J. Appl. Chem. USSR, 1950, 23,   (580)

50WS   W. C. Waggener and R. W. Stoughton, ORNL-795, 1950

51A   S. Ahrland, Acta Chem. Scand., 1951, 5, 1151

51Aa   S. Ahrland, Acta Chem. Scand., 1951, 5, 1271

51Ab   T. V. Arden, J. Chem. Soc., 1951, 350

51B   R. G. Bates, J. Res. Nat. Bur. Stand., 1951, 47, 127

51BB   W. G. Barb, J. H. Baxendale, P. George, and K. R. Hargrave, Trans. Faraday Soc.,
         1951, 47, 591

51BG   R. P. Bell and E. Gelles, J. Chem. Soc., 1951, 2734

51CH   C. E. Crouthamel, A. M. Hayes, and D. S. Martin, J. Amer. Chem. Soc., 1951, 73, 82

51CM   R. E. Connick and S. W. Mayer, J. Amer. Chem. Soc., 1951, 73, 1176

51D   C. W. Davies, J. Chem. Soc., 1951, 1256

51Da   L. P. Ducret, Ann. Chim. (France), 1951, 6, 705

51DB   F. R. Duke and R. F. Bremer, J. Amer. Chem. Soc., 1951, 73, 5179

51DC   P. Deschamps and B. Charreton, Comp. Rend. Acad. Sci. Paris, 1951, 232, 162

51DH   C. W. Davies and B. E. Hoyle, J. Chem. Soc., 1951, 233

51DJ   H. S. Dunsmore and J. C. James, J. Chem. Soc., 1951, 2925

51DM   T. O. Denney and C. B. Monk, Trans. Faraday Soc., 1951, 47, 992

51DP   F. R. Duke and R. C. Pinkerton, J. Amer. Chem. Soc., 1951, 73, 3045

51EU   M. G. Evans and M. Uri, Symp. Soc. Exp. Biol., 1951, 5, 130

51F   S. Fronaeus, Acta Chem. Scand., 1951, 5, 139

51FG   S. C. Furman and C. S. Garner, J. Amer. Chem. Soc., 1951, 73, 4528

51FR   W. Feitknecht and R. Reinmann, Helv. Chim. Acta, 1951, 34, 2255

51HD   G. Harbottle and R. W. Dodson, J. Amer. Chem. Soc., 1951, 73, 2442

51HDC   D. N. Hume, D. D. DeFord, and G. C. B. Cave, J. Amer. Chem. Soc., 1951, 73, 5323

51HM   E. Hayek, F. Mullner, and K. Koller, Monat. Chem., 1951, 82, 959

51HR   T. J. Hardwick and E. Robertson, Canad. J. Chem., 1951, 29, 818

51ID   J. A. Ibers and N. Davidson, J. Amer. Chem. Soc., 1951, 73, 476

51K   N. P. Komar, Uchenye Zapiski Kharkov Univ., 1951, 37, 111

51KL   I. A. Korshunov and L. V. Lipatova, J. Gen. Chem. USSR, 1951, 21,   (615)

51LW   M. Lloyd, V. Wycherley, and C. B. Monk, J. Chem. Soc., 1951, 1786

51M   K. L. Mattern, UCRL-1407, 1951

51Ma   C. B. Monk, Trans. Faraday Soc., 1951, 47, 285

51Mb   C. B. Monk, Trans. Faraday Soc., 1951, 47, 292

51Mc   W. H. McVey, HW-21487, 1951

51MF   T. Moeller and N. Fogel, J. Amer. Chem. Soc., 1951, 73, 4481

51MM   J. Y. Macdonald, K. M. Mitchell, and A. T. S. Mitchell, J. Chem. Soc., 1951, 1574

51MS   S. W. Mayer and S. D. Schwartz, J. Amer. Chem. Soc., 1951, 73, 222

51NK   F. Nelson and K. A. Kraus, J. Amer. Chem. Soc., 1951, 73, 2157

51NL   R. Nasanen and P. Lumme, Acta Chem. Scand., 1951, 5, 13

51PC   S. Peterson and O. W. Cooper, Trans. Kentucky Acad. Sci., 1951, 13, 146

51PN   R. B. Peppler and E. S. Newman, J. Res. Nat. Bur. Stand., 1951, 46, 121
        (see also 64SM)

51Q   M. Quintin, Comp. Rend. Acad. Sci. Paris, 1951, 232, 1303

51RL   S. W. Rabideau and J. F. Lemons, J. Amer. Chem. Soc., 1951, 73, 2895

51S   A. I. Stabrovskii, J. Gen. Chem. USSR, 1951, 21,   (1223)

51Sa   S. Suzuki, Nippon Kagaku Zasshi, 1951, 72, 265

51Sb   D. Stetten, Jr., Anal. Chem., 1951, 23, 1177

51SG   D. F. Swinehart and A. B. Garrett, J. Amer. Chem. Soc., 1951, 73, 507

51SS   G. Saini and C. Sapetti, Atti. Accad. Sci. Torino, Classe Sci. Fis. Mat.-Nat.,
        1951, 86, 247

51SSW   J. A. Schufle, M. F. Stubbs, and R. E. Witman, J. Amer. Chem. Soc., 1951, 73, 1013

51SV   T. H. Siddall, III, and W. C. Vosburgh, J. Amer. Chem. Soc., 1951, 73, 4270

51V   E. N. Vinogradova, Trudy. Komissii Anal. Khim. Akad. Nauk SSSR, 1951, 3, 127, 138

51W   C. A. Wamser, J. Amer. Chem. Soc., 1951, 73, 409

51Y   N. Yui, Sci. Reports Tohoku Univ., 1951, 35, 53

51Z   F. G. Zharovskii, Trudy Komissii Anal. Khim. Akad. Nauk. SSSR, 1951, 3, 101

51ZA   E. L. Zebroski, H. W. Alter, and F. K. Heumann, J. Amer. Chem. Soc., 1951, 73, 5646

52AP   S. Aditya and B. Prasad, J. Indian Chem. Soc., 1952, 29, 293

52B   C. Brosset, Acta Chem. Scand., 1952, 6, 910

52CC   S. Chaberek, R. C. Courtney and A. E. Martell, J. Amer. Chem. Soc., 1952, 74, 5057

52CM   C. A. Colman-Porter and C. B. Monk, J. Chem. Soc., 1952, 1312

52DJ   C. W. Davies, H. W. Jones, and C. B. Monk, Trans. Faraday Soc., 1952, 48, 921

52F   S. Fronaeus, Svensk Kem. Tidskr., 1952, 64, 317

52Fa   P. S. Farrington, J. Amer. Chem. Soc., 1952, 74, 966

52Fb   W. S. Fyfe, J. Chem. Soc., 1952, 2023

52FH   W. Forsling, S. Hietanen, and L. G. Sillen, Acta Chem. Scand., 1952, 6, 901

52GG   K. H. Gayer and A. B. Garrett, J. Amer. Chem. Soc., 1952, 74, 2353

52GGF   J. R. Goates, M. B. Gordon, and N. D. Faux, J. Amer. Chem. Soc., 1952, 74, 835

52GM   J. A. Gledhill and G. M. Malan, Trans. Faraday Soc., 1952, 48, 258

52GMa   C. W. Gibby and C. B. Monk, Trans. Faraday Soc., 1952, 48, 632

52GW   K. H. Gayer and L. Woontner, J. Amer. Chem. Soc., 1952, 74, 1436

52HH   L. G. Hepler and Z. Z. Hugus, Jr., J. Amer. Chem. Soc., 1952, 74, 6115

52HHa   W. Hieber and W. Hubel, Z. Naturforseh, 1952, 7b, 322; Z. Elektrochem., 1953, 57, 235

52HS   S. Hietanen and L. G. Sillen, Acta Chem. Scand., 1952, 6, 747

52J   C. E. Johnson, Jr., J. Amer. Chem. Soc., 1952, 74, 959

52Ja   W. L. Jolly, Thesis, Univ. Calif. Berkeley, 1952; UCRL-1638

52JK   J. S. Johnson and K. A. Kraus, J. Amer. Chem. Soc., 1952, 74, 4436

52JL   W. L. Jolly and W. M. Latimer, J. Amer. Chem. Soc., 1952, 74, 5751

52JM   H. W. Jones and C. B. Monk, Trans. Faraday Soc., 1952, 48, 929

52JMa   J. H. Jonte and D. S. Martin, Jr., J. Amer. Chem. Soc., 1952, 74, 2052

52K   K. E. Kleiner, J. Gen. Chem. USSR, 1952, 22,   (17)

52KF   I. M. Korenman, F. S. Frum, and V. G. Chebakova, J. Gen. Chem. USSR, 1952, 22, (1731)

52KH   Z. Ksandr and M. Hejtmanek, Sbornik I Celostatni Pracovni Konf. Anal. Chem., 1952,
        42; Chem. Abs., 1956, 3150

52KK   E. J. King and G. W. King, J. Amer. Chem. Soc., 1952, 74, 1212

52KP   E. L. King and M. L. Pandow, J. Amer. Chem. Soc., 1952, 74, 1966

52L   I. Leden, Acta Chem. Scand., 1952, 6, 971

52La   I. Leden, Svensk Kem. Tidskr., 1952, 64, 249

52Lb   W. M. Latimer, Oxidation Potentials, 2nd ed., Prentice Hall, New York, 1952

52Lc   M. W. Lister, Canad. J. Chem., 1952, 30, 879

52LH   W. Forsling, S. Hietanen, and L. G. Sillen, Acta Chem. Scand., 1952, 6, 901

52LM   I. Leden and L. E. Marthen, Acta. Chem. Scand., 1952, 6, 1125

52M   C. B. Monk, J. Chem. Soc., 1952, 1314

52Ma   C. B. Monk, J. Chem. Soc., 1952, 1317

52S   K. W. Sykes, J. Chem. Soc., 1952, 124

52Sa   S. Suzuki, Nippon Kagaku Zasshi, 1952, 73, 150

52Sb   S. Suzuki, Nippon Kagaku Zasshi, 1952, 73, 153

52Sc   S. Suzuki, Nippon Kagaku Zasshi, 1952, 73, 278

52Sd   M. B. Shchigol, Zh. Obsh. Khim., 1952, 22, 721

52Se   Sutton, Nature, 1952, 169, 71

52Sf   M. Sahli, Diss., Bern, 1952

52SD   P. Senise and P. Delahay, J. Amer. Chem. Soc., 1952, 74, 6128

52SH   J. C. Sullivan and J. C. Hindman, J. Amer. Chem. Soc., 1952, 74, 6091

52SZ   G. Schwarzenbach, and A. Zobrist, Helv. Chim. Acta, 1952, 35, 1291

52TM   W. E. Trevelyan, P. F. E. Mann, and J. S. Harrison, Arch. Biochem. Biophys., 1952,
        39, 440

52V   C. E. Vanderzee, J. Amer. Chem. Soc., 1952, 74, 4806

52VR   C. E. Vanderzee and D. E. Rhodes, J. Amer. Chem. Soc., 1952, 74, 3552

52WS   W. C. Waggener and R. W. Stoughton, J. Phys. Chem., 1952, 56, 1

52WT   A. S. Wilson and H. Taube, J. Amer. Chem. Soc., 1952, 74, 3509

52YA   K. B. Yatsimirskii and A. A. Astasheva, Zh. Fiz. Khim., 1952, 26, 239

52YG   K. B. Yatsimirskii and Z. M. Grafova, J. Gen. Chem. USSR, 1952, 22,   (1726)

52YV   R. P. Yaffe and A. F. Voigt, J. Amer. Chem. Soc., 1952, 74, 2500

53A   R. Akeret, Diss., Eidg. Techn. Hochschule, Zurich, 1953

53B   G. Biedermann, Arkiv. Kemi, 1953, 5, 441

53Ba   J. Badoz-Lambling, Ann. Chim. (France), 1953, 8, 586

53BD   R. H. Betts and F. S. Dainton, J. Amer. Chem. Soc., 1953, 75, 5721

53BG   R. P. Bell and J. H. B. George, Trans. Faraday Soc., 1953, 49, 619

53BGa   A. K. Babko and A. M. Golub, Sbornik Stat. Obshchei. Khim. Akad. Nauk SSSR, 1953,
        1, 64

53BL   E. Berne and I. Leden, Svensk. Kem. Tidskr., 1953, 65, 88

53BLa   E. Berne and I. Leden, Z. Naturforsch., 1953, 8a, 719

53BP   D. Berg and A. Patterson, Jr., J. Amer. Chem. Soc., 1953, 75, 1484

53BS   C. F. Baes, Jr., and J. M. Schreyer, ORNL-1579, 1953

53BSL   C. F. Baes, Jr., J. M. Schreyer, and J. M. Lesser, ORNL-Y-12, ORNL-1577, 1953

53CH   C. C. B. Cave and D. N. Hume, J. Amer. Chem. Soc., 1953, 75, 2893

53CM   R. E. Connick and W. H. McVey, J. Amer. Chem. Soc., 1953, 75, 474

53CP   H. Chateau and J. Pouradier, Sci. Ind. Phot., 1953, 24, 129

53CS   J. W. Cobble, W. T. Smith, Jr., and G. E. Boyd, J. Amer. Chem. Soc., 1953, 75, 5777

53CT   R. E. Connick and M. S. Tsao, 123rd Meeting Amer. Chem. Soc., 1953, p. 4p

53DH   C. W. Davies and B. E. Hoyle, J. Chem. Soc., 1953, 4134

53E   L. Eriksson, Acta Chem. Scand., 1953, 7, 1146

53Ea   J. O. Edwards, J. Amer. Chem. Soc., 1953, 75, 6151

53Eb   J. O. Edwards, J. Amer. Chem. Soc., 1953, 75, 6154

53EM   A. J. Everett and G. Minkoff, Trans. Faraday Soc., 1953, 49, 410

53F   S. Fronaeus, Svensk Kem. Tidskr., 1953, 65, 1

53Fa   W. Feitknecht, "Loslichkeiten von Hydroxyden", report to Analytical Section,
        IUPAC, July 1953

53FH   R. E. Frank and D. N. Hume, J. Amer. Chem. Soc., 1953, 75, 1736

53G   A. M. Golub, Ukr. Khim. Zh., 1953, 19, 205, 467

53Ga   R. M. Garrels, Amer. Min., 1953, 38, 1251

53GC   R. E. Gosselin and E. R. Coghlan, Arch. Biochem. Biophys., 1953, 45, 301

53GJ   G. A. Gamlen and D. O. Jordan, J. Chem. Soc., 1953, 1435

53GM   J. A. Gledhill and G. M. Malan, Trans. Faraday Soc., 1953, 49, 166

53H   B. O. A. Hedstrom, _Arkiv Kemi_, 1953, _5_, 457

53Ha   B. O. A. Hedstrom, _Arkiv Kemi_, 1953, _6_, 1

53HJ   L. G. Hepler, W. L. Jolly, and W. M. Latimer, _J. Amer. Chem. Soc._, 1953, _75_, 2809

53HS   H. M. Hershenson, M. E. Smith, and D. N. Hume, _J. Amer. Chem. Soc._, 1953, _75_, 507

53HW   J. Hudis and A. C. Wahl, _J. Amer. Chem. Soc._, 1953 _75_, 4153

53IY   T. Ito and N. Yui, _Sci. Reports Tohoku Univ._, 1953, _37_, 19, 185

53J   W. Jaenicke, _Z. Elektochem._, 1953, _57_, 843

53KF   I. M. Korenman, F. S. Frum, and A. I. Kudinova, _Sbornik Statei Obsh. Khim._, 1953, _1_, 83

53KP   M. Kilpatrick and L. Pokras, _J. Electrochem. Soc._, 1953, _100_, 85

53KZ   J. W. Kury, A. J. Zielen, and W. M. Latimer, _J. Electrochem. Soc._, 1953, _100_, 468

53L   I. Leden, "Solubilities of Sulfates", Report to Anal. Sec., IUPAC, July 1953

53La   M. W. Lister, _Canad. J. Chem._, 1953, _31_, 638

53LJ   W. M. Latimer and W. L. Jolly, _J. Amer. Chem. Soc._, 1953, _75_, 1548

53LK   D. L. Leussing and I. M. Kolthoff, _J. Amer. Chem. Soc._, 1953, _75_, 2476

53LU   A. J. Levin and E. A. Ukshe, _Sbornik Stat. Obsh. Khim._, 1953, _2_, 798

53M   L. Meites, _J. Amer. Chem. Soc._, 1953, _75_, 6059

53MK   T. Moeller and G. L. King, _J. Amer. Chem. Soc._, 1953, _75_, 4852

53N   R. Nasanen, _Suomen Kem._, 1953, _B26_, 37

53Na   R. Nasanen, _Suomen Kem._, 1953, _B26_, 67

53NA   T. W. Newton and G. M. Arcand, _J. Amer. Chem. Soc._, 1953, _75_, 2449

53P   D. Peschanski, _J. Chim. Phys._, 1953, _50_, 640

53Pa   A. M. Posner, _Nature_, 1953, _171_, 519

53Pb   F. M. Page, _J. Chem. Soc._, 1953, 1719

53PH   K. Pan and T. M. Hseu, _Bull. Chem. Soc. Japan_, 1953, _26_, 126

53R   A. Ringbom, "Solubilities of Sulfides", Report to Anal. Sect., IUPAC, July, 1953
       (see 64SM)

53RL   R. L. Rebertus, H. A. Laitinen, and J. C. Bailar, Jr., _J. Amer. Chem. Soc._, 1953, _75_, 3051

53RS   E. G. Rochow and D. Seyferth, _J. Amer. Chem. Soc._, 1953, _75_, 2877

53S   S. Suzuki, _Sci. Reports Res. Inst. Tohoku Univ._, 1953, _A5_, 311

53Sa   S. Suzuki, _Nippon Kagaku Zasshi_, 1953, _74_, 219

53Sb   S. Suzuki, _Nippon Kagaku Zasshi_, 1953, _74_, 269

53Sc   G. Saini, _Gazz. Chim. Ital._, 1953, _83_, 677

53Sd   C. G. Spike, Thesis, Univ. Michigan, 1953

53Se   S. I. Sobol, _J. Gen. Chem. USSR_, 1953, _23_, 945 (906)

53SH   P. Souchay and A. Hessaby, _Bull. Soc. Chim. France_, 1953, 599, 614

53SK   N. M. Selivanova and A. F. Kapustinskii, _Zh. Fiz. Khim._, 1953, _27_, 265

53SKa   N. M. Selivanova and A. F. Kapustinskii, _Zh. Fiz. Khim._, 1953, _27_, 565

53SL   S. A. Shchukarev, L. S. Lilich, and V. A. Latysheva, _Dokl. Akad. Nauk SSSR_, 1953,
        _91_, 273

53SP   C. G. Spike and R. W. Parry, _J. Amer. Chem. Soc._, 1953, _75_, 2726

53SPa   C. G. Spike and R. W. Parry, _J. Amer. Chem. Soc._, 1953, _75_, 3770

53TK   J. Y. Tong and E. L. King, _J. Amer. Chem. Soc._, 1953, _75_, 6180

53VD   C. E. Vanderzee and H. J. Dawson, Jr., _J. Amer. Chem. Soc._, 1953, _75_, 5659

53VT   N. K. Vitchenko and A. S. Tikonov, _Trudy Voronezh Univ._, 1953, _32_, 129

53WA   J. I. Watters and A. Aaron, _J. Amer. Chem. Soc._, 1953, _75_, 611

53WD   R. A. Whiteker and N. Davidson, _J. Amer. Chem. Soc._, 1953, _75_, 3081

53WS   E. J. Wheelwright, F. H. Spedding, and G. Schwarzenbach, _J. Amer. Chem. Soc._, 1953,
        _75_, 4196

53Y   K. B. Yatsimirskii, _Sbornik Stat. Obsh. Khim._, 1953, _1_, 97

53Ya   K. B. Yatsimirskii, _Sbornik Stat. Obsh. Khim._, 1953, _1_, 193

54AHI   G. B. Alexander, W. M. Heston, and R. K. Iler, _J. Phys. Chem._, 1954, _58_, 453

54AHS   S. Ahrland, S. Hietanen, and L. G. Sillen, _Acta Chem. Scand._, 1954, _8_, 1907

54AL   S. Ahrland and R. Larsson, _Acta Chem. Scand._, 1954, _8_, 137

54ALa   S. Ahrland and R. Larsson, _Acta Chem. Scand._, 1954, _8_, 354

54BB   R. D. Brown, W. B. Bunger, W. L. Marshall, and C. H. Secoy, _J. Amer. Chem. Soc._,
        1954, _76_, 1532

54BBa   R. D. Brown, W. B. Bunger, W. L. Marshall, and C. H. Secoy, _J. Amer. Chem. Soc._,
        1954, _76_, 1580

54BBS   C. Brosset, G. Biedermann, and L. G. Sillen, _Acta Chem. Scand._, 1954, _8_, 1917

54BR   J. Benkenkamp, W. Rieman, III, and S. Lindenbaum, _Anal. Chem._, 1954, _26_, 505

54C   R. Cohen-Adad, _Compt. Rend. Acad. Sci. Paris_, 1954, _238_, 810

54CC   H. B. Clarke, D. G. Cusworth, and S. P. Datta, Biochem. J., 1954, 58, 146

54CI   B. G. F. Carleson and H. Irving, J. Chem. Soc., 1954, 4390

54CT   R. E. Connick and M. S. Tsao, J. Amer. Chem. Soc., 1954, 76, 5311

54CV   D. Cozzi and S. Vivarelli, Z. Elektrochem., 1954, 58, 907

54D   G. N. Dobrokhotov, J. Appl. Chem. USSR, 1954, 27,   (1056)

54Da   G. F. Davidson, J. Chem. Soc., 1954, 1649

54DP   R. A. Day, Jr., and R. M. Powers, J. Amer. Chem. Soc., 1954, 76, 3895

54DQ   F. R. Duke and P. R. Quinney, J. Amer. Chem. Soc., 1954, 76, 3800

54EL   D. H. Everett and D. A. Landsman, Trans. Faraday Soc., 1954, 50, 1221

54F   J. Faucherre, Bull. Soc. Chim. France, 1954, 253

54Fa   D. C. Feay, Thesis, Univ. Calif. Berkeley, 1954; UCRL-2547

54FH   W. Feitknecht and L. Hartmann, Chimia (Switz.), 1954, 8, 95 (see also 63FS)

54FS   W. Feitknecht and M. Sahli, Helv. Chim. Acta, 1954, 37, 1431

54FSa   J. W. Fulton and D. F. Swinehart, J. Amer. Chem. Soc., 1954, 76, 864

54G   G. W. Goward, Thesis, Princeton Univ., 1954

54GC   J. D. McGilvery and J. P. Crowther, Canad. J. Chem., 1954, 32, 174

54GL   K. H. Gayer and H. Leider, J. Amer. Chem. Soc., 1954, 76, 5938

54GM   J. A. Gledhill and G. M. Malan, Trans. Faraday Soc., 1954, 50, 126

54GMa   F. G. R. Gimblett and C. B. Monk, Trans. Faraday Soc., 1954, 50, 965

54H   S. Hietanen, Acta Chem. Scand., 1954, 8, 1626

54HK   L. G. Hepler, J. W. Kury, and Z. Z. Hugus, Jr., J. Phys. Chem., 1954, 58, 26

54HP   L. E. Holt, Jr., J. A. Pierce, and C. N. Kajdi, J. Colloid Sci., 1954, 9, 409

54HR   G. C. Hood, O. Redlich, and C. A. Reilly, J. Chem. Phys. 1954, 22, 2067

54IA   I. M. Issa and S. A. Awad, J. Phys. Chem., 1954, 58, 948

54IK   I. M. Issa and H. Khalifa, J. Indian Chem. Soc., 1954, 31, 91

54JK   J. S. Johnson, K. A. Kraus, and T. F. Young, J. Amer. Chem. Soc., 1954, 76, 1436

54K   I. M. Korenman, J. Gen. Chem. USSR, 1954, 24,   (1910)

54KH   K. A. Kraus and R. W. Holmberg, J. Phys. Chem., 1954, 58, 325

54KP   M. Kilpatrick and L. Pikras, J. Electrochem. Soc., 1954, 101, 39

54KT   J. Kratohvil, B. Tezak, and V. B. Vouk, Arhiv Kemi, 1954, 26, 191

54L   R. Lloyd, Thesis, Temple Univ., Phila., Penn., 1954

54LN   I. Leden and R. Nilsson, Svensk Kem. Tidskr., 1954, 66, 126

54LP   K. S. Lyalikov and V. N. Piskunova, Russ. J. Phys. Chem., 1954, 28,   (127),
        (595)

54LS   I. Leden and N. H. Schoon, Trans. Chalmers Univ. Technol. Gothenburg, 1954, No. 144
        (see 64SM)

54M   G. Mattock, J. Amer. Chem. Soc., 1954, 76, 4835

54Ma   W. Miedreich, Thesis, Frankfurt am Main, 1954

54N   R. Nasanen, Acta Chem. Scand., 1954, 8, 1587

54NK   R. Nasanen and B. Klaile, Suomen Kem., 1954, B27, 50

54NR   M. S. Novakovskii and A. P. Ryazantseva, Uchenye Zap. Kharkov Univ., 1954, 54, 277

54NS   M. S. Novakovskii and T. M. Shmaeva, Ukr. Khim. Zh., 1954, 20, 615

54P   F. M. Page, Trans. Faraday Soc., 1954, 50, 120

54Pa   K. Pan, J. Chinese Chem. Soc., 1954, 1, 1

54Pb   K. Pan, J. Chinese Chem. Soc., 1954, 1, 16

54Pc   K. Pan, J. Chinese Chem. Soc., 1954, 1, 26

54Pd   D. Peschanski, Compt. Rend. Acad. Sci. France, 1954, 238, 2077

54PB   K. G. Poulsen, J. Bjerrum, and I. Poulsen, Acta Chem. Scand., 1954, 8, 921

54PM   R. A. Paris and J. C. Merlin, Colloquium IUPAC Munster, 1954, 237

54PV   J. Pouradier, A. M. Venet, and H. Chateau, J. Chim. Phys., 1954, 51, 375

54R   R. L. Rebertus, Diss., Univ. Illinois, 1954 (see 64SM)

54Ra   F. W. Rakowsky, Thesis, Ohio State Univ., 1954

54S   N. Sunden, Svensk Kem. Tidskr., 1954, 66, 20

54Sa   N. Sunden, Svensk Kem. Tidskr., 1954, 66, 50

54Sb   N. Sunden, Svensk Kem. Tidskr., 1954, 66, 173

54Sc   N. Sunden, Svensk Kem. Tidskr., 1954, 66, 345

54Sd   R. P. Seward, J. Amer. Chem. Soc., 1954, 76, 4850

54Se   K. W. Sykes, The Kinetics and Mechanism of Inorganic Reactions in Solution, Spec.
        Publ. No. 1, Chem. Soc., London, 1954, p. 64, (see 64SM)

54SD   W. B. Schaap, J. A. Davies, and W. H. Nebergall, J. Amer. Chem. Soc., 1954, 76,
        5226

54SE   J. A. Schufle and H. M. Eiland, J. Amer. Chem. Soc., 1954, 76, 960

54SH   J. C. Sullivan and J. C. Hindman, J. Amer. Chem. Soc., 1954, 76, 5931

54SJ   F. H. Spedding and S. Jaffe, J. Amer. Chem. Soc., 1954, 76, 882

54SS   M. Sato and J. Sato, J. Elektrochem. Soc., Japan, 1954, 22, 411

54ST   R. K. Schofield and A. W. Taylor, J. Chem. Soc., 1954, 4445

54T   V. T. Toropova, J. Gen. Chem. USSR, 1954, 24,   (423)

54TK   Y. P. Tong and E. L. King, J. Amer. Chem. Soc., 1954, 76, 2132

54TO   Sh. T. Talipov and P. F. Obelchenko, Trudy Sredneasiat Gosud. Univ., 1954, 55, 3,
        87

54UL   E. A. Ukshe and A. I. Levin, J. Gen. Chem. USSR, 1954, 24,   (775); ibid., 1956,
        26, 2963 (2657)

54W   R. J. P. Williams, J. Phys. Chem., 1954, 58, 121

54YS   K. B. Yatsimirskii and A. A. Shutov, Zh. Fiz. Khim., 1954, 28, 30

55A   A. Agren, Acta Chem. Scand., 1955, 9, 49

55Aa   E. L. Anderson, Jr., Thesis, Washington State Coll., 1955

55Ab   P. J. Antikainen, Suomen Kem., 1955, B28, 159; Acta Chem. Scand., 1956, 10, 756

55B   R. H. Betts, Canad. J. Chem., 1955, 33, 1775

55BPP   A. I. Biggs, M. H. Panckhurst, and H. N. Parton, Trans. Faraday Soc., 1955, 51,
        802

55BPR   A. I. Biggs, H. N. Parton, and R. A. Robinson, J. Amer. Chem. Soc., 1955, 77, 5844

55BS   T. N. Bondareva and A. G. Stromberg, J. Gen. Chem. USSR, 1955, 25,   (666)

55C   J. S. Clark, Canad. J. Chem., 1955, 33, 1696

55CD   M. Cher and N. Davidson, J. Amer. Chem. Soc., 1955, 77, 793

55CS   D. Cohen, J. C. Sullivan, and J. C. Hindman, J. Amer. Chem. Soc., 1955, 77, 4964

55D   J. A. Davis, Thesis, Indiana Univ., 1955

55DG   M. E. Dry and J. A. Gledhill, Trans. Faraday Soc., 1955, 51, 1119

55DJ   C. W. Davies and A. L. Jones, Trans. Faraday Soc., 1955, 51, 812

55DP   W. G. Davies and J. E. Prue, Trans. Faraday Soc., 1955, 51, 1045

55DW   R. A. Day, Jr., R. N. Wilhite, and F. D. Hamilton, J. Amer. Chem. Soc., 1955, 77,
        3180

55ER   E. Eichler and S. Rabideau, J. Amer. Chem. Soc., 1955, 77, 5501

55F   S. N. Flengas, Trans. Faraday Soc., 1955, 51, 62

55Fa   W. S. Fyfe, J. Chem. Soc., 1955, 1347

55Fb   V. V. Fomin, Zh. Fiz. Khim., 1955, 29, 1728

55GE   L. F. Grantham, T. S. Elleman, and D. S. Martin, Jr., J. Amer. Chem. Soc., 1955,
        77, 2965

55GG   R. J. Gross and J. W. Gryder, J. Amer. Chem. Soc., 1955, 77, 3695

55GL   K. H. Gayer and H. Leider, J. Amer. Chem. Soc., 1955, 77, 1448

55GM   F. G. R. Gimblett and C. B. Monk, Trans. Faraday Soc., 1955, 51, 793

55HB   L. J. Heidt and J. Berestecki, J. Amer. Chem. Soc., 1955, 77, 2049

55IR   H. Irving and F. J. C. Rossotti, J. Chem. Soc., 1955, 1946

55K   P. Kivalo, Suomen Kem., 1955, B28, 155

55Ka   J. Kenttamaa, Suomen Kem., 1955, B28, 172; Ann. Acad. Sci. Fenn., 1955, A2, No. 67

55Kb   A. Krawetz, Thesis, Univ. Chicago, 1955 (see 60HR)

55Kc   I. M. Korenman, J. Gen. Chem. USSR, 1955, 25,   (1859)

55Kd   D. M. H. Kern, J. Amer. Chem. Soc., 1955, 77, 5458

55KE   Y. Kanko and S. Eyubi, Tekn. Foren. Finland Forhandl., 1955, 263; Chem.-Ztg.,
        1956, 80, 130; Oster. Chem. Z., 1957, 58, 259; Finska Kem. Medd., 1960, 69,
        104.

55KJ   J. A. Kiltrick and M. L. Jackson, Proc. Soil Sci. Soc. Amer., 1955, 19, 455

55KN   K. A. Kraus and F. Nelson, J. Amer. Chem. Soc., 1955, 77, 3721

55L   M. Lourijsen-Teyssedre, Bull. Soc. Chim. France, 1955, 1111

55La   M. Lourijsen-Teyssedre, Bull. Soc. Chim. France, 1955, 1118

55Lb   E. D. Loughran, Thesis, Ohio State Univ., 1955

55LC   M. J. LaSalle and J. W. Cobble, J. Phys. Chem., 1955, 59, 519

55LN   I. Leden and R. Nilsson, Z. Naturforsch, 1955, 10a, 67

55LR   M. W. Lister and D. E. Rivington, Canad. J. Chem., 1955, 33, 1603

55M   J. C. McCoubrey, Trans. Faraday Soc., 1955, 51, 743

55MB   L. A. McClaine, E. P. Bullwinkel, and J. C. Huggins, Proc. Int. Conf. Geneva, 1955,
        8, 26

55MS   L. N. Mulay and I. W. Selwood, J. Amer. Chem. Soc., 1955, 77, 2693

55MV   R. M. Milburn and W. C. Vosburgh, J. Amer. Chem. Soc., 1955, 77, 1352

55N   C. J. Nyman, J. Amer. Chem. Soc., 1955, 77, 1371

55Na   G. H. Nancollas, J. Chem. Soc., 1955, 1458

55P   A. D. Paul, UCRL-2926, 1955

55PH   K. Pan and T. M. Hseu, Bull. Chem. Soc. Japan, 1955, 28, 162

55PK   C. Postmus and E. L. King, J. Phys. Chem., 1955, 59, 1208, 1216

55PP   M. H. Panckhurst and H. N. Parton, Trans. Faraday Soc., 1955, 51, 806

55R   J. Rydberg, Arkiv Kemi, 1955, 8, 113

55Ra   O. Redlich, Monat. Chem., 1955, 86, 329

55Rb   S. Rydholm, Svensk Papperstidn, 1955, 58, 273

55Rc   R. Nasanen, Suomen Kem., 1955, B28, 111

55RC   S. W. Rabideau and H. D. Cowan, J. Amer. Chem. Soc., 1955, 77, 6145

55RR   F. J. C. Rossotti and H. S. Rossotti, Acta Chem. Scand., 1955, 9, 1177

55RU   I. G. Ryss and P. V. Ustyanova, Ukr. Khim. Zh., 1955, 21, 6

55S   D. Singh, J. Sci. Res. Banaras Hindu Univ., 1955, 6, 131

55SB   N. M. Selivanova and R. Ya. Boguslavskii, Zh. Fiz. Khim., 1955, 29, 128

55SC   Z. G. Szabo, L. J. Csanyi, and M. Kavai, Z. Anal. Chem., 1955, 146, 401

55SG   S. K. Siddhanta and M. P. Guha, J. Indian Chem. Soc., 1955, 32, 355

55SL   S. A. Shchukarev, L. S. Lilich, and V. A. Latysheva, J. Gen. Chem. USSR,   1955, 25,
        1389 (1444)

55T   L. C. Thompson, Thesis, Wayne Univ., Detroit, Mich., 1955

55TB   V. F. Toropova and E. A. Belaya, Uchenye Zapiski Kazanskogo Univ., 1955, 115,
        No. 3, 61

55TSL   V. F. Toropova, I. A. Sirotina, and T. I. Lisova; Uchenye Zapiski Kazanskogo Univ.,
        1955, 115, No. 3, 43

55TSR   V. F. Toropova, I. A. Sirotina, and V. B. Rotanova, Uchenye Zapiski Kazanskogo
        Univ., 1955, 115, No. 3, 53

55V   A. A. Vlcek, Coll. Czech. Chem. Comm., 1955, 20, 400

55WW   T. D. Waugh, H. F. Walton, and J. A. Laswick, J. Phys. Chem., 1955, 59, 396

55Y   L. M. Yates, Thesis, Washington State Coll., 1955

55ZD   N. de Zoubov, E. Deltombe, and M. Pourbaix, Cebelcor Rapp. Tech., No. 27, 1955

56A   J. Angeli, Comp. Rend. Acad. Sci. Paris, 1956, 242, 1021

56Aa   A. R. Amell, J. Amer. Chem. Soc., 1956, 78, 6234

56AL   S. Ahrland, R. Larsson, and K. Rosengren, Acta Chem. Scand., 1956, 10, 705

56AR   S. Ahrland and K. Rosengren, Acta Chem. Scand., 1956, 10, 727

56B   C. Berecki-Biedermann, Arkiv Kemi, 1956, 9, 175

56Ba   G. Biedermann, _Arkiv Kemi_, 1956, _9_, 277

56Bb   C. F. Baes, Jr., _J. Phys. Chem._, 1956, _60_, 878 (see 64SM)

56BC   C. A. Blake, C. F. Coleman, K. B. Brown, D. G. Hill, R. S. Lowrie, and J. M.
        Schmitt, _J. Amer. Chem. Soc._, 1956, _78_, 5978

56BD   W. D. Bale, E. W. Davies, and C. B. Monk, _Trans. Faraday Soc._, 1956, _52_, 816

56BK   G. Biedermann, M. Kilpatrick, L. Pokras, and L. G. Sillen, _Acta Chem. Scand._,
        1956, _10_, 1327

56BL   A. K. Babko and G. S. Lisetskaya, _J. Inorg. Chem. USSR_, 1956, _1_, No. 5, 95 (969)

56BP   R. P. Bell and M. H. Panckhurst, _Rec. Trav. Chim._, 1956, _75_, 725

56BPa   R. P. Bell and M. H. Panckhurst, _J. Chem. Soc._, 1956, 2836

56C   B. Charreton, _Bull. Soc. Chim. France_, 1956, 323, 337, 347

56Ca   L. G. Carpenter, Thesis, Columbia Univ., 1956

56CD   H. Chateau, M. Durante, and B. Hervier, _Sci. Ind. Photo._, 1956, _27_, 81, 257

56CH   R. E. Connick, L. G. Hepler, Z. Z. Hugus, Jr., J. W. Kury, W. M. Latimer, and M. S.
        Tsao, _J. Amer. Chem. Soc._, 1956, _78_, 1827

56CK   F. Cuta, Z. Ksandr, and M. Hejtmanek, _Chem. Listy_, 1956, _50_, 1064; _Coll. Czech._
        Chem. Comm., 1956, _21_, 1388

56CS   V. G. Chukhlantsev and S. I. Stepanov, _J. Inorg. Chem. USSR_, 1956, _1_, No. 3, 135
        (478)

56DP   E. Deltombe and M. Pourbaix, _Cebelcor Rapp. Tech._, No. 42, 1956

56DZ   E. Deltombe, N. de Zoubov, and M. Pourbaix, _Cebelcor Rapp. Tech._, No. 31, 1956

56DZa   E. Deltombe, N. de Zoubov, and M. Pourbaix, _Cebelcor Rapp. Tech._, No. 32, 1956

56DZb   E. Deltombe, N. de Zoubov, and M. Pourbaix, _Cebelcor Rapp. Tech._, No. 33, 1956

56DZc   E. Deltombe, N. de Zoubov, and M. Pourbaix, _Cebelcor Rapp. Tech._, No. 35, 1956

56DZd   E. Deltombe, N. de Zoubov, and M. Pourbaix, _Cebelcor Rapp. Tech._, No. 41, 1956

56FM   V. V. Fomin and E. P. Maiorova, _J. Inorg. Chem. USSR_, 1956, _1_, No. 8, 7 (1703);
        No. 12, 109 (2749)

56FP   A. Fava and G. Pajaro, _J. Amer. Chem. Soc._, 1956, _78_, 5203

56G   H. S. Gates, Thesis, Univ. Wisconsin, 1956

56GG   R. A. Gilbert and A. B. Garrett, _J. Amer. Chem. Soc._, 1956, _78_, 5501

56GW   P. Gray and T. C. Waddington, _Proc. Roy. Soc., London_, 1956, _A235_, 106

56GWa   K. H. Gayer and L. Woontner, _J. Phys. Chem._, 1956, _60_, 1569

56GWb   K. H. Gayer and L. Woontner, _J. Amer. Chem. Soc._, 1956, _78_, 3944

56H   S. Hietanen, Acta Chem. Scand., 1956, 10, 1531; Rec. Trav. Chim., 1956, 75, 711

56HD   R. E. Huffman and N. Davidson, J. Amer. Chem. Soc., 1956, 78, 4836

56HS   S. Hietanen and L. G. Sillen, Suomen Kem., 1956, B29, 31; Arkiv Kemi., 1956, 10,
       103

56IA   A. N. Ivanov and S. N. Aleshin, Dokl. Moskov. Selskokhoz Akad. Nauch. Konf., 1956,
       22, 386

56J   C. K. Jorgensen, Acta Chem. Scand., 1956, 10, 500, 518

56JP   P. K. Jena and B. Prasad, J. Indian Chem. Soc., 1956, 33, 122

56K   P. Kivalo, Suomen Kem., 1956, B29, 8

56Ka   J. Kenttamaa, Suomen Kem., 1956, B29, 59

56Kb   P. Kivalo, Suomen Kem., 1956, B29, 189

56Kc   M. Kobayashi, Nippon Kagaku Zasshi, 1956, 77, 279

56Kd   P. N. Kovalenko, J. Inorg. Chem. USSR, 1956, 1, No. 8, 22 (1717)

56Ke   P. N. Kovalenko, Ukr. Khim. Zh., 1956, 22, 801

56Kf   H. Kubota, Diss., Univ. Wisc. 1956, Diss. Abs., 1956, 16, 864

56Kg   G. B. Kauffman, Thesis, Univ. Florida, 1956

56KE   P. Kivalo and A. Ekman, Suomen Kem., 1956, B29, 139

56KF   I. M. Korenman, F. S. Frum, and G. A. Tsyganova, J. Gen. Chem. USSR, 1956, 26,
       1945.   (1558)

56KR   P. Kivalo and A. Ringbom, Suomen Kem., 1956, B29 109

56KS   H. Kakihana and L. G. Sillen, Acta Chem. Scand., 1956, 10, 985

56KSa I. M. Kolthoff and J. T. Stock, J. Amer. Chem. Soc., 1956, 78, 2081

56KT   C. M. Kelley and H. V. Tartar, J. Amer. Chem. Soc., 1956, 78, 5752

56L   G. S. Lawrence, Trans. Faraday Soc., 1956, 52, 236

56La   I. Leden, Acta Chem. Scand., 1956, 10, 540, 812

56LP   I. Leden and C. Parck, Acta Chem. Scand., 1956, 10, 535

56LPa A. V. Lapitskii and V. A. Pchelkin, Vestnik Moskov. Univ., 1956, 11, No. 5; Ser.
       Fiz. Mat., No. 3, 69

56LS   G. W. Leonard, M. E. Smith, and D. N. Hume, J. Phys. Chem., 1956, 60, 1493

56M   H. A. C. McKay, Trans. Faraday Soc., 1956, 52, 1568

56MH   C. N. Muldrow, Jr., and L. G. Hepler, J. Amer. Chem. Soc., 1956, 78, 5989

56MS   A. E. Martell and G. Schwarzenbach, Helv. Chim. Acta, 1956, 39, 653; J. Phys.
       Chem., 1958, 62, 886

56N   L. Newman, Diss., Mass. Inst. Tech., 1956

56NM   M. S. Novakovskii and M. G. Mushkina, Ukr. Khim. Zh., 1956, 22, 313

56OB   E. Orban, M. K. Barnett, J. S. Boyle, J. R. Heiks, and L. V. Jones, J. Phys. Chem.,
       1956, 60, 413

56P   P. Papoff, Suomen Kem., 1956, B29, 97

56PC   H. M. Papee, W. J. Canady, and K. J. Laidler, Canad. J. Chem., 1956, 34, 1677

56PJ   R. A. Penneman and L. H. Jones, J. Chem. Phys., 1956, 24, 293

56PV   D. Peschanski and S. Valladas-Dubois, Bull. Soc. Chim. France, 1956, 1170

56GP   M. Quintin and S. Pelletier, J. Chim. Phys., 1956, 53, 226

56R   R. W. Ramette, J. Chem. Educ., 1956, 33, 610

56RM   J. J. Renier and D. S. Martin, J. Amer. Chem. Soc., 1956, 78, 1833

56RR   F. J. C. Rossotti and H. Rossotti, Acta Chem. Scand., 1956, 10, 779

56RRa F. J. Rossotti and H. S. Rossotti, J. Inorg. Nucl. Chem., 1956, 2, 201; Acta Chem.
       Scand., 1956, 10, 957

56SA   R. M. Smith and R. A. Alberty, J. Phys. Chem., 1956, 60, 180

56SAa R. M. Smith and R. A. Alberty, J. Amer. Chem. Soc., 1956, 78, 2376

56SL   S. A. Shchukarev, L. S. Lilich, and V. A. Latysheva, J. Inorg. Chem. USSR, 1956, 1,
       No. 2, 36 (225)

56SM   C. C. Stephenson and J. C. Morrow, J. Amer. Chem. Soc., 1956, 78, 275

56SP   S. C. Sircar and B. Prasad, J. Indian Chem. Soc., 1956, 33, 361

56SW   L. H. Sutcliffe and J. R. Weber, Trans. Faraday Soc., 1956, 52, 1225

56SZ   N. M. Selivanova and G. A. Zubova, Trudy Moskov. Khim. Tek. Inst., 1956, 22, 38

56T   V. F. Toropova, J. Inorg. Chem. USSR, 1956, 1, No. 2, 56 (243)

56Ta   Ya. I. Turyan, J. Anal. Chem. USSR, 1956, 11,   (71)

56Tb   Ya. I. Turyan, J. Inorg. Chem. USSR, 1956, 1, No. 10, 171 (2337)

56TG   I. V. Tananaev, M. A. Glushkova, and G. B. Seifer, J. Inorg. Chem. USSR, 1956, 1,
       No. 1, 72 (66)

56TK   Sh. T. Talipov and O. F. Kutumova, Dokl. Akad. Nauk Uzbek. SSR, 1956, No. 8, 23

56TS   V. N. Tolmachev and L. N. Serpukhova, Zh. Fiz. Khim., 1956, 30, 134

56UL   E. A. Ukshe and A. I. Levin, J. Gen. Chem. USSR, 1956, 26, 2963 (2657)

56VS   C. E. Vanderzee and W. E. Smith, J. Amer. Chem. Soc., 1956, 78, 721

56WL   J. I. Watters, E. D. Longhran, and S. M. Lambert, J. Amer. Chem. Soc., 1956, 78,
       4855

56WW   M. Ward and G. A. Welch, J. Inorg. Nucl. Chem., 1956, 2, 395

56Y   K. B. Yatsimirskii, J. Anal. Chem. USSR, 1956, 10, (344)

56YV   K. B. Yatsimirskii and V. P. Vasilev, Zh. Fiz. Khim., 1956, 30, 28

56YVa K. B. Yatsimirskii and V. P. Vasilev, Zh. Fiz. Khim., 1956, 30, 901

56YVb K. B. Yatsimirskii and V. P. Vasilev, J. Anal. Chem. USSR, 1956, 11,   (536)

56ZC   A. J. Zielen and R. E. Connick, J. Amer. Chem. Soc., 1956, 78, 5785

56ZK   V. L. Zolotavin and V. K. Kuznetsova, J. Inorg. Chem. USSR, 1956, 1, No. 10, 87
          (2257)

57A   G. Anderegg, Helv. Chim. Acta, 1957, 40, 1022

57Aa   P. J. Antikainen, Suomen Kem., 1957, B30, 22, 201

57Ab   P. J. Antikainen, Suomen Kem., 1957, B30, 74

57Ac   P. J. Antikainen, Ann. Acad. Sci. Fenn., 1957, AII, No. 56

57Ad   G. M. Arcand, J. Amer. Chem. Soc., 1957, 79, 1865

57AG   S. Ahrland and I. Grenthe, Acta Chem. Scand., 1957, 11, 1111; 1961, 15, 932

57B   S. Broersma, J. Chem. Phys., 1957, 26, 1405

57BG   B. D. Blaustein and J. W. Gryder, J. Amer. Chem. Soc., 1957, 79, 540

57BH   E. A. Burns and D. N. Hume, J. Amer. Chem. Soc., 1957, 79, 2704

57BL   S. A. Brown and J. E. Land, J. Amer. Chem. Soc., 1957, 79, 3015

57BLa A. Basinski and U. Ledzinska, Rocz. Chem., 1957, 31, 457

57BM   C. H. Brubaker, Jr., and J. P. Mickel, J. Inorg. Nucl. Chem., 1957, 4, 55

57BP   A. Basinski and S. Poczopko, Rocz. Chem., 1957, 31, 449

57BPa T. A. Bak and E. L. Praestgaard, Acta Chem. Scand., 1957, 11, 901

57BS   G. Biedermann and P. Schindler, Acta Chem. Scand., 1957, 11, 731

57BW   E. A. Burns and R. A. Whiteker, J. Amer. Chem. Soc., 1957, 79, 866

57BWa J. H. Baxendale and C. F. Wells, Trans. Faraday Soc., 1957, 53, 800

57C   R. Caramazza, Gazz. Chim. Ital., 1957, 87, 1507

57Ca   B. Charreton, Comp. Rend. Acad. Sci. Paris, 1957, 244, 1208

57CB   F. Cuta, E. Beranek, and J. Pisecky, Chem. Listy, 1957, 51, 1614; Coll. Czech.
          Chem. Comm., 1958, 23, 1496

57CH   H. Chateau and B. Hervier, J. Chim. Phys., 1957, 54, 637

57CHP H. Chateau, B. Hervier, and J. Pouradier, J. Phys. Chem., 1957, 61, 250; J. Chim.
          Phys., 1957, 54, 246

57CJ   S. C. Chang and M. L. Jackson, Proc. Soil Sci. Soc. Amer., 1957, 21, 265

57CP   S. R. Cohen and R. A. Plane, J. Phys. Chem., 1957, 61, 1096

57D   G. D'Amore, Atti. Soc. Peloritana Sci. Fis. Mat. Nat., 1956-7, 3, 95

57DB   Y. Doucet and S. Bugnon, J. Chem. Phys., 1957, 54, 155

57DBF   H. A. Droll, B. P. Block, and W. C. Fernelius, J. Phys. Chem., 1957, 61, 1000

57DM   E. W. Davies and C. B. Monk, Trans. Faraday Soc., 1957, 53, 442

57DS   J. Durell and J. M. Sturtevant, Biochim. Biophys. Acta, 1957, 26, 282

57E   D. F. Evans, J. Chem. Soc., 1957, 4013

57ET   F. Ender, W. Teltschik, and K. Schafer, Z. Elektrochem., 1957, 61, 775

57F   N. R. Fetter, Thesis, Univ. Oregon, 1957

57FH   L. J. Frolen, W. S. Harris, and D. F. Swinehart, J. Phys. Chem., 1957, 61, 1672

57FK   R. M. Fuoss and C. A. Kraus, J. Amer. Chem. Soc., 1957, 79, 3304

57GL   K. H. Gayer and H. Leider, Canad. J. Chem., 1957, 35, 5

57GP   S. A. Greenberg and E. W. Price, J. Phys. Chem., 1957, 61, 1539

57GS   A. M. Golub and V. M. Samoilenko, Ukr. Khim. Zh., 1957, 23, 17

57GW   K. H. Gayer and L. Woontner, J. Phys. Chem., 1957, 61, 364

57H   P. Hagenmuller, Compt. Rend. Acad. Sci. Paris, 1957, 244, 2061

57HH   K. H. Hsu and S. H. Ho, Kexue Tongbao, 1957, 433; Chem. Abs., 1961, 21953b

57HN   J. R. Howard and G. H. Nancollas, Trans. Faraday Soc., 1957, 53, 1449

57HW   J. A. Hearne and A. G. White, J. Chem. Soc., 1957, 2168

57I   A. Iwase, Nippon Kagaku Zasshi, 1957, 78, 1659

57IL   N. Ingri, G. Lagerstrom, M. Frydman, and L. G. Sillen, Acta Chem. Scand., 1957, 11,
        1034

57JS   J. Jortner and G. Stein, Bull. Res. Council Israel, 1957, 6A, 239

57K   J. Kenttamaa, Suomen Kem., 1957, B30, 9

57Ka   M. Kobayashi, Nippon Kagaku Zasshi, 1957, 78, 611

57Kb   A. Kleiber, Thesis, Univ. Strasbourg, 1957 (see also 64SM)

57Kc   P. N. Kovalenko, J. Appl. Chem. USSR, 1957, 30,   (52)

57Kd   P. N. Kovalenko, Khim. Nauka Prom., 1957, 2, 533

57KC   E. I. Krylov and V. G. Chukhlantsev, J. Anal. Chem. USSR, 1957, 12,   (451)

57KD   Y. Kauko and S. Doger, Acta Chem. Scand., 1957, 11, 804

V. BIBLIOGRAPHY

57KE   P. Kivalo and P. Ekari, Suomen Kem., 1957, B30, 116

57KH   J. W. Kury, Z. Z. Hugus, Jr., and W. M. Latimer, J. Phys. Chem., 1957, 61, 1021

57KL   P. Kivalo and R. Luoto, Suomen Kem., 1957, B30, 163

57KM   F. Ya. Kulba and V. E. Mironov, J. Inorg. Chem. USSR, 1957, 2, No. 8, 46 (1741)

57KP   I. A. Korshunov, A. P. Pochinailo, and V. M. Tikhomirova, J. Inorg. Chem. USSR, 1957, 2, No. 1, 101 (68).

57KS   J. J. Katz and G. T. Seaborg, The Chemistry of the Actinide Elements, Methuen & Co. Ltd., London, 1957, 264, 305, 313, 363

57L   K. H. Lieser, Z. Anorg. Allg. Chem., 1957, 292, 97

57La   J. Lefebvre, J. Chim. Phys., 1957, 74, 567

57Lb   S. M. Lambert, Thesis, Ohio State Univ., 1957

57LW   S. M. Lambert and J. I. Watters, J. Amer. Chem. Soc., 1957, 79, 4262

57LWa   S. M. Lambert and J. I. Watters, J. Amer. Chem. Soc., 1957, 79, 5606

57M   Y. Marcus, Acta Chem. Scand., 1957, 11, 599

57Ma   R. M. Milburn, J. Amer. Chem. Soc., 1957, 79, 537

57Mb   T. C. MacAvoy, Thesis, Univ. Cincinnati, 1957

57MF   K. P. Mishchenko and I. E. Flis, J. Appl. Chem. USSR, 1957, 30,   (665)

57MN   G. Mahapatra, C. Nanda, and D. Patnaik, J. Indian Chem. Soc., 1957, 34, 457

57MO   N. N. Mironov and A. I. Odnosevtsev, J. Inorg. Chem. USSR, 1957, 2, No. 9, 342 (2202)

57MP   R. N. Misra and S. Pani, J. Indian Chem. Soc., 1957, 34, 393

57MR   S. S. Muhammad and T. N. Rao, J. Indian Chem. Soc., 1957, 34, 250

57MRa   S. S. Muhammad and T. N. Rao, J. Chem. Soc., 1957, 1077

57MRH   S. S. Muhammad, D. H. Rao, and M. A. Haleem, J. Indian Chem. Soc., 1957, 34, 101

57MS   J. Meier and G. Schwarzenbach, Helv. Chim. Acta, 1957, 40, 907

57MT   I. S. Morozov and G. M. Toptygina, J. Inorg. Chem. USSR, 1957, 2, No. 7 , 286 (1629)

57MV   H. C. Moser and A. F. Voigh, J. Inorg. Nucl. Chem., 1957, 4, 354

57N   R. O. Nilsson, Arkiv Kemi, 1957, 10, 363

57NB   T . W. Newton and F. B. Baker, J. Phys. Chem., 1957, 61, 934

57NBC   M. Nardelli, A. Braibanti, and I. Chierici, Gazz. Chim. Ital., 1957, 87, 510

57NM   M. S. Novakovskii, M. G. Mushkina, and E. G. Vorobeva, Trudy Khim. Fak. Kharkov Univ., 1957, 16, 107 (see 64SM)

57NN   V. S. K. Nair and G. N. Nancollas, J. Chem. Soc., 1957, 318

57O   A. Olin, Acta Chem. Scand., 1957, 11, 1445

57P   J. J. Podesta, Rev. Fac. Cienc. Quim. LaPlata, 1957, 30, 61

57PC   P. Papoff, M. A. Caliumi, and G. Ferrari, Ric. Sci., 1957, 27, suppl. A;
           Polarografia, 3, 131

57PL   A. V. Pamfilov, A. I. Lopushanskaya, and E. B. Gusel, Ukr. Khim. Zh., 1957, 23, 297

57PP   A. L. Pitman, M. Pourbaix, and N. de Zoubov, Cebelcor Rapp. Tech., No. 55, 1957

57PT   F. A. Posey and H. Taube, J. Amer. Chem. Soc., 1957, 79, 255

57R   S. W. Rabideau, J. Amer. Chem. Soc., 1957, 79, 3765

57S   A. S. Solovkin, J. Inorg. Chem. USSR, 1957, 2, No. 3, 216 (611)

57Sa   E. B. Shtemina, J. Inorg. Chem. USSR, 1957, 2, No. 4, 344 (933)

57Sb   J. F. Scaife, Canad. J. Chem., 1957, 35, 1332

57Sc   O. V. Serebrennikova, Trudy Ural. Politekh. Inst., 1957, 58, 57

57Sd   W. M. Smith, Proc. Chem. Soc., 1957, 207

57SL   S. A. Shchukarev, L. S. Lilich, and V. A. Latysheva, Uch. Zap. Leningrad Univ.,
           Ser. Khim. Nauk, 1957, No. 15 (211), 17

57SM   K. Yu, Salnis, K. P. Mishchenko, and I. E. Flis, J. Inorg. Chem. USSR, 1957, 2,
           No. 9, 1 (1985)

57SN   B. S. Strates, W. F. Neuman, and G. J. Levinskas, J. Phys. Chem., 1957, 61, 279

57T   V. F. Toropova, J. Inorg. Chem. USSR, 1957, 2, No. 3, 63 (515)

57Ta   J. Thamer, J. Amer. Chem. Soc., 1957, 79, 4298

57TE   V. M. Tarayan and L. A. Eliazyan, Izvest. Akad. Nauk. Armyan SSR, Khim., 1957, 10,
           189

57TH   H. G. Tsiang and   K. H. Hsu, Acta Chim. Sinica, 1957, 23, 196

57TK   E. Thilo and G. Kruger, Z. Elektrochem., 1957, 61, 24

57TM   N. Tanaka and T. Murayama, Z. Phys. Chem. (Frankfurt), 1957, 11, 366

57TMa R. C. Turner and K. E. Miles, Canad. J. Chem., 1957, 35, 1002

57TMb T. A. Tumanova, K. P. Mishchenko, and I. E. Flis, J. Inorg. Chem. USSR, 1957, 2,
           No. 9, 9 (1990)

57TS   Ya. I. Turyan and G. F. Serova, J. Inorg. Chem. USSR, 1957, 2, No. 2, 165 (336)

57TSa T. Takahashi and K. Sasaki, J. Electrochem. Soc. Japan, 1957, 25, 58, 118

57TV   I. V. Tananaev and A. D. Vinogradova, J. Inorg. Chem. USSR, 1957, 2, No. 10, 276
           (2455)

57V   V. P. Vasilev, Zh. Fiz. Khim., 1957, 31, 692

57W   R. H. Wood, Thesis, 1957, Univ. California Berkeley; UCRL-3751

57WL   J. I. Watters, S. M. Lambert, and E. D. Loughran, J. Amer. Chem. Soc., 1957, 79,
        3651

57YG   K. B. Yatsimirskii and L. V. Guskova, J. Inorg. Chem. USSR, 1957, 2, No. 9, 91
        (2039)

57YM   K. B. Yatsimirskii and P. M. Milyukov, J. Inorg. Chem. USSR, 1957, 2, No. 5, 98
        (1046)

57YMa   K. B. Yatsimirskii and P. M. Milyukov, Zh. Fiz. Khim., 1957, 31, 842

57YT   K. B. Yatsimirskii and V. D. Tetyushkina, J. Inorg. Chem. USSR, 1957, 2, No. 2,
        141 (320)

57ZM   N. de Zoubov, J. van Muylder, and M. Pourbaix, Cebelcor Rapp. Tech., No. 60, 1957

58A   T. Ackerman, Z. Elektrochem., 1958, 62, 411

58Aa   K. A. Allen, J. Amer. Chem. Soc., 1958, 80, 4133

58Ab   F. Achenza, Ann. Chim. (Italy), 1958, 48, 565

58AK   M. Akabane and A. Kurosawa, Kogyo Kagaku Zasshi, 1958, 61, 303

58AN   S. Ahrland and B. Noren, Acta Chem. Scand., 1958, 12, 1595

58AS   N. V. Akselrud and V. B. Spivakovskii, J. Inorg. Chem. USSR, 1958, 3, No. 2, 38
        (269)

58ASP   G. Anderegg, G. Schwarzenbach, M. Padmoyo, and O. F. Borg, Helv. Chim. Acta,
        1958, 41, 988

58AT   M. Anbar and H. Taube, J. Amer. Chem. Soc., 1958, 80, 1073

58B   S. A. Brusilovskii, Dokl. Chem., 1958, 120, 343 (305)

58BB   P. B. Barton, Jr., and P. M. Bethke, Econ. Geol., 1958, 53, 914

58BC   A. Braibanti and I. Chierici, Gazz. Chim. Ital., 1958, 88, 793

58BCC   J. F. Below, Jr., R. E. Connick, and C. P. Coppel, J. Amer. Chem. Soc., 1958, 80,
        2961

58BS   A. Basinski and W. Szymanski, Rocz. Chem., 1958, 32, 23

58C   R. Caramazza, Gazz. Chim. Ital., 1958, 88, 308

58Ca   Y. TiChia, UCRL-8311, 1958

58Cb   J. W. Collat, Anal. Chem., 1958, 30, 1726

58CP   R. E. Connick and A. D. Paul, J. Amer. Chem. Soc., 1958, 80, 2069

58CPC   R. Cernatescu, I. Popescu, A. Craciun, M. Bostan, and N. Iorga, Stud. Cercetari,
        Chim., Filiala Iasi, 1958, 9, 1

58D   D. G. Davis, <u>Anal. Chem.</u>, 1958, <u>30</u>, 1729

58Da   R. J. Dietz, Jr., Thesis, <u>Mass. Inst. Tech.</u>, 1958 (see 64SM)

58DT   C. Dragulescu and P. Tribunescu, <u>Studii Cercetari Sti. Chim.</u>, <u>Timisoara</u>, 1958, <u>5</u>,
         No. 3-4, 19

58E   C. L. P. van Eck, Thesis, Leiden, 1958

58ES   J. S. Elliot, R. F. Sharp, and L. Lewis, <u>J. Phys. Chem.</u>, 1958, <u>62</u>, 686

58F   M. L. Freedman, <u>J. Amer. Chem. Soc.</u>, 1958, <u>80</u>, 2072

58Fa   I. E. Flis, <u>J. Appl. Chem. USSR</u>, 1958, <u>31</u>,   (1194)

58FK   V. V. Fomin, R. E. Kartushova, and T. I. Rudenko, <u>J. Inorg. Chem. USSR</u>, 1958, <u>3</u>,
         No. 9, 167 (2117)

58FM   I. E. Flis, K. P. Michchenko, and N. V. Pakhomova, <u>J. Inorg. Chem. USSR</u>, 1958, <u>3</u>,
         No. 8, 72 (1772)

58FN   M. Frydman, G. Nilsson, T. Rengemo, and L. G. Sillen, <u>Acta Chem. Scand.</u>, 1958, <u>12</u>,
         878

58G   S. A. Greenberg, <u>J. Amer. Chem. Soc.</u>, 1958, <u>80</u>, 6508

58Ga   A. K. Grzybowski, <u>J. Phys. Chem.</u>, 1958, <u>62</u>, 555

58Gb   F. I. Golovin, <u>Trudy Molod. Nauch Rabot. Akad. Nauk. Kirgiz SSR</u>, 1958, 119

58GH   R. L. Graham and L. G. Hepler, <u>J. Amer. Chem. Soc.</u>, 1958, <u>80</u>, 3538

58GK   H. S. Gates and E. L. King, <u>J. Amer. Chem. Soc.</u>, 1958, <u>80</u>, 5011

58GT   K. H. Gayer and L. C. Thompson, <u>Canad. J. Chem.</u>, 1958, <u>36</u>, 1649

58GTZ   K. H. Gayer, L. C. Thompson, and O. T. Zajicek, <u>Canad. J. Chem.</u>, 1958, <u>36</u>, 1268

58H   R. A. Horne, <u>J. Inorg. Nucl. Chem.</u>, 1958, <u>6</u>, 338

58Ha   R. A. Horne, <u>Nature</u>, 1958, <u>181</u>, 410

58Hb   L. G. Hepler, <u>J. Amer. Chem. Soc.</u>, 1958, <u>80</u>, 6181

58HN   J. R. Howard, V. S. K. Nair, and G. H. Nancollas, <u>Trans. Faraday Soc.</u>, 1958, <u>54</u>,
         1034

58HT   K. H. Hsu and H. G. Tsiang, <u>Acta Chim. Sinica</u>, 1958, <u>24</u>, 277

58HW   W. Hieber and G. Wagner, <u>Z. Naturforsch</u>, 1958, <u>13b</u>, 339

58I   A. Indelli, <u>Ann. Chim.</u> (Italy), 1958, <u>48</u>, 345

58J   M. B. Jensen, <u>Acta Chem. Scand.</u>, 1958, <u>12</u>, 1657

58Ja   L. Jager, <u>Chem. Listy</u>, 1958, <u>52</u>, 734

58Jb   C. K. Jorgensen, DA-91, 508-EUC-247, 1958

58JB   E. Jorgensen and J. Bjerrum, <u>Acta Chem. Scand.</u>, 1958, <u>12</u>, 1047; <u>J. Inorg. Nucl.
         Chem.</u>, 1958, <u>8</u>, 313

58K   P. N. Kovalenko, J. Inorg. Chem. USSR, 1958, 3, No. 5, 1 (1065)

58KB   P. N. Kovalenko and K. N. Bagdasarov, Uchenye Zapiski Univ. Rostov-Na Donu, 1958,
         No. 41 (=Trudy Khim. Fak., 9,), 107

58KC   I. K. Krotova and M. L. Chepelevetskii, Soobschch Nauch. Inst. Udobren
         Insektofungisid, 1958, No. 10, 58

58KG   P. N. Kovalenko and O. I. Geiderovich, Nauch. Dokl. Vys. Shk., Khim., 1958, 294

58KGL   I. M. Korenman, V. G. Ganina, and N. P. Lebedeva, J. Inorg. Chem. USSR, 1958, 3,
         No. 5, 281 (1265)

58KK   O. I. Khotsyanovskii and O. K. Kudra, Izv. Vyssh. Ucheb. Zaved., Khim, 1958, No. 1,
         43

58KT   G. Kruger and E. Thilo, Z. Phys. Chem. (Leipzig), 1958, 209, 190

58KV   S. K. Kor and G. S. Verma, J. Chem. Phys., 1958, 29, 9

58L   R. Larsson, Acta Chem. Scand., 1958, 12, 708

58La   O. N. Lapteva, J. Appl. Chem. USSR, 1958, 31,   (1210)

58LG   P. E. Lake and J. M. Goodings, Canad. J. Chem., 1958, 36, 1089

58M   R. A. Myers, Thesis, Univ. Nebr., 1958 (see 64SM)

58Ma   V. E. Mironov, Diss. (Kand), Leningrad Tech. Inst., 1958 (see 64SM)

58Mb   Y. Marcus, Proc. 2nd. Inter. Conf. Geneva, 1958, 3, 465

58Mc   P. M. Mader, J. Amer. Chem. Soc., 1958, 80, 2634

58MF   E. P. Maiorova and V. V. Formin, J. Inorg. Chem. USSR, 1958, 3, No. 8, 295 (1937)

58MG   A. I. Moskvin and A. D. Gelman, J. Inorg. Chem. USSR, 1958, 3, No. 4, 198 (962)

58MH   C. N. Muldrow, Jr., and L. G. Hepler, J. Phys. Chem., 1958, 62, 982

58ML   W. L. Marshall, F. J. Loprest, and C. H. Secoy, J. Amer. Chem. Soc., 1958, 80,
         5646

58MP   J. van Muylder and M. Pourbaix, Cebelcor Rapp. Tech., No. 59, 1958

58MW   P. E. Martin and A. G. White, J. Chem. Soc., 1958, 2490

58N   R. O. Nilsson, Arkiv Kemi, 1958, 12, 219, 337

58Na   R. O. Nilsson, Arkiv Kemi, 1958, 12, 371

58NC   L. Newman, J. De O. Cabral, and D. N. Hume, J. Amer. Chem. Soc., 1958, 80, 1814

58NL   L. Newman, W. J. LaFleur, F. J. Brousaides, and A. M. Ross, J. Amer. Chem. Soc.,
         1958, 80, 4491

58NN   V. S. K. Nair and G. H. Nancollas, J. Chem. Soc., 1958, 3706

58NNa   V. S. K. Nair and G. H. Nancollas, J. Chem. Soc., 1958, 4144

58NR   G. Nilsson, T. Regnemo, and L. G. Sillen, Acta Chem. Scand., 1958, 12, 868

58O   R. K. Osterheld, J. Phys. Chem., 1958, 62, 1133

58P   D. D. Perrin, J. Amer. Chem. Soc., 1958, 80, 3852

58PD   F. Pantani and P. Desideri, Gazz. Chim. Ital., 1958, 88, 1183

58PL   D. Pavlov and D. Lazarov, J. Inorg. Chem. USSR, 1958, 3, No. 9, 142 (2099)

58PS   V. I. Paramonova and A. N. Sergeev, J. Inorg. Chem. USSR, 1958, 3, No. 1, 331 (215)

58PT   V. P. Persiantseva and P. S. Titov, Nauch. Dokl. Vysshei Shkoly, Khim., 1958, 584

58PW   M. H. Panckhurst and K. G. Woolmington, Proc. Roy, Soc. (London), 1958, A244, 124

58R   D. R. Rosseinsky, Trans. Faraday Soc., 1958, 54, 116

58RA   S. W. Rabideau, L. B. Asprey, T. K. Keenan, and T. W. Newton, Proc. Geneva Conf.
        Peaceful Uses of Atomic Energy, 1958, 28, p. 361

58RB   T. Rengemo, U. Brune, and L. G. Sillen, Acta Chem. Scand., 1958, 12, 873

58S   P. Schindler, Helv. Chim. Acta, 1958, 41, 527

58Sa   N. M. Selivanova, Zh. Fiz. Khim., 1958, 32, 1277

58Sb   J. F. B. Silman, Thesis, Harvard Univ., 1958

58Sc   P. Sakellaridis, Chim. Chronika, 1958, 23, 263

58Sd   P. Sakellaridis, Compt. Rend. Acad. Sci. Paris, 1958, 247, 2367

58SG   G. Schwarzenbach, O. Gubeli, and H. Zust, Chimia (Switz.), 1958, 12, 84

58SK   K. Schug and E. L. King, J. Amer. Chem. Soc., 1958, 80, 1089

58SM   G. Schwarzenbach and J. Meier, J. Inorg. Nucl. Chem., 1958, 8, 302

58SMa   R. Schwarz and W. D. Muller, Z. Anorg. Allg. Chem., 1958, 296, 273

58SO   G. Saini and G. Ostracoli, J. Inorg. Nucl. Chem., 1958, 8, 346

58SP   P. Senise and M. Perrier, J. Amer. Chem. Soc., 1958, 80, 4194

58SPa   G. Schwarzenbach and G. Parissakis, Helv. Chim. Acta, 1958, 41, 2425

58SPS   R. W. Stromatt, R. M. Peekema, and F. A. Scott, HW-58212, 1958

58SS   V. I. Spitsyn and I. A. Savich, J. Inorg. Chem. USSR, 1958, 3, No. 8, 351 (1979)

58SSa   N. M. Selivanova and V. A. Shneider, Nauch. Dokl. Vysshei Shkoly, Khim., 1958, 216

58ST   D. B. Scaife and H. J. V. Tyrrell, J. Chem. Soc., 1958, 392; J. Inorg. Nucl.
        Chem., 1958, 8, 353

58SW   J. M. Smithson and R. J. P. Williams, J. Chem. Soc., 1958, 457

58SWa   E. A. Simpson and G. M. Waind, J. Chem. Soc., 1958, 1746

58T   R. S. Tobias, Acta Chem. Scand., 1958, 12, 198

58TF   W. B. Treumann and L. M. Ferris, J. Amer. Chem. Soc., 1958, 80, 5048

58TG   R. S. Tobias and A. B. Garrett, J. Amer. Chem. Soc., 1958, 80, 3532

58TW   T. A. Turney and G. A. Wright, J. Chem. Soc., 1958, 2415

58VE   E. G. Vassian and W. H. Eberhardt, J. Phys. Chem., 1958, 62, 84

58VG   G. I. Volkov and V. I. Grinevich, J. Inorg. Chem. USSR, 1958, 3, No. 8, 337 (1968)

58VP   C. Vanleugenhaghe and M. Pourbaix, Cebelcor Rapp. Tech., No. 74, 1958

58VPa   C. Vanleugenhaghe and M. Pourbaix, Cebelcor Rapp. Tech., No. 75, 1958

58VR   J. Vaid and T. L. Ramachar, Bull. India Sect. Elecktrochem. Soc., 1958, 7, 5

58VRa   A. Valvassori and R. Riccardi, Boll. Sci. Fac. Chim. Ind. Bologna, 1958, 16, 80

58VS   C. Vanleugenhaghe, K. Schwabe, and M. Pourbaix, Cebelcor Rapp. Tech., No. 76, 1958

58W   R. H. Wood, J. Amer. Chem. Soc., 1958, 80, 1559

58YA   K. B. Yatsimirskii and I. I. Alekseeva, Izv. Vyssh. Ucheb. Zaved., Khim., 1958,
         No. 1, 53

58YF   K. B. Yatsimirskii and T. I. Fedorova, Izv. Vyssh. Ucheb. Zaved, Khim., 1958,
         No. 3, 40

58YK   K. B. Yatsimirskii and V. D. Korableva, J. Inorg. Chem. USSR, 1958, 3, No. 2, 139
         (339)

58YM   K. B. Yatsimirskii and P. M. Milyukov, Trudy Ivanovsk. Khim-Tekh. Inst., 1958,
         7, 16

58ZB   A. I. Zelyanskaya, N. V. Bausova, and L. Ya. Kukalo, Trudy Inst. Met. Akad. Nauk.
         SSSR, Ural. Filial., 1958, No. 2, 263

59A   F. Achenza, Ann. Chim. (Italy), 1959, 49, 624, 848

59Aa   F. Achenza, Rend. Seminar Fac. Sci. Univ. Cagliari, 1959, 29, 52

59Ab   K. P. Ang, J. Chem. Soc., 1959, 3822

59AS   N. V. Akselrud and V. B. Spivakovskii, Russ. J. Inorg. Chem., 1959, 4, 22 (56)

59ASa   N. V. Akselrud and V. B. Spivakovskii, Russ. J. Inorg. Chem., 1959, 4, 449 (989)

59ASb   N. V. Akselrud and V. B. Spivakovskii, Ukr. Khim. Zh., 1959, 25, 14

59ASc   L. P. Adamovich and G. S. Shupenko, Ukr. Khim. Zh., 1959, 25, 155

59B   E. Bock, Canad. J. Chem., 1959, 37, 1888

59Ba   M. Bronnimann, Diss., Bern, 1959 (see 63FS)

59BBC   R. G. Bates, V. E. Bower, R. G. Canham, and J. E. Prue, Trans. Faraday Soc., 1959,
         55, 2062

59BBD   A. Basinski, F. Burnicki, and W. Dzierza, Rocz. Chem., 1959, 33, 177

59BC   E. A. Burns and F. D. Chang, J. Phys. Chem., 1959, 63, 1314

59BE   J. Besson and W. Eckert, Bull. Soc. Chim. France, 1959, 1676

59BK   A. I. Busev and N. A. Kanaev, Vestnik. Moskov. Univ., Ser. Mat., 1959, No. 1, 135

59BS   A. Basinski and W. Szczerba, Rocz. Chem., 1959, 33, 283

59BSB   A. Basinski, W. Szymanski, and T. Betto, Rocz. Chem., 1959, 33, 289

59C   G. K. Czamanske, Econ. Geol., 1959, 54, 57

59CH   H. C. Chiang and K. H. Hsu, Kexue Tongbao, 1959, 397

59CN   H. Coll, R. V. Nauman, and P. W. West, J. Amer. Chem. Soc., 1959, 81, 1284

59D   W. H. Dumbaugh, Jr., Thesis, Penn. State Univ., 1959

59DP   P. G. Desideri and F. Pantani, Gazz. Chim. Ital., 1959, 89, 1349

59DT   C. Dragulescu and P. Tribunescu, Stud. Cercetari Sti. Chim., Timisoara, 1959, 6, 59

59E   A. J. Ellis, Amer. J. Sci., 1959, 257, 217, 354

59Ea   A. J. Ellis, Amer. J. Sci., 1959, 257, 287

59EF   J. E. Earley, D. Fortnum, A. Wojcicki, and J. O. Edwards, J. Amer. Chem. Soc.,
        1959, 81, 1295

59EG   K. Emerson and W. M. Graven, J. Inorg. Nucl. Chem., 1959, 11, 309

59EGa   A. J. Ellis and R. M. Golding, J. Chem. Soc., 1959, 127

59ES   H. K. El-Shamy and F. G. Sherif, Egypt. J. Chem., 1959, 2, 217

59FB   J. Faucherre and Y. Bonnaire, Compt. Rend. Acad. Sci. Paris, 1959, 248, 3705

59FS   Ya. D. Fridman and Dzh. S. Sarbaev, Russ. J. Inorg. Chem., 1959, 4, 835 (1849)

59FSa   H. Freund and C. R. Schneider, J. Amer. Chem. Soc., 1959, 81, 4780

59G   A. M. Golub, Russ. J. Inorg. Chem., 1959, 4, 711 (1577)

59GJ   L. O. Gilpatrick, H. R. Jolley, M. J. Kelly, M. D. Silverman, and G. M. Watson,
        CF-59-10-121, 1959

59GK   A. M. Golub and Yu. V. Kosmatyi, Russ. J. Inorg. Chem., 1959, 4, 606 (1347)

59GR   R. P. H. Gasser and R. E. Richards, Molec. Phys., 1959, 2, 357

59GS   J. Gandeboeuf and P. Souchay, J. Chim. Phys., 1959, 56, 358

59HJ   G. C. Hood, A. C. Jones, and C. A. Reilly, J. Phys. Chem., 1959, 63, 101

59HS   S. Hietanen and L. G. Sillen, Acta Chem. Scand., 1959, 13, 533

59HSa   S. Hietanen and L. G. Sillen, Acta Chem. Scand., 1959, 13, 1828

59I   N. Ingri, Acta Chem. Scand., 1959, 13, 758

59IB   N. Ingri and F. Brito, Acta Chem. Scand., 1959, 13, 1971

59K   J. Kenttamaa, Suomen Kem., 1959, B32, 55

59Ka   S. K. Kor, Z. Phys. Chem., 1959, 210, 288

59Kb   J. R. Kramer, J. Sediment. Petrol., 1959, 29, 465

59Kc   J. Kenttamaa, Suomen Kem., 1959, B32, 68

59Kd   E. V. Komarov, Russ. J. Inorg. Chem., 1959, 4, 591 (1313)

59KB   I. M. Korenman and V. N. Burova, Trudy Khim. Khim. Tekh. (Univ. Gorkii), 1959, 2, 366

59KG   E. L. King and P. K. Gallagher, J. Phys. Chem., 1959, 63, 1073

59KGa   P. N. Kovalenko and O. I. Geiderovich, Russ. J. Inorg. Chem., 1959, 4, 895 (1974)

59KGb   K. E. Kleiner and G. I. Gridchina, Russ. J. Inorg. Chem., 1959, 4, 915 (2020)

59KK   P. Kivalo and R. Kurkela, Suomen Kem., 1959, B32, 39

59KL   P. N. Kovalenco and T. V. Lindorf, Russ. J. Inorg. Chem., 1959, 4, 868, (1919)

59KP   J. W. Kury, A. D. Paul, L. G. Hepler, and R. E. Connick, J. Amer. Chem. Soc., 1959, 81, 4185

59KS   A. E. Klygin and I. D. Smirnova, Russ. J. Inorg. Chem., 1959, 4, 16 (42)

59KSN   A. E. Klygin, I. D. Smirnova, and N. A. Nikolskaya, Russ. J. Inorg. Chem., 1959, 4, 754 (1674)

59L   G. Lagerstrom, Acta Chem. Scand., 1959, 13, 722

59LP   W. L. Lindsay, M. Peech, and J. S. Clark, Proc. Soil Sci. Soc. Amer., 1959, 23, 357

59M   B. N. Mattoo, Z. Phys. Chem. (Frankfurt), 1959, 19, 156

59Ma   Y. Marcus, J. Phys. Chem., 1959, 63, 1000

59Mb   J. S. Mendez-Schalchi, Thesis, Mass. Inst. Tech. 1959

59Mc   Y. Marcus, Bull. Res. Council Israel, 1959, A8, 17

59Md   G. Maronny, Electrochim. Acta, 1959, 1, 58

59Me   T. T. Mityurova, Dopovidi Akad. Nauk Ukr. SSR, 1959, 166

59Ma   H. Matsuda and Y. Ayabe, Z. Elektrochem., 1959, 63, 1164

59MC   Y. Marcus and C. D. Coryell, Bull. Res. Council Israel, 1959, A8, 1

59MG   P. K. Migal, N. Kh. Grinberg, and Ya. I. Turyan, Russ. J. Inorg. Chem., 1959, 4, 833 (1844)

59MV   C. C. Meloche and F. Vratny, Anal. Chim. Acta, 1959, 20, 415 (see also 64SM)

59NH   L. Newman and D. N. Hume, J. Amer. Chem. Soc., 1959, 81, 5901

59NN   V. S. K. Nair and G. H. Nancollas, J. Chem. Soc., 1959, 3934

59NQ   L. Newman and K. P. Quinlan, J. Amer. Chem. Soc., 1959, 81, 547

59O   A. Olin, Acta Chem. Scand., 1959, 13, 1791

59OH   J. W. Olver and D. N. Hume, Anal. Chim. Acta, 1959, 20, 559

59P   D. D. Perrin, J. Chem. Soc., 1959, 1710

59Pa   I. Popescu, Stud. Cercetari, Chim., Filiala Iasi, 1959, 10, 25

59Pb   M. Pryszczewska, Roca. Chem., 1959, 33, 755

59Pc   C. N. Polydoropoulos, Chim. Chronika, 1959, 24, 147; Chem. Abs., 1960, 15048c

59PD   F. Pantani and P. G. Desideri, Gazz. Chim. Ital., 1959, 89, 1360

59PL   K. Pan and J. L. Lin, J. Chinese Chem. Soc., 1959, 6, 1

59R   R. W. Ramette, J. Chem. Educ., 1959, 36, 191

59S   M. Spiro, Trans. Faraday Soc., 1959, 55, 1746

59Sa   K. W. Sykes, J. Chem. Soc., 1959, 2473

59Sb   J. L. Schultz, Thesis, Univ. Minnesota, 1959

59Sc   P. C. Scott, Thesis, Univ. Minnesota, 1959

59Sd   B. M. Shchigol, Russ. J. Inorg. Chem., 1959, 4, 913 (2014)

59Se   G. B. Seifer, Russ. J. Inorg. Chem., 1959, 4, 1311 (2832)

59Sf   P. Schindler, Helv. Chim. Acta, 1959, 42, 2736

59SD   R. L. Seth and A. K. Dey, Z. Phys. Chem. (Leipzig), 1959, 210, 108

59SH   J. C. Sullivan and J. C. Hindman, J. Phys. Chem., 1959, 63, 1332

59SHC   J. C. Hindman, J. C. Sullivan, and D. Cohen, J. Amer. Chem. Soc., 1959, 81, 2316

59SK   N. M. Selivanova, A. F. Kapustinskii, and G. A. Zubova, Bull. Acad. Sci. USSR,
        1959, 174 (187)

59SL   S. A. Shchukarev, L. S. Lilich, V. A. Latysheva, and D. K. Andreeva, Russ. J.
        Inorg. Chem., 1959, 4, 1001 (2198)

59SLa   S. A. Shchukarev, L. S. Lilich, V. A. Latysheva, and I. I. Chuburkova, Vestnik
        Leningrad. Univ., 1959, 14, No. 10, 66

59SS   N. M. Selivanova and V. A. Shneider, Izv. Vyssh. Ucheb. Zaved., Khim., 1959, 2,
        475, 651

59ST   V. B. Shevchenko, V. G. Timoshev, and A. A. Volkova, Soviet J. Atom. Ener., 1959,
        6, 293 (426)

59SV   C. D. Schmulbach, J. R. Van Wazer, and R. R. Irani, J. Amer. Chem. Soc., 1959, 81,
        6347

59SY   S. Saito and N. Yui, <u>Nippon Kagaku Zasshi</u>, 1959, <u>80</u>, 139

59SZ   N. M. Selivanova and G. A. Zubova, <u>Zh. Fiz. Khim.</u>, 1959, <u>33</u>, 141

59SZF   N. M. Selivanova, G. A. Zubova, and F. I. Finkelshtein, <u>Russ. J. Phys. Chem.</u>,
         1959, <u>33</u>, 430 (2365)

59T   L. O. Tuazon, Thesis, Iowa State Coll., 1959

59Ta   I. V. Tananaev, <u>Acta Chim. Sinica</u>, 1959, <u>25</u>, 391

59TC   Ya. I. Turyan and N. G. Chebotar, <u>Russ. J. Inorg. Chem.</u>, 1959, <u>4</u>, 273 (599)

59TH   H. G. Tsiang and K. H. Hsu, <u>Kexue Tongbao</u>, 1959, 331

59TL   I. V. Tananaev and C. D. Lu, <u>Russ. J. Inorg. Chem.</u>, 1959, <u>4</u>, 961 (2122)

59TS   Ya. I. Turyan and R. Ya. Shtipelman, <u>Russ. J. Inorg. Chem.</u>, 1959, <u>4</u>, 366 (808)

59TT   N. Tanaka and T. Takamura, <u>J. Inorg. Nucl. Chem.</u>, 1959, <u>9</u>, 15

59U   R. Uggla, <u>Acta Acad. Sci. Fenn.</u>, 1959, <u>AII</u>, No. 97

59VK   D. G. Vartak and M. B. Kabadi, <u>Proc. Symp. Chem. Coord. Comp.</u>, Allahabad, 1959, <u>1</u>,
         127

59VN   V. G. Voden, G. P. Nikitina, and M. F. Pushlenkov, <u>Radiokhim.</u>, 1959, <u>1</u>, 121

59W   H. v. Wartenberg, <u>Z. Anorg. Allg. Chem.</u>, 1959, <u>299</u>, 227

59WL   J. I. Watters and S. M. Lambert, <u>J. Amer. Chem. Soc.</u>, 1959, <u>81</u>, 3201

59WO   J. A. Wolhoff and J. T. G. Overbeek, <u>Rec. Trav. Chim.</u>, 1959, <u>78</u>, 759

59WP   J. L. Weaver and W. C. Purdy, <u>Anal. Chim. Acta</u>, 1959, <u>20</u>, 376

59YD   T. Yamane and N. Davidson, <u>J. Amer. Chem. Soc.</u>, 1959, <u>81</u>, 4438

59Z   A. J. Zielen, <u>J. Amer. Chem. Soc.</u>, 1959, <u>81</u>, 5022

60A   G. Atkinson, <u>J. Amer. Chem. Soc.</u>, 1960, <u>82</u>, 818

60Aa   N. V. Akselrud, <u>Russ. J. Inorg. Chem.</u>, 1960, <u>5</u>, 928 (1910)

60Ab   H. L. Conley, Thesis, 1960; UCRL-9332

60AD   P. J. Antikainen and D. Dyrssen, <u>Acta Chem. Scand.</u>, 1960, <u>14</u>, 86

60AH   G. Atkinson and C. J. Hallada, <u>J. Phys. Chem.</u>, 1960, <u>64</u>, 1487

60AHS   P. J. Antikainen, S. Hietanen, and L. G. Sillen, <u>Acta Chem. Scand.</u>, 1960, <u>14</u>, 95

60AM   D. W. Anderson, G. N. Malcolm, and H. N. Parton, <u>J. Phys. Chem.</u>, 1960, <u>64</u>, 494

60AR   P. J. Antikainen and V. M. K. Rossi, <u>Suomen Kem.</u>, 1960, <u>B33</u>, 210

60AS   N. V. Akselrud and V. B. Spivakovskii, <u>Russ. J. Inorg. Chem.</u>, 1960, <u>5</u>, 158 (327)

60ASa   N. V. Akselrud and V. B. Spivakovskii, <u>Russ. J. Inorg. Chem.</u>, 1960, <u>5</u>, 163 (340)

60ASb   N. V. Akselrud and V. B. Spivakovskii, Russ. J. Inorg. Chem., 1960, 5, 168 (348)

60ASc   J. Angeli and P. Souchay, Comp. Rend. Acad. Sci. Paris, 1960, 250, 713

60AT   P. J. Antikainen and K. Tevanen, Suomen Kem., 1960, B33, 59

60B   B. Baysal, Act. 2 Congr. Internat. Catalyse, Paris, 1960, 1, 559

60Ba   S. A. Brusilovskii, Trudy Inst. Geol. Rud. Mestorozhdenii etc., (Akad. Nauk
        SSSR), 1960, 42, 58

60Bb   H. L. Barnes, Bull. Geol. Soc. Amer., 1960, 71, 1821

60BB   P. B. Barton, Jr., and P. M. Bethke, Amer. J. Sci., 1960, A258, 21

60BC   N. Bailey, A. Carrington, K. A. K. Lott, and M. C. R. Symons, J. Chem. Soc., 1960,
        290

60BG   G. Bianucci and L. Ghiringhelli, Ann. Chim. (Italy), 1960, 50, 99

60BH   G. Biedermann and S. Hietanen, Acta. Chem. Scand., 1960, 14, 711

60BI   F. Brito and N. Ingri, An. Fis. Quim., Ser. B, 1960, 56, 165

60BK   A. K. Babko and V. S. Kodenskaya, Russ. J. Inorg. Chem., 1960, 5, 1241 (2568)

60BN   F. B. Baker, T. W. Newton, and M. Kahn, J. Phys. Chem., 1960, 64, 109

60BR   A. Basinski and M. Rozwadowski, Rocz. Chem., 1960, 34, 47

60BS   G. Biedermann and L. G. Sillen, Acta Chem. Scand., 1960, 14, 717

60BT   A. I. Busev, V. G. Tiptsova, and T. A. Sokolova, Vestnik Moskov. Univ., Ser. II
        Khim, 1960, No. 6, 42

60C   J. M. Creeth, J. Phys. Chem., 1960, 64, 920

60Ca   F. Chauveau, Bull. Soc. Chim. France, 1960, 810, 819

60Cb   F. Chauveau, Bull. Soc. Chim. France, 1960, 834

60Cc   M. Cola, Gazz. Chim. Ital., 1960, 90, 1037

60CC   V. Caglioti, L. Ciavatta, and A. Liberti, J. Inorg. Nucl. Chem., 1960, 15, 115

60CE   M. M. Crutchfield and J. O. Edwards. J. Amer. Chem. Soc., 1960, 82, 3533

60CL   C. Chen-ping and L. Lien-sen, Acta Chim. Sinica, 1960, 26, 148 (see 64VG)

60CLa   L. Ciavatta and A. Liberti, Ric. Sci., 1960, 30, 1186

60CO   B. Carell and A. Olin, Acta Chem. Scand., 1960, 14, 1999

60D   J. Danon, J. Inorg. Nucl. Chem., 1960, 13, 112

60Da   G. Daniele, Gazz. Chim. Ital., 1960, 90, 1371

60DF   J. L. Dye, M. P. Faber, and D. J. Karl, J. Amer. Chem. Soc., 1960, 82, 314

60DM   R. G. Denotkina, A. I. Moskvin, and V. B. Shevchenko, Russ. J. Inorg. Chem., 1960,
        5, 387 (805), 731 (1509)

60EK   J. H. Espenson and E. L. King, J. Phys. Chem., 1960, 64, 380

60FB   D. H. Fortnum, C. J. Battaglia, S. R. Cohen, and J. O. Edwards, J. Amer. Chem.
            Soc., 1960, 82, 778

60FSA   P. Franzosini, C. Sinistri, and G. Ajroldi, Ric. Sci., 1960, 30, 1707

60FSS   Ya. D. Fridman, D. S. Sarbaev, and R. I. Sorochan, Russ. J. Inorg. Chem., 1960,
            5, 381 (791)

60FT   F. M. Filinov, E. N. Tekster, A. A. Kolpakova, and E. P. Panteleeva, Russ. J.
            Inorg. Chem., 1960, 5, 552 (1149)

60G   E. Giesbrecht, J. Inorg. Nucl. Chem., 1960, 15, 265

60Ga   R. M. Golding, J. Chem. Soc., 1960, 3711

60Gb   C. J. Garrigues, Publ. Sci. Tech. Ministere de L'Air (Paris), NT-93, 1960

60GB   G. Gordon and C. H. Brubaker, Jr., J. Amer. Chem. Soc., 1960, 82, 4448

60GC   S. A. Greenberg, T. N. Chang, and E. Anderson, J. Phys. Chem., 1960, 64, 1151

60GG   A. A. Grinberg and M. I. Gelfman, Proc. Acad. Sci. USSR, 1960, 133, 895 (1081)

60GK   P. K. Gallagher and E. L. King, J. Amer. Chem. Soc., 1960, 82, 3510

60GL   P. K. Glasoe and F. A. Long, J. Phys. Chem., 1960, 64, 188

60GN   I. Grenthe and B. Noren, Acta Chem. Scand., 1960, 14, 2216

60GR   R. L. Gustafson, C. Richard, and A. E. Martell, J. Amer. Chem. Soc., 1960, 82,
            1526

60GS   A. A. Grinberg and G. A. Shagisultanova, Russ. J. Inorg. Chem., 1960, 5, 134 (280),
            920 (1895)

60GSa   A. M. Golub and V. V. Skopenko, Russ. J. Inorg. Chem., 1960, 5, 961 (1973)

60H   E. Hogfeldt, Acta Chem. Scand., 1960, 14, 1597

60HR   G. C. Hood and C. A. Reilly, J. Chem. Phys., 1960, 32, 127

60HT   K. H. Hsu, T. C. Tan, and C. M. Yen, Scientia Sinica, 1960, 9, 232; Acta Chim.
            Sinica, 1959, 25, 229

60IC   R. R. Irani and C. F. Callis, J. Phys. Chem., 1960, 64, 1398

60JP   B. Jeszowska-Trzebiatowska and L. Pajdowski, Rocz. Chem., 1960, 34, 787

60K   T. Kumai, J. Chem. Soc., Japan, 1960, 81, 1687

60Ka   K. E. Kleiner, Russ. J. Phys. Chem., 1960, 34, 194 (416)

60KG   K. E. Kleiner and G. I. Gridchina, Russ. J. Inorg. Chem., 1960, 5, 96 (202)

60KM   F. Ya. Kulba and V. E. Mironov, Russ. J. Inorg. Chem., 1960, 5, 138 (287)

60KMa   F. Ya. Kulba and V. E. Mironov, Russ. J. Inorg. Chem., 1960, 5, 922 (1898)

60KV   K. E. Kleiner and V. T. Vasilenko, Russ. J. Inorg. Chem., 1960, 5, 53 (112)

60L   K. H. Lieser, Z. Anorg. Allg. Chem., 1960, 304, 296

60LC   A. Lodzinska and H. Cichocka, Rocz. Chem., 1960, 34, 297

60LP   I. A. Lebedev, S. V. Pirozhkov, and G. N. Yakovlev, Radiokhim., 1960, 2, 549

60LR   M. W. Lister and P. Rosenblum, Canad. J. Chem., 1960, 38, 1827

60LS   M. H. Lietzke and R. W. Stoughton, J. Phys. Chem., 1960, 64, 816

60LSV   A. V. Lapitskii, B. V. Strizhkov, and L. G. Vlasov, Vesinik Moskov. Univ., Ser. II
        Khim., 1960, No. 4, 25

60LY   M. W. Lister and Y. Yoshino, Canad. J. Chem., 1960, 38, 45

60LYa   M. W. Lister and Y. Yoshino, Canad. J. Chem., 1960, 38, 2342

60M   C. B. Monk, J. Amer. Chem. Soc., 1960, 82, 5762

60Ma   Y. Marcus, J. Inorg. Nucl. Chem., 1960, 12, 287

60Mb   S. Matsuo, Nippon Kagaku Zasshi, 1960, 81, 833

60MB   E. C. Moreno, W. E. Brown, and G. Osborn, Proc. Soil Sci. Soc. Amer., 1960, 24, 94,
        99

60MJ   R. L. McCullough, L. H. Jones, and R. A. Penneman, J. Inorg. Nucl. Chem., 1960,
        13, 286

60MM   R. W. Mooney and G. J. Meisenhelter, J. Chem. Eng. Data, 1960, 5, 373

60MT   M. Mori and R. Tsuchiya, Bull. Chem. Soc. Japan, 1960, 33, 841

60MTF   K. P. Mishchenko, T. A. Tumanova, and I. E. Flis, J. Anal. Chem. USSR, 1960, 15,
        241 (211)

60N   R. Nasanen, Suomen Kem., 1960, B33, 47

60NA   C. J. Nyman and G. S. Alberts, Anal. Chem., 1960, 32, 207

60NM   R. Nasanen and P. Merilainen, Suomen Kem., 1960, B33, 149, 197

60O   A. Olin, Acta Chem. Scand., 1960, 14, 126, 814

60Oa   J. Olivard, Arch. Biochem. Biophys., 1960, 88, 382

60P   D. D. Perrin, J. Chem. Soc., 1960, 3189

60PB   M. G. Panova, N. E. Brezhneva, and V. I. Levin, Radiokhim., 1960, 2, 208

60PN   M. F. Pushlenkov, G. P. Nikitina, and V. G. Voden, Radiokhim., 1960, 2, 215

60R   J. Raaflaub, Helv. Chim. Acta, 1960, 43, 629

60RK   S. W. Rabideau and R. J. Kline, J. Phys. Chem., 1960, 64, 650

60RKa   B. vant Riet and I. M. Kolthoff, J. Phys. Chem., 1960, 64, 1045

60RS   D. H. Richards and K. W. Sykes, J. Chem. Soc., 1960, 3626

60S   J. Stary, Coll. Czech. Chem. Comm., 1960, 25, 890

60Sa   L. Schoepp, Diss., Tech. Univ. Berlin, 1960

60SB   A. W. Savage, Jr., and J. C. Browne, J. Amer. Chem. Soc., 1960, 82, 4817

60SF   G. Schwarzenbach and A. Fisher, Helv. Chim. Acta, 1960, 43, 1365

60SG   P. Schindler and A. B. Garrett, Helv. Chim. Acta, 1960, 43, 2176

60SL   R. W. Stoughton and M. H. Lietzke, J. Phys. Chem., 1960, 64, 133

60SS   V. B. Shevchenko, I. V. Shilin, and Yu. F. Zhdanov, Russ. J. Inorg. Chem., 1960,
        5, 1367 (2832)

60SW   F. Seel and R. Winkler, Z. Phys. Chem. (Frankfurt), 1960, 25, 217

60SWa   L. H. Suthcliffe and J. R. Weber, J. Inorg. Nucl. Chem., 1960, 12, 281

60T   C. C. Templeton, J. Chem. Eng. Data, 1960, 5, 514

60Ta   T. A. Turney, J. Chem. Soc., 1960, 4263

60Tb   R. S. Tobias, J. Amer. Chem. Soc., 1960, 82, 1070

60TM   Ya. I. Turyan and Yu. S. Milyavskii, Russ. J. Inorg. Chem., 1960, 5, 1086 (2242)

60TV   I. V. Tananaev and A. D. Vinogradova, Russ. J. Inorg. Chem., 1960, 5, 155 (321)

60TZ   Ya. I. Turyan and B. P. Zhantalai, Russ. J. Inorg. Chem., 1960, 5, 848 (1748)

60YD   T. Yamane and N. Davidson, J. Amer. Chem. Soc., 1960, 82, 2123

60ZK   S. S. Zavodnov and P. A. Kryukov, Bull. Acad. Sci. USSR, 1960, 1583 (1704)

61A   N. V. Akselrud, Ukr. Khim. Zh., 1961, 27, 431

61AE   N. V. Akselrud and V. I. Ermolenko, Russ. J. Inorg. Chem., 1961, 6, 397 (777)

61AH   K. Aurivillius and O. von Heidenstam, Acta Chem. Scand., 1961, 15, 1993

61AT   P. J. Antikainen and K. Tevanen, Suomen Kem., 1961, B34, 135

61B   A. Bolzan, Rev. Fac. Cienc. Quim. Univ. Nacl. La Plata, 1961, 33, 67; Chem. Abs.,
        1963, 58, 3091d

61BC   G. Biedermann and L. Ciavatta, Acta Chem. Scand., 1961, 15, 1347

61BD   D. Bunn, F. S. Dainton, and S. Duckworth, Trans. Faraday Soc., 1961, 57, 1131

61BL   G. Biedermann, N. C. Li, and J. Yu, Acta Chem. Scand., 1961, 15, 555

61BM   M. Boulet and J. R. Marier, Arch. Biochem. Biophys., 1961, 93, 157

61BN   S. Bruckenstein and D. C. Nelson, J. Chem. Eng. Data, 1961, 6, 605

61BP   A. A. Bezzubenko and B. I. Pechchevitskii, Izv. Sibir. Otdel. Akad. Nauk. SSSR,
        1961, No. 8, 62

61BS   A. Basinski, W. Szymanski, and Z. Bebnista, Rocz. Chem., 1961, 35, 59

61BT   D. Banerjea and K. K. Tripathi, J. Inorg. Nucl. Chem., 1961, 18, 199

61CA   V. G. Chukhlantsev and K. V. Alyamovskaya, Russ. J. Inorg. Chem., 1961, 6, 223
       (443)

61CAa V. G. Chukhlantsev and K. V. Alyamovskaya, Izv Vyssh. Ucheb. Zayed., Khim., 1961,
       4, 359, 706

61CF   R. E. Connick and D. A. Fine, J. Amer. Chem. Soc., 1961, 83, 3414

61CO   B. Carell and A. Olin, Acta Chem. Scand., 1961, 15, 727

61COa B. Carell and A. Olin, Acta Chem. Scand., 1961, 15, 1875

61CP   R. E. Connick and A. D. Paul, J. Phys. Chem., 1961, 65, 1216

61D   T. F. Dyachenko, Nauch. Trudy Dnepropetrovsk Khim. Tech. Inst., 1961, 12, No. 2,
       87

61Da   G. A. Dean, Canad. J. Chem., 1961, 39, 1174

61DK   N. S. Drozdov and V. P. Krylov, Russ. J. Phys. Chem., 1961, 35, 1264 (2557)

61DT   D. Dyrssen and V. Tyrrell, Acta Chem. Scand., 1961, 15, 393, 1622

61EA   A. J. Ellis and D. W. Anderson, J. Chem. Soc., 1961, 1765

61EAa A. J. Ellis and D. W. Anderson, J. Chem. Soc., 1961, 4678

61F   Ya. D. Fridman, Russ. J. Inorg. Chem., 1961, 6, 771 (1501)

61GG   A. A. Grinberg and M. I. Gelfman, Proc. Acad. Sci. USSR, 1961, 137, 257 (87)

61GM   A. D. Gelman, A. I. Moskvin, L. M. Zaitsev, and M. P. Mefodeva, Kompleksye
       Soedineniya Transuranovykh Elementov, Moscow, 1961, p. 90, 98

61GO   A. M. Golub and S. S. Ognyanik, Ukr. Khim. Zh., 1961, 27, 283

61GS   A. M. Golub and E. P. Skorobogatko, Ukr. Khim. Zh., 1961, 27, 16

61GSa A. M. Golub and V. V. Skopenko, Russ. J. Inorg. Chem., 1961, 6, 69 (140)

61GSb A. M. Golub and V. V. Skopenko, Dokl. Chem., 1961, 138, 517 (601)

61GT   R. M. Garrels, M. E. Thompson, and R. Siever, Amer. J. Sci., 1961, 259, 24

61H   G. P. Height, Jr., Acta Chem. Scand., 1961, 15, 2012

61Ha   E. Hogfeldt, J. Inorg. Nucl. Chem., 1961, 17, 302

61Hb   T. Hurlen, Acta Chem. Scand., 1961, 15, 1231

61HA   C. J. Hallada and G. Atkinson, J. Amer. Chem. Soc., 1961, 83, 3759

61HG   C. J. Hardy, B. F. Greenfield, and D. Scargill, J. Chem. Soc., 1961, 174

61I   R. R. Irani, J. Phys. Chem., 1961, 65, 1463

V. BIBLIOGRAPHY

61IC   R. R. Irani and C. F. Callis, J. Phys. Chem., 1961, 65, 296

61ICa   R. R. Irani and C. F. Callis, J. Phys. Chem., 1961, 65, 934

61K   V. I. Karpov, Russ. J. Inorg. Chem., 1961, 6, 271 (531)

61Ka   P. N. Kovalenko, Russ. J. Inorg. Chem., 1961, 6, 275 (539)

61Kb   A. G. Kozlov, Russ. J. Inorg. Chem., 1961, 6, 668 (1302)

61Kc   P. Krumholz, Proc. 6th Int. Conf. Coord. Chem., 1961, 564

61Kd   A. F. Kuteinikov, Radiochem (AEC-TR-4581), 1961, 3, No. 6, 202 (706)

61KB   P. N. Kovalenko and K. N. Bagdasarov, J. Appl. Chem. USSR, 1961, 34, 759 (789)

61KBa   P. N. Kovalenko and K. N. Bagdasarov, Russ. J. Inorg. Chem., 1961, 6, 272 (534)

61KMF   F. Ya. Kulba, V. E. Mironov, and V. A. Fedorov, Russ. J. Inorg. Chem., 1961, 6, 813 (1586)

61KMT   F. Ya. Kulba, V. E. Mironov, G. S. Troitskaya, and N. G. Maksimova, Russ. J. Inorg. Chem., 1961, 6, 952 (1865)

61KT   G. Kruger and E. Thilo, Z. Anorg. Allg. Chem., 1961, 308, 242

61KY   G. A. Krestov and K. B. Yatsimirskii, Russ. J. Inorg. Chem., 1961, 6, 1170 (2304)

61KZ   A. E. Klygin, D. M. Zavrazhnova, and N. A. Nikolskaya, J. Anal. Chem. USSR, 1961, 16, 311 (297)

61L   M. W. Lister, Canad. J. Chem., 1961, 39, 2330

61LH   H. L. Loy and D. M. Himmelblau, J. Phys. Chem., 1961, 65, 264

61LP   I. Leden and G. Persson, Acta Chem. Scand., 1961, 15, 1141

61LPa   J. L. Lin and K. Pan, J. Chinese Chem. Soc., 1961, 8, 14

61LW   M. W. Lister and D. W. Wilson, Canad. J. Chem., 1961, 39, 2606

61M   V. E. Mironov, Russ. J. Inorg. Chem., 1961, 6, 205 (405)

61Ma   V. E. Mironov, Russ. J. Inorg. Chem., 1961, 6, 336 (659)

61Mb   T. Murayama, Sci. Repts. Tohoku Univ., Ser. I, 1961, 45, 84

61Mc   S. Matsuo, Nippon Nagaku Zasshi, 1961, 82, 1330, 1334

61MA   D. S. Martin, Jr., and R. J. Adams, Proc. 6th Int. Conf. Coord. Chem., 1961, 579

61MB   D. W. Margerum, T. J. Bydalek and J. J. Bishop, J. Amer. Chem. Soc., 1961, 83, 1791

61MC   I. N. Marov and M. K. Chmutova, Russ. J. Inorg. Chem., 1961, 6, 1340 (2654)

61MD   P. P. Mohapatra, R. C. Das, and S. Aditya, J. Indian Chem. Soc., 1961, 38, 845

61MF   C. J. Mandleberg, K. E. Francis, and R. Smith, J. Chem. Soc., 1961, 2464

61ML   C. H. Muendel, H. B. Linford, and W. A. Selke, J. Amer. Inst. Chem. Eng., 1961, 7, 133

61MN   L. N. Mulay and M. C. Naylor, Proc. 6th Int. Conf. Coord. Chem., 1961, 520

61MP   G. N. Malcolm, H. N. Parton, and I. D. Watson, J. Phys. Chem., 1961, 65, 1900

61MS   D. F. C. Morris and E. L. Short, J. Chem. Soc., 1961, 5148

61MSa   S. S. Muhammad and E. V. Sundaram, J. Sci. Ind. Res. (India), 1961, B20, 16

61NL   N. S. Nikolaev and Yu. A. Lukyanychev, Atom. Ener., 1961, 11, 67 (see also 64SM)

61NM   R. Nasanen, P. Merilainen, and K. Leppanen, Acta. Chem. Scand., 1961, 15, 913

61NP   B. K. S. Nair, L. H. Prabhu, and D. G. Vartak, J. Sci. Ind. Res. (India), 1961,
        B20, 489

61NR   C. J. Nyman, D. K. Roe, and R. A. Plane, J. Amer. Chem. Soc., 1961, 83, 323

61NS   C. J. Nyman and T. Salazar, Anal. Chem., 1961, 33, 1467

61P   A. Peterson, Acta Chem. Scand., 1961, 15, 101

61PB   K. S. Pitzer and L. Breuer, Thermodynamics, rev. ed. by G. N. Lewis and M. Randall,
        McGraw-Hill, New York, 1961, pp. 400, 676

61PF   A. Patterson, Jr., and H. Freitag, J. Elecktrochem. Soc., 1961, 108, 529

61PG   A. D. Paul, L. S. Gallo, and J. B. Van Camp, J. Phys. Chem. 1961, 65, 441

61PK   N. S. Poluektov and L. I. Kononenko, Russ. J. Inorg. Chem., 1961, 6, 938 (1837)

61PM   V. M. Peshkova, N. V. Melchakova, and S. G. Zhemchuzhin,, Russ. J. Inorg. Chem.,
        1961, 6, 630 (1233)

61PP   I. V. Pyatnitskii and E. S. Pilipyuk, Ukr. Khim. Zh., 1961, 27, 247

61PS   S. S. Parikh and T. R. Sweet, J. Phys. Chem., 1961, 65, 1909

61PY   C. N. Polydoropoulos and Th. Yannapoulos, J. Inorg. Nucl. Chem., 1961, 19, 107

61RB   R. A. Robinson and V. E. Bower, J. Phys. Chem., 1961, 65, 1279

61RK   A. L. Rotinyan, V. L. Kheifets, and S. A. Nikolaeva, Russ. J. Inorg. Chem., 1961,
        6, 10 (21)

61RM   S. W. Rabideau and R. H. Moore, J. Phys. Chem., 1961, 65, 371

61RMa   S. W. Rabideau and B. J. Masters, J. Phys. Chem., 1961, 65, 1256

61RS   R. W. Ramette and R. F. Stewart, J. Phys. Chem., 1961, 65, 243

61RSS   R. A. Robinson, J. M. Stokes, and R. H. Stokes, J. Phys. Chem., 1961, 65, 542

61RT   I. G. Ryss and V. B. Tulchinskii, Russ. J. Inorg. Chem., 1961, 6, 947 (1856)

61RW   T. E. Rogers and G. M. Waind, Trans. Far. Soc., 1961, 83, 3373

61S   S. Sobkowski, J. Inorg. Nucl. Chem., 1961, 23, 81

61Sa   R. B. Simpson, J. Amer. Chem. Soc., 1961, 83, 4711

61Sb   M. B. Shchigol, Russ. J. Inorg. Chem., 1961, 6, 1361 (2693)

61Sc   K. Schlyter, Trans. Roy. Inst. Tech. Stockholm, 1961, No. 182 (see 64SM)

61Sd   J. P. Schwing, Thesis, Univ. Strasbourg, 1961

61SA   F. G. Sherif and A. M. Awad, J. Inorg. Nucl. Chem., 1961, 19, 94

61SB   M. B. Shchigol and N. B. Burchinskaya, Russ. J. Inorg. Chem., 1961, 6, 1267 (2504);
        (see also A. K. Babko, ibid., 1962, 7, 1373 (2642)

61SF   R. W. Soughton, A. J. Fry, and J. E. Barney, II, J. Inorg. Nucl. Chem., 1961, 19,
        286

61SM   E. L. Short and D. F. C. Morris, J. Inorg. Nucl. Chem., 1961, 18, 192

61SMa   C. I. Sanders and D. S. Martin, Jr., J. Amer. Chem. Soc., 1961, 83, 807

61SN   P. Senise and E. F. A. Neves, J. Amer. Chem. Soc., 1961, 83, 4146

61SR   P. Schindler and M. Reinert, Chimia (Switz.), 1961, 15, 534

61SRD   N. Sutin, J. K. Rowley, and R. W. Dodson, J. Phys. Chem., 1961, 65, 1248

61T   Ya. I. Turyan, Russ. J. Inorg. Chem., 1961, 6, 80 (162)

61Ta   G. A. Tsigdinos, Thesis, Boston Univ., 1961

61Tb   J. Y. Tong, TID-13337, 1961

61Tc   I. A. Taub, Thesis, Univ. Minnesota, 1961; Diss. Abs., 1961, 22, 1401

61TB   V. F. Toropova and F. M. Batyrshina, Izv. Vyssh. Ucheb. Zaved., Khim., 1961, 4, 11

61TD   I. V. Tananaev and E. N. Deichman, Radiochem. (AEC-TR-4581), 1961, 3, No. 6, 208
        (712)

61TG   A. W. Taylor and E. L. Gurney, J. Phys. Chem., 1961, 65, 1613

61TH   R. S. Tobias and Z. Z. Hugus, Jr., J. Phys. Chem., 1961, 65, 2165

61TJ   J. F. Tate and M. M. Jones, J. Phys. Chem., 1961, 65, 1661

61TO   N. Tanaka and H. Ogino, Bull. Chem. Soc., Japan, 1961, 34, 1040

61V   V. P. Vasilev, Izv. Vyssh. Ucheb. Zaved., Khim., 1961, 4, 936

61VQ   C. E. Vanderzee and A. S. Quist. J. Phys. Chem., 1961, 65, 118

61WD   R. M. Wallace and E. K. Dukes, J. Phys. Chem., 1961, 65, 2094

61WK   J. M. White, P. Kelly, and N. C. Li, J. Inorg. Nucl. Chem., 1961, 16, 337

61WL   J. M. Wright, W. T. Lindsay, Jr., and T. R. Druga, TID 4500=WADD-TM-204, 1961

61Y   R. G. Yalman, J. Amer. Chem. Soc., 1961, 83, 4142

62A   I. Ahlberg, Acta Chem. Scand., 1962, 16, 887

62AM   H. Asai and M. Morales, Arch. Biochem. Biophys., 1962, 99, 383

62AMa   J. M. Austin and A. D. Mair, J. Phys. Chem., 1962, 66, 519

62AP   V. I. Altynov and B. V. Ptitsyn, Russ. J. Inorg. Chem., 1962, 7, 1088 (2103)

62AY   G. Atkinson and M. Yokoi, J. Phys. Chem., 1962, 66, 1520

62B   H. Bode, Z. Anorg. Allg. Chem., 1962, 317, 3

62Ba   Yu. A. Buslaev, Russ. J. Inorg. Chem., 1962, 7, 619 (1204)

62Bb   G. Biedermann, Proc. 7th Int. Conf. Coord. Chem., Stockholm–Uppsala, 1962, p. 159

62Bc   L. A. Blatz, J. Phys. Chem., 1962, 66, 160

62BA   J. A. Bolzan and A. J. Arvia, Electrochim. Acta, 1962, 7, 589

62BB   A. Bukowska and H. Basinska, Chem. Anal., 1962, 7, 559

62BBa   A. Bukowska and H. Basinska, Chem. Anal., 1962, 7, 563

62BC   G. Biedermann and L. Ciavatta, Acta Chem. Scand., 1962, 16, 2221

62BD   O. D. Bonner, H. Dolyniuk, C. F. Jordan, and G. B. Hanson, J. Inorg. Nucl. Chem.,
         1962, 24, 689

62BG   A. K. Babko, G. I. Gridchina, and B. I. Nabivanets, Russ. J. Inorg. Chem., 1962,
         7, 66 (132)

62BK   T. A. Belyavskaya and I. F. Kolosova, Vestnik Moskov. Univ., Ser. II, 1962, No. 5,
         55

62BM   C. F. Baes, Jr., and N. J. Meyer, Inorg. Chem., 1962, 1, 780

62BS   D. Banerjea and I. P. Singh, J. Indian Chem. Soc., 1962, 39, 353

62BW   B. Behr and H. Wendt, Z. Elektrochem., 1962, 66, 223

62CI   J. J. Christensen and R. M. Izatt, J. Phys. Chem., 1962, 66, 1030

62CM   J. R. Cooke and M. J. Minski, J. Appl. Chem., 1962, 12, 123

63CMa   R. L. Causey and R. M. Mazo, Anal. Chem., 1962, 34, 1630

62CO   B. Carell and A. Olin, Acta Chem. Scand., 1962, 16, 2350

62COa   B. Carell and A. Olin, Acta Chem. Scand., 1962, 16, 4

62DC   G. D'Amore, G. Calabro, and P. Curro, Atti Soc. Peloritana, Sci. Fis. Mat. Nat.,
         1962, 8, 265

62DG   M. K. Dorfman and J. W. Gryder, Inorg. Chem., 1962, 1, 799

62DGa   K. P. Dubey and S. Ghosh, Z. Anorg. Allg. Chem., 1962, 319, 204

62DK   J. F. Duncan and D. L. Kepert, J. Chem. Soc., 1962, 205

62DL   D. Dyrssen and P. Lumme, Acta Chem. Scand., 1962, 16, 1785

63DS   G. G. Davis and W. M. Smith, Canad. J. Chem., 1962, 40, 1836

62EE   H. R. Ellison, J. O. Edwards, and E. A. Healy, J. Amer. Chem. Soc., 1962, 84, 1820

62EK   N. P. Ermolaev and N. N. Krot, Soviet Radiochem., 1962, 4, 600 (678)

62ET   A. M. Egorov and Z. P. Titova, Russ. J. Inorg. Chem., 1962, 7, 141 (275)

62F   M. H. Frere, Proc. Soil Sci. Soc. Amer., 1962, 26, 48

62FE   T. D. Farr and K. L. Elmore, J. Phys. Chem., 1962, 66, 315

62FP   C. R. Frink and M. Peech, Soil Sci., 1962, 20, 346; Inorg. Chem., 1963, 2, 473

62FR   R. O. Fournier and J. J. Rowe, Amer. Mineralogist, 1962, 47, 897

62FSD   Ya. D. Fridman, R. I. Sorochan, and N. V. Dolgashova, Russ. J. Inorg. Chem., 1962,
        7, 1100 (2127)

62FSV   Ya. D. Fridman, Dzh. S. Sarbaev, and R. A. Veresova, Russ. J. Inorg. Chem., 1962,
        7, 15b (305)

62FT   N. Fogel, J. M. J. Tai, and J. Yarborough, J. Amer. Chem. Soc., 1962, 84, 1145

62G   I. Grenthe, Acta Chem. Scand., 1962, 16, 2300

62GA   A. M. Golub and O. E. Andreichenko, Russ. J. Inorg. Chem., 1962, 7, 279 (549)

62GG   H. Gnepf, O. Gubeli, and G. Schwarzenbach, Helv. Chim. Acta., 1962, 45, 1171

62GM   A. D. Gelman, A. I. Moskvin, and V. P. Zaitseva, Radiokhim., 1962, 4, 154

62GT   R. M. Garrels and M. E. Thompson, Amer. J. Sci., 1962, 260, 57

62H   F. Halla, Monat. Chem., 1962, 93, 948

62Ha   G. P. Haight, Jr., Proc. 7th Int. Conf. Coord. Chem., Stockholm-Uppsala, 1962, 318

62Hb   R. E. Hester, Thesis, Cornell Univ.; 1962; Diss. Abs., 1962, 23, 1510

62HZ   G. P. Haight, Jr., J. Zoltewicz, and W. Evans, Acta Chem. Scand., 1962, 16, 311

62I   N. Ingri, Acta Chem. Scand., 1962, 16, 439

62IC   R. M. Izatt, J. J. Christensen, R. T. Pack, and R. Bench, Inorg. Chem., 1962, 1,
        828

62IN   B. N. Ivanov-Emin, L. A. Niselson, and L. E. Larionova, Russ. J. Inorg. Chem.,
        1962, 7, 266 (522)

62JE   J. Jordan and G. J. Ewing, Inorg. Chem., 1962, 1, 587

62JP   P. K. Jena and B. Prasad, J. Indian Chem. Soc., 1962, 39, 33

62KB   P. N. Kovalenko and K. N. Bagdasarov, Russ. J. Inorg. Chem., 1962, 7, 380 (739),
        913 (1765), 915 (1769)

62KC   F. Ya. Kulba and N. N. Chernova, Russ. J. Inorg.. Chem., 1962, 7, 824 (1595)

62KG   P. N. Kovalenko and O. I. Geiderovich, Izv. Vyssh. Ucheb. Zaved., Khim, 1962, 5, 58

62L   V. K. LaMer, J. Phys. Chem., 1962, 66, 973

62La   W. T. Lindsay, Jr., J. Phys. Chem., 1962, 66, 1341

62LG   V. A. Latysheva and L. P. Goryanina, Russ. J. Inorg. Chem., 1962, 7, 377 (732)

62LL   P. Lumme and H. Lumme, Suomen Kem., 1962, B35, 120

62LN   Yu. A. Lukyanychev and N. S. Nikolaev, Soviet J. Atom. Ener., 1962, 13, 779 (179)

62LY   I. A. Lebedev and G. N. Yakovlev, Radiokhim., 1962, 4, 304

62LYa   M. W. Lister and Y. Yoshino, Canad. J. Chem., 1962, 40, 1490

62M   K. Mizumachi, Nippon Kagaku Zasshi, 1962, 83, 61, 67

62MF   G. W. Morey, R. O. Fournier, and J. J. Rowe, Geochim. Cosmochim. Acta, 1962, 26, 1029

62ML   A. O. McDougall and F. A. Long, J. Phys. Chem., 1962, 66, 429

62MM   P. G. Manning and C. B. Monk, Trans. Faraday Soc., 1962, 58, 938

62MR   I. N. Marov and D. I. Ryalbchikov, Russ. J. Inorg. Chem., 1962, 7, 533 (1036)

62MS   D. F. C. Morris and E. L. Short, Electrochim. Acta, 1962, 7, 385

62MSa   D. F. C. Morris and E. L. Short, J. Chem. Soc., 1962, 2672

62MZ   A. I. Moskvin and V. P. Zaitseva, Radiokhim., 1962, 4, 73

62N   B. I. Nabivanets, Russ. J. Inorg. Chem., 1962, 7, 212 (417)

62NA   R. K. Nanda and S. Aditya, Z. Phys. Chem. (Frankfurt), 1962, 35, 139

62NF   V. A. Nazarenko, G. V. Flyantikova, and N. V. Lebedeva, Ukr. Khim. Zh., 1962, 28, 266; AEC-TR-6192, No. 2, p. 309

62NL   N. S. Nikolaev and Yu. A. Lukyanychev, Soviet J. Atom. Ener., 1962, 12, 356 (334)

62NM   M. S. Novakovskii and M. G. Mushkina, Russ. J. Inorg. Chem., 1962, 1, 549 (1068)

62NP   G. P. Nikitina and M. F. Pushlenkov, Radiokhim. 1962, 4, 137

62P   A. D. Paul, J. Phys. Chem., 1962, 66, 1248

62Pa   D. D. Perrin, J. Chem. Soc., 1962, 2197

62Pb   D. D. Perrin, J. Chem. Soc., 1962, 4500

62Pc   H. A. Pohl, J. Chem. Eng. Data, 1962, 7, 295

62PA   V. M. Peshkova and P. Ang, Russ. J. Inorg. Chem., 1962, 7, 1091 (2110)

62PB   N. P. Prokhorova and N. E. Brezhneva, Russ. J. Inorg. Chem., 1962, 7, 953 (1846)

62PF   R. L. Pecsok and A. N. Fisher, Inorg. Chem., 1962, 1, 155

62PM   D. F. Peppard, G. W. Mason, and I. Hucker, J. Inorg. Nucl. Chem., 1962, 24, 881

62PN   V. I. Paramonova and N. M. Nikolaevna, Radiokhim. 1962, 4, 84

62PNN   V. I. Paramonova, B. P. Nikolskii, and N. M. Nikolaeva, Russ. J. Inorg. Chem.,
        1962, 7, 528 (1028)

62PO   L. Pajdowski and A. Olin, Acta Chem. Scand., 1962, 16, 983

62PPC   J. A. Perez-Bustamante, J. B. Polonio, and R. F. Cellini, An. Soc., Esp. Fis.
        Quim., 1962, B48, 677

62PPL   B. I. Peshchevitskii, B. V. Ptitsyn, and N. M. Leskova, Izv. Sibir. Otdel. Akad.
        Nauk. SSSR, 1962, No. 11, 143

62RB   R. E. Reeves and P. Bragg, J. Amer. Chem. Soc., 1962, 84, 2491

62RD   H. M. Rootare, V. R. Deitz, and F. G. Carpenter, J. Colloid Sci., 1962, 17, 179

62RE   D. I. Ryabchikov, A. N. Ermakov, V. K. Belyaeva, I. N. Marov, and K. M. Yao, Russ.
        J. Inorg. Chem., 1962, 7, 34 (69)

62RJ   R. M. Rush, J. S. Johnson, and K. A. Kraus, Inorg. Chem., 1962, 1, 378

62RK   A. Roppongi and T. Kato, Bull. Chem. Soc. Japan, 1962, 35, 1086

62RKa   A. Roppongi and T. Kato, Bull. Chem. Soc. Japan, 1962, 35, 1092

62S   P. Sakellaridis, Compt. Rend. Acad. Sci. Paris, 1962, 255, 127

62Sa   Y. Sasaki, Acta Chem. Scand., 1962, 16, 719

62Sb   D. N. Sokolv, Trudy Po Khim. Khim. Tekh., 1962, No. 1, 55; Chem. Abs., 1963, 58,
        8453a

62Sc   K. Schlyter, Trans. Roy. Inst. Tech. Stockholm, 1962, No. 195

62Sd   K. Schlyter, Trans. Roy. Inst. Tech. Stockholm, 1962, No. 196

62Se   G. F. Smith, Trans. Faraday Soc., 1962, 58, 350

62SA   P. Schindler, H. Althaus, A. Schurch, and W. Feitkuecht, Chimia (Switz.), 1962,
        16, 42

62SD   A. B. Scott, R. G. Dartau, and S. Sapsoonthorn, Inorg. Chem., 1962, 1, 313

62SE   S. I. Syshlyaev and N. P. Edeleva, Izv. Vyssh. Ucheb. Zaved., Khim., 1962, 5,
        871

62SB   G. Schwarzenbach, G. Geier, and J. Littler, Helv. Chim. Acta, 1962, 45, 2601

62SH   E. R. Segnit, H. D. Holland, and C. J. Biscardi, Geochim. Cosmochim. Acta, 1962,
        26, 1301

62SK   Z. A. Sheka and E. E. Kriss, Radiokhim. 1962, 4, 720

62SLK   N. M. Selivanova, Z. L. Leshchinskaya, and T. V. Klushina, Russ. J. Phys. Chem.,
        1962, 36, 719 (1349)

62SLW   P. E. Sturrock, E. D. Loughran, and J. I. Watters, Inorg. Chem., 1962, 1, 457

62ST   K. W. Sykes and B. L. Taylor, Proc. 7th Int. Conf. Coord. Chem., Stockholm-Uppsala,
        1962, p. 31

62TE   N. Tanaka, K. Ebata and T. Murayama, Bull. Chem. Soc. Japan, 1962, 35, 124

62TR   Ya. I. Turyan and V. F. Romanov, Russ. J. Inorg. Chem., 1962, 7, 558 (1087)

62TS   J. Terpilowski and R. Staroscik, Chemia Analityczna, 1962, 7, 629

62TZ   S. Tribalat and C. Zeller, Bull. Soc. Chim. France, 1962, 2041

62V   V. P. Vasilev, Russ. J. Inorg. Chem., 1962, 7, 283 (555)

62Va   V. P. Vasilev, Russ. J. Inorg. Chem., 1962, 7, 924 (1788)

62VF   L. P. Varga and H. Freund, J. Phys. Chem., 1962, 66, 21

62W   T. Williams, J. Inorg. Nucl. Chem., 1962, 24, 1215

62WG   M. J. M. Woods, P. K. Gallagher, and E. L. King, Inorg. Chem., 1962, 1, 55

62WS   L. J. Wittenberg and R. H. Steinmeyer, J. Inorg. Nucl. Chem., 1962, 24, 1015

62YI   T. F. Young and D. E. Irish, Ann. Rev. Phys. Chem., 1962, 13, 448

63A   N. V. Akselrud, Russ. Chem. Rev., 1963, 32, 353 (800)

63AG   S. Ahrland, I. Grenthe, L. Johansson, and B. Noren, Acta Chem. Scand., 1963, 17,
        1567

63AK   S. Ahrland, D. Karipides, and B. Noren, Acta Chem. Scand., 1963, 17, 411

63AM   K. A. Allen and W. J. McDowell, J. Phys. Chem., 1963, 67, 1138

63AS   A. M. Azzam and I. A. W. Shimi, Z. Anorg. Allg. Chem., 1963, 321, 284

63BA   J. A. Bolzan and A. Arvia, Electrochim. Acta, 1963, 8, 375

63BF   H. Bilinski, H. Furedi, M. Branica, and B. Tezak, Croat. Chim. Acta, 1963, 35, 19

63BJ   J. A. Bolzan, E. A. Jauregui, and A. J. Arvia, Electrochem. Acta, 1963, 8, 841

63BL   K. Bachmann and K. H. Lieser, Ber. Bunsengesel. Phys. Chem., 1963, 67, 802

63BLN   A. K. Babko, V. V. Lukachina, and B. I. Nabivanets, Russ. J. Inorg. Chem., 1963,
        8, 957 (1939)

63BP   R. L. Benoit and J. Place, Canad. J. Chem., 1963, 41, 1170

63BPa   J. R. Buckholz and R. E. Powell, J. Amer. Chem. Soc., 1962, 85, 509

63BPA   J. A. Bolzan, J. J. Podesta, and A. J. Arvia, An. Asoc. Quim. Argentina, 1962, 51,
        43

63BS   M. Bjorkman and L. G. Sillen, Trans. Roy. Inst. Technol. Stockholm, 1963, No. 199

63BW   N. S. Bayliss and D. W. Watts, Aust. J. Chem., 1963, 16, 927

63C   L. Ciavatta, Arkiv Kemi, 1963, 21, 129

63Ca   P. L. Cloke, Geochim. Cosmochim. Acta, 1963, 27, 1264 1299

63CI   J. J. Christensen, R. M. Izatt, J. D. Hale, R. T. Pack and G. D. Watt, Inorg.
         Chem., 1963, 2, 337

63CK   M. L. Chepelevetskii and K. F. Kharitonovich, J. Anal. Chem. USSR, 1963, 18,
         314 (357)

63CU   G. R. Choppin and P. J. Unrein, J. Inorg. Nucl. Chem., 1963, 25, 387

63DD   J. Douphin, S. Dauphin, D. Chatonier, and G. Andraud, Bull. Soc. Chim. France,
         1963, 2751; J. Dauphin, S. Dauphin, D. Chatonier, and M. T. Vialatte, ibid.,
         2754

63DH   H. S. Dunsmore, S. Hietanen, and L. G. Sillen, Acta Chem. Scand., 1963, 17, 2644

63DK   H. S. Dunsmore, T. R. Kelly, and G. H. Nancollas, Trans. Faraday Soc., 1963
         59, 2606

63DL   A. J. Dedman, T. J. Lewis, and D. H. Richards, J. Chem. Soc., 1963, 2456

63DS   H. S. Dunsmore and L. G. Sillen, Acta Chem. Scand., 1963, 17, 2657

63DW   L. A. D'Orazio and R. H. Wood, J. Phys. Chem., 1963, 67, 1435

63E   A. J. Ellis, Amer. J. Sci., 1963, 261, 259

63Ea   A. J. Ellis, J. Chem. Soc., 1963, 4300

63EK   J. H. Espenson and E. L. King, J. Amer. Chem. Soc., 1963, 85, 3328

63EM   I. Eliezer and Y. Marcus, J. Inorg. Nucl. Chem., 1963, 25, 1465

63EMK   T. Erdey-Gruz, L. Majthenyi, and E. Kugler, Acta Chim. Acad. Sci. Hung., 1963, 67,
         393

63F   M. L. Freedman, J. Inorg. Nucl. Chem., 1963, 25, 575

63FD   Ya. D. Fridman, R. K. Drachevskoya, and V. A. Shestkova, in Redkozemelnye
         Elementy, Izd. Nauka, Moskva, 1963, p. 166

63FL   P. Flood, T. J. Lewis, and D. H. Richards, J. Chem. Soc., 1963, 2446

63FP   V. Frei and J. Podlahova, Chem. Z., 1963, 87, 47

63FS   W. Feitknecht and P. Schindler, Pure Appl. Chem., 1963, 6, 130

63FU   V. Frei and A. Ustyanovichova, Russ. J. Phys. Chem., 1963, 37, 612 (1153)

63G   I. Greenwald, J. Phys. Chem., 1963, 67, 2853

63GK   A. M. Golub and R. A. Kostrova, Ukr. Khim. Zh., 1963, 29, 128

63GKa   G. Gattow and B. Krebs, Z. Anorg. Allg. Chem., 1963, 323, 13

63GKG   A. A. Grinberg, N. V. Kiseleva, and M. I. Gelfman, Doklady Chem., 1963, 153, 1025
         (1327)

63GL   P. Gerding, I. Leden and S. Sunner, Acta Chem. Scand., 1963, 17, 2190

63GR   J. F. Goodman and P. Robson, J. Chem. Soc., 1963, 2871

63GS   H. Galal-Gorchev and W. Stumm, J. Inorg. Nucl. Chem., 1963, 25, 567

63H   J. D. Hem, J. Chem. Eng. Data. 1963, 8, 99

63Ha   P. B. Hostetler, J. Phys. Chem., 1963, 67, 720

63Hb   P. B. Hostetler, Amer. J. Sci., 1963, 261, 238

63Hc   E. Hogfeldt, Acta Chem. Scand., 1963, 17, 785

63Hd   S. D. Hamann, J. Phys. Chem., 1963, 67, 2233

63HC   R. N. Heistand and A. Clearfield, J. Amer. Chem. Soc., 1963, 85, 2566

63HI   L. D. Hansen, R. M. Izatt, and J. J. Christensen, Inorg. Chem., 1963, 2, 1243

63HIa   J. D. Hale, R. M. Izatt and J. J. Christensen, J. Phys. Chem., 1963, 67, 2605

63HR   S. Hietanen, B. R. L. Row, and L. G. Sillen, Acta Chem. Scand., 1963, 17, 2735

63HS   M. N. Hughes and G. Stedman, J. Chem. Soc., 1963, 1239

63I   N. Ingri, Acta Chem. Scand., 1963, 17, 573, 581

63Ia   N. Ingri, Acta Chem. Scand., 1963, 17, 597

63JW   A. Johansson and E. Wanninen, Talanta, 1963, 10, 769

63K   E. I. Kolesnikova, Russ. J. Inorg. Chem., 1963, 8, 641 (1239)

63Ka   S. K. Kundra, Indian J. Chem., 1963, 1, 362

63Kb   N. P. Komar, Uch. Zap. Kharkov Univ., 1963, 133, 189

63KB   I. F. Kolosova and T. A. Belyauskaya, Vestnik Moskov. Univ., Khim., 1963, 18,
         No. 1, 52

63KBa   P. N. Kovalenko and K. N. Bagdasarov, Izv. Vyssh. Ucheb. Zaved., Khim., 1963, 6,
         546

63KM   F. Ya. Kulba, V. E. Mironov, V. A. Fedorov, and V. A. Baevskii, Russ. J. Inorg.
         Chem., 1963, 8, 1012 (1945)

63KS   H. Koch and H. Schmidt, Z. Naturforsch., 1963, 18b, 936

63KV   D. Konrad and A. A. Vlcek, Coll. Czech. Chem. Comm., 1963, 28, 595

63L   D. Lewis, Acta Chem. Scand., 1963, 17, 1891

63LC   A. Liberti, V. Chiantella, and F. Corigliano, J. Inorg. Nucl. Chem., 1963, 25, 415

63LK   V. I. Levin, G. V. Korpusov, N. M. Manko, E. N. Patrusheva, N. P. Prokhorova,
         and G. F. Platnov, Soviet Atom. Ener. 1963, 15, 828 (138)

63LL   P. Lumme and H. Lumme, Suomen Kem., 1963, B36, 176

63LLa   P. Lumme and H. Lumme, Suomen Kem., 1963, B36, 192

63LM   S. H. Laurie and C. B. Monk, J. Chem. Soc., 1963, 3343

63LMa   J. Lefebvre and H. Maria, Compt. Rend. Acad. Sci. Paris, 1963, 256, 3121

63LR   T. J. Lewis, D. H. Richards, and D. A. Salter, J. Chem. Soc., 1963, 2434

63LZ   L. I. Lebedeva and N. A. Zhukova, Russ. J. Inorg. Chem., 1963, 8, 841 (1634)

63M   V. E. Mironov, Russ. J. Inorg. Chem., 1963, 8, 388 (764)

63Ma   M. Cl. Musikas, Radiochim. Acta, 1963, 1, 92

63MF   V. E. Mironov, V. A. Fedorov, and V. A. Nazarov, Russ. J. Inorg. Chem., 1963, 8, 1102 (2109)

63MG   A. I. Moskvin, I. Geletseanu, and A. V. Lapitskii, Doklady Chem., 1963, 149, 264 (611)

63MH   S. S. Mesaric and D. N. Hume, Inorg. Chem., 1963, 2, 788

63MHa   S. S. Mesaric and D. N. Hume, Inorg. Chem., 1963, 2, 1063

63MK   V. E. Mironov, F. Ya. Kulba, and V. A. Fedorov, Russ. J. Inorg. Chem. 1963, 8, 601 (1161)

63MKa   V. E. Mironov, F. Ya. Kulba, V. A. Fedorov, and T. F. Nikitenko, Russ. J. Inorg. Chem., 1963, 8, 964 (1852)

63MKb   V. E. Mironov, F. Ya. Kulba, V. A. Fedorov, and O. B. Tikhomirov, Russ. J. Inorg. Chem., 1963, 8, 1322 (2524)

63MKc   V. E. Mironov, F. Ya. Kulba, V. A. Fedorov, and O. B. Tikhomirov, Russ. J. Inorg. Chem., 1963, 8, 1328 (2536)

63MKN   V. E. Mironov, F. Ya. Kulba, and V. A. Nazarov, Russ. J. Inorg. Chem., 1963, 8, 470 (916)

63MKT   V. E. Mironov, F. Ya. Kulba, and O. I. Trifonov, Russ. J. Inorg. Chem., 1963, 8, 1104 (2113)

63MM   Y. Marcus and D. Maydan, J. Phys. Chem., 1963, 67, 979

63MMa   D. Maydan and Y. Marcus, J. Phys. Chem., 1963, 67, 987

63ND   D. A. Netzel and H. A. Droll, Inorg. Chem., 1963, 2, 412

63NP   G. H. Nancollas and N. Purdie, Trans. Faraday Soc., 1963, 59, 735

63NPa   J. Nowikow and G. Pfrepper, Z. Naturforsch., 1963, 18b, 993

63P   L. Pajdowski, Rocz. Chem., 1963, 37, 1351, 1363

63Pa   K. N. Polydoropoulos, Diss., Univ. Athens, 1963

63Pb   A. J. Panson, J. Phys. Chem., 1963, 67, 2177

63PA   V. M. Peshkova and P. An. Vestnik Moskov. Univ., Khim., 1963, 18, No. 1, 40

63PB   R. A. Penneman, R. Bain, G. Gilbert, L. H. Jones, R. S. Nyholm, and G. K. N. Reddy, J. Chem. Soc., 1963, 2266

63PE   L. Pinto, K. Egger, and P. Schindler, Helv. Chim. Acta, 1963, 46, 425

63PF   D. Peschanski and J. M. Fruchart, Compt. Rend. Acad. Sci. Paris, 1963, 257, 1853

63PG   R. C. Phillips, P. George, and F. J. Rutman, Biochem. 1963, 2, 501

63PL   G. Popa, C. Luca, and E. Iosif, Z. Phys. Chem. (Leipzig) 1963, 222, 49

63PS   B. G. Pozharskii, T. N. Sterlingova, and A. E. Petrova, Russ. J. Inorg. Chem.,
        1963, 8, 831 (1594)

63R   B. N. Ryzhenko, Geochem., 1963, 151 (137)

63RB   R. W. Ramette and R. F. Broman, J. Phys. Chem., 1963, 67, 942

63RC   E. F. C. H. Rohwer and J. J. Cruywagen, J. S. African Chem. Inst., 1963, 16, 26

63RD   R. W. Ramette and E. A. Dratz, J. Phys. Chem., 1963, 67, 940

63RF   W. L. Reynolds and S. Fukushima, Inorg. Chem., 1963, 2, 176

63RH   C. L. Rulfs, R. F. Hirsch, and R. A. Pacer, Nature, 1963, 199, 66

63RJ   R. M. Rush and J. S. Johnson, J. Phys. Chem., 1963, 67, 821

63RS   T. P. Radhakrishnan and A. K. Sundaram, J. Electroanal. Chem., 1963, 5, 124

63RSa   R. W. Ramette and J. B. Spencer, J. Phys. Chem., 1963, 67, 944

63S   I. Szilard, Acta Chem. Scand., 1963, 17, 2674

63Sa   M. B. Shchigol, Russ. J. Inorg. Chem., 1963, 8, 707 (1361)

63Sb   M. Schellenberg, Diss., ETH, Zurich, 1963

63Sc   E. Schumann, Diss., Tech. Hoch., Karsruhe, 1963

63Sd   P. Schindler, Chimia (Switz.), 1963, 17, 313

63SA   P. Schindler, H. Althaus, and W. Feitknecht, Gazz. Chim. Ital., 1963, 93, 168

63SB   G. Schorsch and J. Bye, Compt. Rend. Acad. Sci. Paris, 1963, 257, 2833

63SBa   U. V. Seshaiah and S. N. Banerji, Proc. Nat. Acad. Sci. India, 1963, 33, 61

63SD   J. Shankar and B. V. DeSouza, Aust. J. Chem., 1963, 16, 1119

63SG   G. Schwarzenbach and G. Geier, Helv. Chim. Acta. 1963, 46, 906

63SI   I. E. Starik and L. I. Ilmenkova, Soviet Radiochem., 1963, 5, 209 (236)

63SK   V. P. Shvedov and K. Kotegov, Soviet Radiochem., 1963, 5, 342 (374)

63SL   N. M. Selivanova and Z. L. Leshchinskaya, Russ. J. Inorg. Chem., 1963, 8, 286 (563)

63SM   P. Schindler, W. Michaelis, and W. Feitknecht, Helv. Chim. Acta, 1963, 46, 444

63SR   J. W. Stout and R. A. Robie, J. Phys. Chem., 1963, 67, 2248

63SRM   P. E. Sturrock, J. D. Ray, J. McDowell, and H. R. Hunt, Jr., Inorg. Chem., 1963,
        2, 649

V. BIBLIOGRAPHY

63SS   O. E. Schupp, III, P. E. Sturrock, and J. I. Watters, Inorg. Chem., 1963, 2, 106

63SW   H. Strehlow and H. Wendt, Inorg. Chem., 1963, 2, 6

63SWa   G. Schwarzenbach and M. Widmer, Helv. Chim. Acta, 1963, 46, 2613

63SZ   Yu. I. Sannikov, V. L. Zolotavin, and I. Ya. Bezrukov, Russ. J. Inorg. Chem., 1963, 8, 474 (923)

63TC   S. Tribalat and J. M. Caldero, J. Electroanal. Chem., 1963, 5, 176; Compt. Rend. Acad. Sci. Paris, 1962, 255, 925

63TF   A. W. Taylor, A. W. Frazier, E. L. Gurney and J. P. Smith, Trans. Faraday Soc., 1963, 59, 1585

63TN   I. N. Tananaev, N. S. Nikolaev, Yu. A. Lukyanychev, and I. F. Alenchikova, Khimiya Ftoristykh Sodinenii Aktinidov, Izd. Akad. Nauk SSSR, 1963

63TS   N. Tanaka, Y. Saito, and H. Ogino, Bull. Chem. Soc. Japan, 1963, 36, 794

63TU   R. Tsuchiya and A. Umayahara, Bull. Chem. Soc. Japan; 1963, 36, 554

63TV   I. V. Tananaev and V. P. Vasileva, Russ. J. Inorg. Chem., 1963, 8, 555 (1070)

63UK   A. I. Ulyanov and T. I. Kazakova, Bull. Acad. Sci. USSR, 1963, 355 (393)

63VH   L. P. Varga and D. N. Hume, Inorg. Chem., 1963, 2, 201

63VM   V. P. Vasilev and P. S. Mukhina, Russ. J. Inorg. Chem., 1963, 8, 986 (1895)

63VR   V. M. Vdovenko, G. A. Ramanov, and V. A. Shcherbakov, Soviet Radiochem., 1963, 5, 538 (581)

63VRa   V. M. Vdovenko, G. A. Romanov, and V. A. Shcherbakov, Soviet Radiochem., 1963, 5, (664)

63VS   C. E. Vanderzee and J. A. Swanson, J. Phys. Chem., 1963, 67, 2608

63VV   V. P. Vasilev, V. N. Vasileva, N. A. Klindukhova, and A. N. Parfenova, Izv. Vyssh. Ucheb. Zaved., Khim., 1963, 6, 339

63W   R. L. S. Willix, Trans. Faraday Soc., 1963, 59, 1315

63WS   J. I. Watters, P. E. Sturrock, and R. E. Simonaitis, Inorg. Chem., 1963, 2, 765

63YR   K. B. Yatsimirskii and L. P. Raizman, Russ. J. Inorg. Chem., 1963, 8, 574 (1107)

64A   J. Aveston, Inorg. Chem., 1964, 3, 98

64Aa   F. Achenza, Ann. Chim. (Italy), 1964, 54, 240

64AA   J. Aveston, E. W. Anacker, and J. S. Johnson, Inorg. Chem., 1964, 3, 735

64AF   G. P. Arkhipova, I. E. Flis, and K. P. Mishchenko, J. Appl. Chem. USSR, 1964, 37, 2275 (2306)

64AJ   S. Ahrland and L. Johansson, Acta Chem. Scand., 1964, 18, 2125

64AP   V. I. Altynov and B. V. Ptitsyn, Russ. J. Inorg. Chem., 1964, 9, 1301 (2407)

64B   J. C. Barnes, J. Chem. Soc., 1964, 3880·

64Ba   K. Burger, Acta Chim. Acad. Sci. Hung., 1964, 40, 261

64Bb   G. Biedermann, Svensk Kem. Tidskr., 1964, 76, 362

64BC   G. Biedermann and L. Ciavatta, Arkiv Kemi. 1964, 22, 253

64BI   F. Brito, N. Ingri, and L. G. Sillen, Acta Chem. Scand., 1964, 18, 1557

64BK   R. J. Baltisberger and E. L. King, J. Amer. Chem. Soc., 1964, 86, 795

64BL   F. Becker and H. M. Luschow, Proc. 8th Int. Conf. Coord. Chem., Vienna, 1964, 334

64BM   J. C. Barnes and C. B. Monk, Trans. Faraday Soc., 1964, 60, 578

64BMa   E. Ya. Benyash and T. G. Maslakova, Russ. J. Inorg. Chem., 1964, 9, 1472 (2731)

64BN   G. Biedermann and L. Newman, Arkiv Kemi, 1964, 22, 303

64BP   B. M. L. Bausal, S. K. Patil, and H. D. Sharma, J. Inorg. Nucl. Chem., 1964, 26, 993

64BR   L. I. Budarin, T. A. Rumyantseva, and T. T. Sherinva, Izv. Vyssh. Ucheb. Zaved., Khim., 1964, 7, 715

64BS   A. A. Birynkov and V. I. Shlenskaya, Russ. J. Inorg. Chem., 1964, 9, 450 (813)

64BSa   M. Bartusek and L. Sommer, Z. Phys. Chem. (Leipzig), 1964, 226, 309

64BU   E. A. Buketov, M. Z. Ugorets, and A. S. Pashinkin, Russ. J. Inorg. Chem., 1964, 9, 292 (526)

64C   D. M. Czakis-Sulikowska, Rocz. Chem., 1964, 38, 533

64Ca   E. H. P. Cordfunke, J. Phys. Chem., 1964, 68, 3353

64CI   J. J. Christensen, R. M. Izatt, L. D. Hansen, and J. D. Dale, Inorg. Chem., 1964, 3, 130

63CL   C. Chen-ping and L. Lien-sen, Scientia Sinica, 1964, 13, 1334

64DB   W. Davis, Jr., and H. J. de Bruin, J. Inorg. Nucl. Chem., 1964, 26, 1069

64DG   F. Z. Dzhabarov and S. V. Gorbachev, Russ. J. Inorg. Chem., 1964, 9, 1297 (2399)

64DR   D. Deveze and R. Rumpf, Compt. Rend. Acad. Sci. Paris, Sec. C, 1964, 258, 6135

64DRC   V. R. Deitz, H. M. Rootare, and F. G. Carpenter, J. Colloid Sci., 1964, 19, 87

64DS   D. Dyrssen and T. Sekine, J. Inorg. Nucl. Chem., 1964, 26, 981

64DSa   H. Diebler and N. Sutin, J. Phys. Chem., 1964, 68, 174

64EH   A. J. Eve and D. N. Hume, Inorg. Chem., 1964, 3, 276

64EM   H. Ellison and A. E. Martell, J. Inorg. Nucl. Chem., 1964, 26, 1555

64F   A. N. Fletcher, J. Inorg. Nucl. Chem., 1964, 26, 955

64FB   R. Fischer and J. Bye, Bull. Soc. Chim. France, 1964, 2920

64FC   J. P. Fackler, Jr., and I. D. Chawla, Inorg. Chem., 1964, 3, 1130

64FD   Ya. D. Fridman and N. V. Dolgashova, Russ. J. Inorg. Chem., 1964, 9, 345 (623)

64FF   T. L. Fabry and R. M. Fuoss, J. Phys. Chem., 1964, 68, 974

64FM   I. E. Flis, K. P. Mishchenko, and G. I. Pusenok, Izv. Vyssh. Ucheb. Zaved., Khim.,
        1964, 7, 764

64FP   V. Frei, J. Podlahova and J. Podlaha, Coll. Czech. Chem. Comm., 1964, 29, 2587

64FR   H. N. Farrer and F. J. C. Rossotti, J. Inorg. Nucl. Chem., 1964, 26, 1959

64FW   J. S. Fritz and H. Waki, J. Inorg. Nucl. Chem., 1964, 26, 865

64G   E. Sh. Ganelina, J. Appl. Chem. USSR, 1964, 37, 1350 (1358)

64GA   A. M. Golub and R. Akmyradov, Ukr. Khim. Zh., 1964, 30, 1016

64GH   F. H. Guzzetta and W. B. Hadley, Inorg. Chem., 1964, 3, 259

64GK   G. Gordon and D. M. H. Kern, Inorg. Chem., 1964, 3, 1055

64GL   R. S. Guzairov, V. A. Leitsin, and S. D. Grekov, Russ. J. Inorg. Chem., 1964, 9,
        10 (20)

64GM   R. Geyer and H. Mucke, Z. Anal. Chem., 1964, 200, 210

64GS   I. Gainar and K. W. Sykes, J. Chem. Soc., 1964, 4452

64H   R. Hugel, Bull. Soc. Chim. France, 1964, 1462

64HK   J. Haas, N. Konopik, F. Mark, and A. Neckel, Monat. Chem., 1964, 95, 1141, 1166,
        1173

64HM   G. G. Hammes and M. L. Morrell, J. Amer. Chem. Soc., 1964, 86, 1497

64HMF   R A. Horne, B. R. Myers, and G. R. Frysinger, Inorg. Chem., 1964, 3, 452

64HP   R. E. Hester and R. A. Plane, J. Chem. Phys., 1964, 40, 411

64HPS   S. D. Hamann, P. J. Pearce, and W. Strauss, J. Phys. Chem., 1964, 68, 375

64HR   G. P. Haight, Jr., D. C. Richardson, and N. H. Coburn, Inorg. Chem., 1964, 3,
        1777

64HS   S. Hietanen and L. G. Sillen, Acta Chem. Scand., 1964, 18, 1015

64HSa   S. Hietanen and L. G. Sillen, Acta Chem. Scand., 1964, 18, 1018

64ID   B. Z. Iofa and G. M. Dakar, Soviet Radiochem., 1964, 6, 396 (411)

64K   V. N. Kumok, Russ. J. Inorg. Chem., 1964, 9, 198 (362)

64KB   P. N. Kovalenko and K. N. Bagdasarov, Russ. J. Inorg. Chem., 1964, 9, 296 (534);
        J. Appl. Chem. USSR, 1964, 37, 739 (735)

64KS   V. N. Kumok and V. V. Serebrennikov, Russ. J. Inorg. Chem., 1964, 9, 1160 (2148)

64KY   F. Ya. Kulba, Yu. B. Yakovlev, and V. E. Mironov, <u>Russ. J. Inorg. Chem.</u>, 1964, <u>9</u>,
            1390 (2573)

64L   K. H. Lieser, <u>J. Inorg. Nucl. Chem.</u>, 1964, <u>26</u>, 1571

64La   V. A. Latysheva, <u>Khimiya Redkikh Elementov</u>, Izd. Leningrad Univ., 1964, p. 133
            (see 64SM)

64LA   S. C. Lahiri and S. Aditya, <u>J. Indian Chem. Soc.</u>, 1964, <u>41</u>, 517

64LD   Ch. Ch. Liang and Yu. M. Tu, <u>Russ. J. Inorg. Chem.</u>, 1964, <u>9</u>, 727 (1333)

64LP   T. S. Laximinarayanan, S. K. Patil, and H. D. Sharma, <u>J. Inorg. Nucl. Chem.</u>, 1964,
            <u>26</u>, 1001

64LR   I. Leden and T. Ryhl, <u>Acta Chem. Scand.</u>, 1964, <u>18</u>, 1196

64LRC   S. Lynn, R. G. Rinker, and W. H. Corcoran, <u>J. Phys. Chem.</u>, 1964, <u>68</u>, 2363

64LW   S. H. Laurie, J. M. Williams, and C. J. Nyman, <u>J. Phys. Chem.</u>, 1964, <u>68</u>, 1311

64MK   V. E. Mironov, F. Ya. Kulba, and V. A. Fedorov, <u>Russ. J. Inorg. Chem.</u>, 1964, <u>9</u>,
            888 (1641)

64MKa   V. E. Mironov, F. Ya. Kulba, A. V. Fokina, V. S. Golubeva, and V. A. Nazarov,
            <u>Russ. J. Inorg. Chem.</u>, 1964, <u>9</u>, 1152 (2133)

64MKb   V. E. Mironov, F. Ya. Kulba, V. A. Fedorov, and A. V. Fedorova, <u>Russ. J. Inorg.
            Chem.</u>, 1964, <u>9</u>, 1155 (2138)

64MKc   J. C. Mason and A. D. Kowalak, <u>Inorg. Chem.</u>, 1964, <u>3</u>, 1248

64MKI   V. E. Mironov, F. Ya. Kulba, and Yu. E. Ivanov, <u>Russ. J. Inorg. Chem.</u>, 1964, <u>9</u>,
            884 (1633)

64MP   A. I. Moskvin and V. F. Peretrukhin, <u>Soviet Radiochem.</u>, 1964, <u>6</u>, 198 (206)

64MR   D. F. C. Morris, G. L. Reed, and K. J. Sutton, <u>J. Inorg. Nucl. Chem.</u>, 1964, <u>26</u>,
            1461

64MW   H. A. C. McKay and J. L. Woodhead, <u>J. Chem. Soc.</u>, 1964, 717

64N   J. Nassler, <u>Coll. Czech. Chem. Comm.</u>, 1964, <u>29</u>, 174

64Na   V. S. K. Nair, <u>J. Inorg. Nucl. Chem.</u>, 1964, <u>26</u>, 1911

64Nb   G. Neumann, <u>Acta Chem. Scand.</u>, 1964, <u>18</u>, 278

64NH   A. W. Naumann and C. J. Hallada, <u>Inorg. Chem.</u>, 1964, <u>3</u>, 70

64NK   B. I. Nabivanets and L. N. Kudritskaya, <u>Ukr. Khim. Zh.</u>, 1964, <u>30</u>, 891

64NKa   B. I. Kabivanets and L. N. Kudritskaya, <u>Ukr. Khim. Zh.</u>, 1964, <u>30</u>, 1007

64NL   B. I. Nabivanets and V. V. Lukachina, <u>Ukr. Khim. Zh.</u>, 1964, <u>30</u>, 1123

64NM   R. I. Novoselov, Z. A. Muzykantova, and B. V. Ptitsyn, <u>Russ. J. Inorg. Chem.</u>,
            1964, <u>9</u>, 442 (799)

64NU   G. Nord (Waind) and J. Ulstrup, <u>Acta Chem. Scand.</u>, 1964, <u>18</u>, 307

64O   H. Ohtaki, <u>Acta Chem. Scand.</u>, 1964, <u>18</u>, 521

64P   D. D. Perrin, <u>J. Chem. Soc.</u>, 1964, 3644

64Pa   R. F. Platford, <u>Canad. J. Chem.</u>, 1964, <u>42</u>, 181

64PB   N. K. Popovicheva, A. A. Biryukov, and V. I. Shlenskaya, <u>Russ. J. Inorg. Chem.</u>, 1964, <u>9</u>, 803 (1482) (See 64SM)

64PH   W. Plumb and G. M. Harris, <u>Inorg. Chem.</u>, 1964, <u>3</u>, 542

64PS   P. I. Protsenko, O. N. Shokina, and N. P. Checkhunova, <u>Russ. J. Phys. Chem.</u>, 1964, <u>38</u>, 1013 (1857)

64RK   I. G. Ryss and N. F. Kulish, <u>Russ. J. Inorg. Chem.</u>, 1964, <u>9</u>, 603 (1103), 752 (1382)

64RP   P. A. Rock and R. E. Powell, <u>Inorg. Chem.</u>, 1964, <u>3</u>, 1593

64RSM   G. L. Reed, K. J. Sutton, and D. F. C. Morris, <u>J. Inorg. Nucl. Chem.</u>, 1964, <u>26</u> 1227

64RSS   T. P. Radhakrishnan, S. C. Saraiya, and A. K. Sundaram, <u>J. Inorg. Nucl. Chem.</u>, 1964, <u>26</u>, 378

64S   A. P. Samodelov, <u>Soviet Radiochem.</u>, 1964, <u>6</u>, 548 (568)

64Sa   G. Schorsch, <u>Bull. Soc. Chim. France</u>, 1964, 1449, 1456

64SA   C. E. Schaffer and P. Andersen in <u>Theory and Structure of Complex Compounds</u> symposium, Wroclaw, 1962, Pergamon, Warszawa, 1964, p. 571

64SAa   I. E. Starik, N. I. Ampelogova, and B. S. Kuznetsov, <u>Soviet Radiochem.</u>, 1964, <u>6</u>, 501 (519) (see 73Ab)

64SAb   I. E. Starik, N. I. Ampelogova, and B. S. Kuznetsov, <u>Soviet Radiochem.</u>, 1964, <u>6</u>, 507 (524) (see 73Ab)

64SAF   P. Schindler, H. Althaus, and W. Feitknecht, <u>Helv. Chim. Acta</u>, 1964, <u>47</u>, 982

64SB   V. I. Shlenskaya and A. A. Bryukov, <u>Vestik Moskov. Univ.</u>, <u>Khim.</u>, 1964, No. 3, 65

64SD   V. I. Spitsyn, R. A. Dyachkova, and V. P. Khlebnikov, <u>Doklady Chem.</u>, 1964, <u>157</u>, 677 (135)

64SH   C. C. Stephenson, H. P. Hopkins, and C. A. Wulff, <u>J. Phys. Chem.</u>, 1964, <u>68</u>, 1427

64SLI   S. A. Shchukarev, O. A. Lobaneva, M. A. Ivanova, and M. A. Kononova, <u>Russ. J. Inorg. Chem.</u>, 1964, <u>9</u>, 1503 (2791)

64SLM   N. M. Selivanova, Z. L. Leshinskaya, A. I. Maier, and E. Yu. Muzalev. <u>Izv. Vyssh. Ucheb. Zaved.</u>, <u>Khim.</u>, 1964, <u>17</u>, 209

64SM   Unpublished values quoted in L. G. Sillen and A. E. Martell, <u>Stability Constants of Metal-Ion Complexes</u>, Special Publication No. 17, The Chemical Soc., London, 1964

64SMa   V. B. Spivakovskii and L. P. Moisa, <u>Russ. J. Inorg. Chem.</u>, 1964, <u>9</u>, 1239 (2287)

64SS   Y. Sasaki and L. G. Sillen, <u>Acta Chem. Scand.</u>, 1964, <u>18</u>, 1014

64SSa   P. S. Shetty and P. R. Subbaraman, Indian J. Chem., 1964, 2, 428

64SSG   P. S. Shetty, P. R. Subbaraman, and J. Gupta, Indian J. Chem., 1964, 2, 8

64SSL   P. Salomaa, L. L. Schaleger, and F. A. Long, J. Amer. Chem. Soc., 1964, 86, 1

64SSW   P. Sellers, S. Sunner, and I. Wadso, Acta Chem. Scand., 1964, 18, 202

64ST   J. Ste-Marie, A. E. Torma, and A. O. Gubeli, Canad. J. Chem., 1964, 42, 662

64T   J. Y. Tong, Inorg. Chem., 1964, 3, 1804

64Ta   G. Thompson, Thesis, Univ. Calif. Berkeley, 1964; UCRL-11410

64TC   I. V. Tananaev and N. N. Chudinova, Russ. J. Inorg. Chem., 1964, 9, 135 (244)

64TCa   S. Tribalat and J. M. Caldero, Comp. Rend. Acad. Sci. Paris, Sec. C, 1964, 258
        2828

64TY   R. S. Tobias and M. Yasuda, J. Phys. Chem., 1964, 68, 1820

64TYa   R. S. Tobias and M. Yasuda, Canad. J. Chem., 1964, 42, 781

64V   V. P. Vasilev, Russ. J. Inorg. Chem., 1964, 9, 354 (647)

64VG   V. P. Vasilev and N. K. Grechina, Russ. J. Inorg. Chem., 1964, 9, 357 (647)

64VH   L. P. Varga and D. N. Hume, Inorg. Chem., 1964, 3, 77

64VHa   A. L. VanGeet and D. N. Hume, Inorg. Chem., 1964, 3, 523

64VM   V. P. Vasilev and P. S. Mikhina, Izv. Vyssh. Ucheb. Zaved., Khim., 1964, 7, 711

64VMa   V. P. Vasilev and P. S. Mikhina, Russ. J. Inorg. Chem., 1964, 9, 620 (1134)

64VR   S. Varva and N. P. Rudenko, Vestnik Moskov. Univ., Khim., 1964, No. 6, 14

64W   E. D. Weed, Thesis, Ohio State Univ., 1964; Diss. Abs., 1964, 25, 795

64Wa   H. Wenger, Diss., Eidgenossische Tech. Hoch., Zurich, 1964

64WD   T. H. Wirth and N. Davidson, J. Amer. Chem. Soc., 1964, 86, 4325

64WE   J. H. Walsh and J. E. Earley, Inorg. Chem., 1964, 3, 343

64WG   M. J. M. Woods, P. K. Gallagher, Z. Z. Hugus, Jr., and E. L. King, Inorg. Chem.,
        1964, 3, 1313

64WS   J. I. Watters and R. A. Simonaitis, Talanta, 1964, 11, 247

64WSa   M. Widmer and G. Schwarzenbach, Helv. Chim. Acta, 1964, 47, 266

64YK   K. B. Yatsimirskii and V. E. Kalinina, Russ. J. Inorg. Chem., 1964, 9, 611 (1117)

64YP   K. B. Yatsimirskii and K. E. Prik, Russ. J. Inorg. Chem., 1964, 9, 995 (1838)

65A   R. Arnek, Arkiv Kemi, 1965, 24, 531

65Aa   G. Anderegg, Helv. Chim. Acta, 1965, 48, 1712

65Ab   J. Aveston, J. Chem. Soc., 1965, 4438

65Ac   J. Aveston, J. Chem. Soc., 1965, 4444

65AB   I. P. Alimarin, T. A. Belyauskaya, and G. D. Brykina, Vestnik Moskov. Univ., Khim.,
         1965, No. 5, 69

65AK   G. Atkinson and S. K. Kor, J. Phys. Chem., 1965, 69, 128, (see also L. S.
         Jackopin and E. Yeagen; G. Atkinson and S. K. Kor, J. Phys. Chem., 1966,
         70, 313, 314)

65AKP I. P. Alimarin, Sh. A. Khamid, and I. V. Puzdrenkova, Russ. J. Inorg. Chem., 1965,
         10, 209 (389)

65BB   R. Barbieri and J. Bjerrum, Acta Chem. Scand., 1965, 19, 469

65BBa N. N. Baranova and V. L. Barsukov, Geokhim., 1965, 1093

65BC   E. Bottari and L. Ciavatta, J. Inorg. Nucl. Chem., 1965, 27, 133

65BCY J. Botts, A. Chashin, and H. L. Young, Biochem., 1965, 4, 1788

65BE   C. J. Battaglia and J. O. Edwards, Inorg. Chem., 1965, 4, 552

65BG   Yu. A. Buslaev and M. P. Gustyakova, Russ. J. Inorg. Chem., 1965, 10, 831 (1524)

65BL   G. J. Buist and J. D. Lewis, Chem. Comm., 1965, 66

65BLa K. A. Burkov and L. S. Lilich, Vest. Leningrad Univ., 1965, No. 10 (Fiz. Khim.,
         No. 1), 103

65BLS K. A. Burkov, L. S. Lilich, and L. G. Sillen, Acta Chem. Scand., 1965, 19, 14

65BM   C. F. Baes, Jr., N. J. Meyer, and C. E. Roberts, Inorg. Chem., 1965, 4, 518

65BS   M. Bartusek and L. Sommer, J. Inorg. Nucl. Chem., 1965, 27, 2397

65BW   R. H. Boyd and C. H. Wang, J. Phys. Chem., 1965, 69, 3906

65BY   G. A. Bogdanov, G. K. Yurchenko, and L. A. Kuzenko, Russ. J. Phys. Chem., 1965,
         39, 1261 (2359)

65C   J. Chojnacka, Rocz. Chem., 1965, 39, 161

65CD   A. K. Covington and J. V. Dobson, J. Inorg. Nucl. Chem., 1965, 27, 1435

65CG   L. Ciavatta and M. Grimaldi, J. Inorg. Nucl. Chem., 1965, 27, 2019

65CI   J. J. Christensen, R. M. Izatt and D. Eatough, Inorg. Chem., 1965, 4, 1278

65CK   G. R. Choppin and J. Ketels, J. Inorg. Nucl. Chem., 1965, 27, 1335

65CO   J. Chojnacki and B. Oleksyn, Rocz. Chem., 1965, 39, 1141

65CS   G. R. Choppin and W. F. Stazik, Inorg. Chem., 1965, 4, 1250

65CW   J. P. Candlin and R. G. Wilkins, J. Amer. Chem. Soc., 1965, 87, 1490

65D   K. P. Dubey, Z. Anorg. Allg. Chem., 1965, 337, 309

65DB   J. R. Durig, O. D. Bonner, and W. H. Breazeale, J. Phys. Chem., 1965, 69, 3886

65DK   R. G. Deshpande, P. K. Khopkar, C. L. Rao, and H. D. Sharma, J. Inorg. Nucl.
       Chem., 1965, 27, 2171

65F   A. Ferse, Z. Phys. Chem. (Leipzig), 1965, 229, 51

65Fa   R. T. M. Fraser, J. Chem. Soc., 1965, 1747

65Fb   W. Feldmann, Z. Anorg. Allg. Chem., 1965, 338, 235

65FA   I. E. Flis, G. P. Arkhipova, and K. P. Mishchenko, J. Appl. Chem. USSR, 1965, 38,
       1466 (1494)

65FK   A. B. Fasman, G. G. Kutyukov, and D. V. Sokolskii, Russ. J. Inorg. Chem., 1965,
       10, 727 (1338)

65FM   J. R. Fryer and D. F. C. Morris, Electrochim. Acta, 1965, 10, 473

65FMT H. N. Farrer, M. M. McGrady, and R. S. Tobias, J. Amer. Chem. Soc., 1965, 87,
       5019

65FP   V. Frei, J. Podlahova, and J. Podlaha, Russ. J. Inorg. Chem., 1965, 10, 921
       (1690)

65G   R. Guillaumont, Bull. Soc. Chim. France, 1965, 135

65Ga   R. Guillaumont, Compt. Rend. Acad. Sci. Paris, Sec. C, 1965, 260, 1416

65GA   R. W. Green and P. W. Alexander, Aust., J. Chem., 1965, 18, 651

65GC   S. A. Greenberg and T. N. Chang, J. Phys. Chem., 1965, 69, 182

65GCa R. M. Garrels and C. L. Christ, Solutions, Minerals and Equilibria, Harper and
       Row, New York, 1965

65GP   E. Sh. Ganelina and T. N. Pozhidaeva, J. Appl. Chem. USSR, 1965, 38, 2168 (2210)

65GS   T. Goto and M. Smutz, J. Inorg. Nucl. Chem., 1965, 27, 663

65GSS H. Gamsjager, H. U. Stuber, and P. Schindler, Helv. Chim. Acta, 1965, 48, 723

65H   R. Hugel, Bull. Soc. Chim. France, 1965, 968

65Ha   R. Hugel, Bull. Soc. Chim. France, 1965, 971, 2017

65HA   J. D. Hefley and E. S. Amis, J. Phys. Chem., 1965, 69, 2082

65HC   L. D. Hansen, J. J. Christensen, and R. M. Izatt, Chem. Comm., 1965, 36

65HD   R. Haase, K. H. Ducker, and H. A. Kuppers, Ber. Bunsengesell. Phys. Chem., 1965,
       69, 97

65HE   G. P. Haight, Jr. and B. Y. Ellis, Inorg. Chem., 1965, 4, 249

65HI   S. Hamada, Y. Ishikawa, and T. Shirai, Nippon Kagaku Zasshi, 1965, 86, 1042

65HP   G. P. Haight, Jr., and J. R. Peterson, Inorg. Chem., 1965, 4, 1073

65HS   H. E. Hellwege and G. K. Schweitzer, J. Inorg. Nucl. Chem., 1965, 27, 99

65HSE   M. K. Hargreaves, E. A. Stevinson, and J. Evans, J. Chem. Soc., 1965, 4582

65HW   H. P. Hopkins, Jr., and C. A. Wulff, J. Phys. Chem., 1965, 69, 6

65HWa   H. P. Hopkins, Jr., and C. A. Wulff, J. Phys. Chem., 1965, 69, 9

65Wb   H. P. Hopkins, Jr., and C. A. Wulff, J. Phys. Chem., 1965, 69, 1980

65HWH   H. P. Hopkins, C. Wu, and L. G. Hepler, J. Phys. Chem., 1965, 69, 2244

65IC   R. M. Izatt, J. J. Christensen, J. W. Hansen and G. D. Watt, Inorg. Chem., 1965,
            4, 718

65IM   A. Indelli and G. Mantovani, Trans. Faraday Soc., 1965, 61, 909

65JB   K. M. Jones and J. Bjerrum, Acta Chem. Scand., 1965, 19, 974

65JG   D. S. Jain and J. N. Gaur, Bull. Acad. Polon. Sci., Ser. Sci. Chim., 1965, 13, 615

65JL   E. Josefowicz and H. Ladzinska-Kulinska, Rocz. Chem., 1965, 39, 1175

65K   I. V. Kolosov, Russ. J. Inorg. Chem., 1965, 10, 1197 (2200)

65Ka   J. R. Kyrki, Suomen Kem., 1965, B38, 203

65KM   F. Ya. Kulba, V. E. Mironov, and G. Mrnyakova, Russ. J. Inorg. Chem., 1965, 10,
            758 (1393)

65KMa   F. Ya. Kulba, V. E. Mironov, I. F. Mavrin, and Yu. B. Yakovlev, Russ. J. Inorg.
            Chem., 1965, 10, 1117 (2053)

65KMb   F. Ya. Kulba, V. E. Mironov, and I. F. Mavrin, Russ. J. Phys. Chem., 1965, 39,
            1387 (2595)

65KS   S. K. Kundra, P. R. Subbaraman, and J. Gupta, Indian J. Chem., 1965, 3, 60

65KY   F. Ya. Kulba, Yu. B. Yakovlev, and V. E. Mironov, Russ. J. Inorg. Chem., 1965,
            10, 886 (1624)

65KYa   F. Ya. Kulba, Yu. B. Yakovlev, and V. E. Mironov, Russ. J. Inorg. Chem., 1965,
            10, 1113 (1624)

65L   K. H. Lieser, Z. Anorg. Allg. Chem., 1965, 335, 225

65La   O. Lukkari, Suomen Kem., 1965, B38, 121

65Lb   B. J. Levien, Aust. J. Chem., 1965, 18, 1161

65LA   Z. L. Leshchinskaya, M. A. Averbukh, and N. M. Selivanova, Russ. J. Phys. Chem.,
            1965, 39, 1082 (2036)

65LB   K. H. Lieser, G. Beyer, and E. Lakatos, Z. Anorg. Allg. Chem., 1965, 339, 208

65LL   P. Lumme, P. Lahermo, and J. Tummavuori, Acta Chem. Scand., 1965, 19, 2175

65LP   D. Leonesi and G. Piantoni, Ann. Chim. (Italy), 1965, 55, 668

65LS   Z. L. Leshchinskaya and N. M. Selivanova, Russ. J. Phys. Chem., 1965, 39, 1297
            (2430)

65LT   P. Lumme and J. Tummavuori, Acta Chem. Scand., 1965, 19, 617

65LW   D. W. Larsen and A. C. Wahl, Inorg. Chem., 1965, 4, 1281

65M   L. W. Marple, J. Inorg. Nucl. Chem., 1965, 27, 1693

65MB   R. C. Mercier, M. Bonnet, and M. R. Paris, Bull. Soc. Chim. France, 1965, 2926

65MJ   D. F. C. Morris and M. W. Jones, J. Inorg. Nucl. Chem., 1965, 27, 2454

65MK   T. R. Musgrave and R. N. Keller, Inorg. Chem., 1965, 4, 1793

65MKF   V. E. Mironov, F. Ya. Kulba, and V. A. Fedorov, Russ. J. Inorg. Chem., 1965, 10, 495 (914); 755 (1388)

65MR   V. E. Mironov and Yu. I. Rutkovskii, Russ. J. Inorg. Chem., 1965, 10, 580 (1069) 1450 (2670)

65MRI   V. E. Mironov, Yu. I. Rutkovskii, and E. I. Ignatenko, Russ. J. Inorg. Chem., 1965, 10, 1434 (2639)

65MRS   D. F. C. Morris, G. L. Reed, E. L. Short, D. N. Slater, and D. N. Waters, J. Inorg. Nucl. Chem., 1965, 27, 377

65MS   A. S. G. Mazumdar and C. K. Sivaramakrishnan, J. Inorg. Nucl. Chem., 1965, 27, 2423

65MSW   T. D. B. Morgan, G. Stedman, and P. A. E. Whincup, J. Chem. Soc., 1965, 4813

65NH   Y. Narusawa, J. Hashimoto, and H. Hamaguchi, Bull. Chem. Soc. Japan, 1965, 38, 234

65NP   R. I. Novoselov and B. V. Ptitsyn, Russ. J. Inorg. Chem., 1965, 10, 1241 (2282)

65NPG   N. M. Nikolaeva, B. V. Ptitsyn, and I. I. Gorbacheva, Russ. J. Inorg. Chem., 1965, 10, 570 (1051)

65NS   H. Neumann, I. Z. Steinberg, and E. Katchalski, J. Amer. Chem. Soc., 1965, 87, 3841

65NT   T. Nishide and R. Rsuchiya, Bull. Chem. Soc. Japan, 1965, 38, 1398

65P   R. Pottel, Ber. Bunsengesel. Phys. Chem., 1965, 69, 363

65Pa   J. A. Perez-Bustamante, Radiochim. Acta, 1965, 4, 67

65PE   R. Phillips, P. Eisenberg, P. George, and R. J. Rutman, J. Biol. Chem., 1965, 240, 4393

65PF   D. Peschanski and J. M. Fruchart, Compt. Rend. Acad. Sci. Paris, Sec. C, 1965, 260, 3073

65PI   J. A. Partridge, R. M. Izatt, and J. J. Christensen, J. Chem. Soc., 1965, 4231

65PSV   P. Paoletti, J. H. Stern, and A. Vacca, J. Phys. Chem., 1965, 69, 3759

65PSZ   L. Przyborowski, G. Schwarzenbach and T. Zimmerman, Helv. Chim. Acta, 1965, 48, 1556

65PT   P. Papoff, G. Torsi, and P. G. Zambonin, Gazz. Chim. Ital., 1965, 95, 1031

65PY   K. V. Pavlova and K. B. Yatsimirskii, Russ. J. Inorg. Chem., 1965, 10, 557 (1027)

65R   P. A. Rock, Inorg. Chem., 1965, 4, 1667

65Ra   J. K. Ruff, Inorg. Chem., 1965, 4, 1446

65RD   D. E. Ryan, J. R. Dean, and R. M. Cassidy, Canad. J. Chem., 1965, 43, 999

65RP   I. G. Ryss and N. G. Parkhomenko, Ukr. Khim. Zh., 1965, 31, 237

65S   T. Sekine, Acta Chem. Scand., 1965, 19, 1435

65Sa   T. Sekine, Acta Chem. Scand., 1965, 19, 1469

65Sc   T. Sekine, Acta Chem. Scand., 1965, 19, 1519

65Sd   T. Sekine, Acta Chem. Scand., 1965, 19, 1526

65Se   M. B. Shchigol, Russ. J. Inorg. Chem., 1965, 10, 1142 (2097)

65Sf   G. Schorsch, Bull. Soc. Chim. France, 1965, 988

65SAH   P. Schindler, H. Althaus, F. Hofer, and W. Minder, Helv. Chim. Acta, 1965, 48,
        1204

65SAP   A. Kh. Serif, I. P. Alimarin, and I. V. Puzdrenkova, Vestnik Moskov. Univ., Khim.,
        1965, 3, 71

65SG   A. Swinarski and A. Grodzicki, Rocz. Chem., 1965, 39, 3

65SGa   A. Swinarski and A. Grodzicki, Rocz. Chem., 1965, 39, 1155

65SK   T. W. Swaddle and E. L. King, Inorg. Chem., 1965, 4, 532

65SL   S. A. Shchukarev, O. A. Lobaneva, and M. A. Kononova,, Vestnik Leningrad Univ.,
        1965, No. 4, 149

65SM   G. K. Schweitzer and S. W. McCarty, J. Inorg. Nucl. Chem., 1965, 27, 191

65SMa   M. Shiloh and Y. Marcus, Israel J. Chem., 1965, 3, 123

65SMb   R. G. Seys and C. B. Monk, J. Chem. Soc., 1965, 2452

65SMc   V. B. Spivakovskii and G. V. Makovskaya, Russ. J. Inorg. Chem., 1965, 10, 576
        (1062)

65SP   V. I. Spitsyn and N. N. Patsukova, Russ. J. Inorg. Chem., 1965, 10, 1304 (2396)

65SS   G. Schwarzenbach and M. Schellenberg, Helv. Chim. Acta, 1965, 48, 28

65SSa   A. V. Stepanov and V. P. Shvedov, Russ. J. Inorg. Chem., 1965, 10, 541 (1000)

65T   A. Tateda, Bull. Chem. Soc. Japan, 1965, 38, 165

65TF   R. S. Tobias and C. E. Freidline, Inorg. Chem., 1965, 4, 215

65V   G. Vogt, Ber. Bunsengesellschaff Phys. Chem., 1965, 69, 648

65VP   V. Vesely, V. Pekarek, and M. Abbrent, J. Inorg. Nucl. Chem., 1965, 27, 1159

65VW   L. P. Varga, W. D. Wakley, L. S. Nicolson, M. L. Madden, and J. Patterson, Anal.
        Chem., 1965, 37, 1003

65WC   G. D. Watt, J. J. Christensen and R. M. Izatt, Inorg. Chem., 1965, 4, 220

65WS   W. J. Weber, Jr. and W. Stumm, J. Inorg. Nucl. Chem. 1965, 27, 237

65YR   K. B. Yatsimirskii and V. F. Romanov, Russ. J. Inorg. Chem., 1965, 10, 877, (1607)

65ZP   J. Zsako and E. Petri, Rev. Roumaine Chem., 1965, 10, 571

65ZS   D. M. Ziv and I. A. Shestakova, Radiokhim., 1965, 7, 175

66A   J. Aveston, J. Chem. Soc. (A), 1966, 1599

66AB   S. Ahrland and L. Brandt, Acta Chem. Scand., 1966, 20, 328

66AD   R. H. Arntson, F. W. Dickson, and G. Tunell, Science, 1966, 153, 1673

66AM   D. W. Archer and C. B. Monk, J. Chem. Soc. (A), 1966, 1374

66AMa   D. W. Archer and C. B. Monk, Trans. Faraday Soc., 1966, 62, 1583

66AP   G. Atkinson and S. Petrucci, J. Phys. Chem., 1966, 70, 3122

66AS   A. Adin and A. G. Sykes, J. Chem. Soc. (A), 1966, 1518

66ASS   L. P. Andrusenko, Z. A. Sheka, and I. A. Sheka, Russ. J. Inorg. Chem., 1966, 11,
        676 (1266)

66B   F. Brito, An. Fis. Quim., Ser. B, 1966, 62, 123; Acta Chem. Scand., 1967, 21, 1968

66Ba   F. Brito, An. Fis. Quim., Ser. B, 1966, 62, 193

66Bb   F. Brito, An. Fis. Quim., Ser. B, 1966, 62, 197

66BA   T. A. Belyavskaya, I. P. Alimarin, and G. D. Brykina, Moscow Univ. Chem. Bull.,
        1966, 21, 63 (84)

66BB   H. Bilinski and M. Branica, Croat. Chim. Acta, 1966, 38, 263

66BBS   H. Bilinski, M. Branica, and L. G. Sillen, Acta Chem. Scand., 1966, 20, 853

66BE   M. S. Borisov, A. A. Elesin, I. A. Lebedev, V. T. Filimonov, and G. N. Yakovlev,
        Soviet Radiochem., 1966, 8, 40 (42)

66BF   M. N. Bukhsh, J. Flegenheimer, F. M. Hall, A. G. Maddock, and C. Ferreira
        de Miranda, J. Inorg. Nucl. Chem., 1966, 28, 421

66BI   K. A. Burkov and L. V. Ivanova, Vestnik Leningrad Univ., 1966, No. 16 (Fiz. Khim.,
        No. 3), 120

66BK   J. H. Boughton and R. N. Keller, J. Inorg. Nucl. Chem., 1966, 28, 2851

66BM   G. L. Bertrand, F. J. Millero, C. H. Wu, and L. G. Hepler, J. Phys. Chem., 1966,
        70, 699

66BSA   A. A. Biryukov, V. I. Shlenskaya, and I. P. Alimarin, Bull. Acad. Sci. USSR,
        Chem. Sci., 1966, 1 (3)

V. BIBLIOGRAPHY

66BSW   G. L. Bertrand, G. W. Stapleton, C. A. Wulff, and L. G. Hepler, Inorg. Chem., 1966, 5, 1283

66BT   W. S. Broecker and T. Takahashi, J. Geophys. Res., 1966, 71, 1575

66CI   J. J. Christensen, R. M. Izatt, L. D. Hansen, and J. A. Partridge, J. Phys. Chem., 1966, 70, 2003

66CK   N. A. Coward and R. W. Kiser, J. Phys. Chem., 1966, 70, 213

66CL   R. W. Chlebek and M. W. Lister, Canad. J. Chem., 1966, 44, 437

66CM   A. Cassol and L. Magon, Ric. Sci., 1966, 36, 1194

66CN   T. J. Conocchioli, G. H. Nancollas, and N. Sutin, Inorg. Chem., 1966, 5, 1

66CP   A. Cassol, R. Portanova, and L. Magon, Ric. Sci., 1966, 36, 1180; A. Cassol, L. Magon, and R. Barbieri, Inorg. Nucl. Chem. Letters, 1967, 3, 25

66CT   R. L. Carroll and L. B. Thomas, J. Amer. Chem. Soc., 1966, 88, 1376

66CV   R. Cohen-Adad and P. Veys, Bull. Soc. Chim. France, 1966, 1740

66DM   A. V. Davydov, I. N. Marov, and P. N. Palei, Russ. J. Inorg. Chem., 1966, 11, 702 (1316)

66DO   P. R. Danesi, F. Orlandini, and G. Scibona, J. Inorg. Nucl. Chem., 1966, 28, 1047

66DR   E. N. Deichman, G. V. Rodicheva, and L. S. Krysina, Russ. J. Inorg. Chem., 1966, 11, 1199 (2237)

66E   V. I. Ermolenko, Dopovidi Akad. Nauk. Ukr. SSR, Ser. B, 1966, 85

66EB   J. H. Espenson and D. E. Binau, Inorg. Chem., 1966, 5, 1365

66EL   L. I. Elding and I. Leden, Acta Chem. Scand., 1966, 20, 706

66F   T. M. Florence, Aust. J. Chem., 1966, 19, 1343

66Fa   D. S. Flett, Solvent Extr. Chem., Proc. Int. Conf., Gothenburg, Interscience Publ. (1967), 1966, 60

66FK   U. K. Frolova, V. N. Kumok, and V. V. Serebrennikov, Izv. Vyssh. Ucheb. Zaved., Khim., 1966, 9, 176

66FL   Ya. D. Fridman, M. G. Levina, and R. I. Sorochan, Russ. J. Inorg. Chem., 1966, 11, 877 (1641)

66FP   F. H. Firsching and T. R. Paul, J. Inorg. Nucl. Chem., 1966, 28, 2414

66FPS   M. T. Falqui, G. Ponticelli, and F. Sotgiu, Ann. Chim. (Italy), 1966, 56, 464

66FT   C. E. Freidline and R. S. Tobias, Inorg. Chem., 1966, 5, 354

66G   P. Gerding, Acta Chem. Scand., 1966, 20, 79

66Ga   R. Guillaumont, Rev. Chim. Minerale, 1966, 3, 339

66Gb   P. Gerding, Acta Chem. Scand., 1966, 20, 2771

66Gc   M. Givon, Israel J. Chem., 1966, 4, 3p

66GC   S. L. Gupta and M. K. Chatterjee, Indian J. Chem., 1966, 4, 22

66GD   D. W. Gruenwedel and N. Davidson, J. Mol. Biol., 1966, 21, 129

66GK   H. Gamsjager, W. Kraft, and W. Rainer, Monat. Chem., 1966, 97, 833

66GL   S. D. Grekov, V. A. Leitsin, and R. S. Guzairov, Russ. J. Inorg. Chem., 1966, 11, 452 (835)

66GM   M. V. Goloshchapov, B. V. Martynenko, and T. N. Filatova, Russ. J. Inorg. Chem., 1966, 11, 504 (935)

66GS   A. A. Grinberg, A. I. Stetsenko, and G. P. Guryanova, Russ. J. Inorg. Chem., 1966, 11, 1008 (1887)

66GW   G. Gattow and J. Wortmann, Z. Anorg. Allg. Chem., 1966, 345, 172

66HF   W. Holzapfel and E. U. Franck, Ber. Bunsengesell. Phys. Chem., 1966, 70, 1105

66HN   J. Hala, O. Navratil, and V. Nechuta, J. Inorg. Nucl. Chem., 1966, 28, 553

66I   A. A. Ivakin, J. Appl. Chem. USSR, 1966, 39, 252 (277)

66Ia   A. A. Ivakin, J. Appl. Chem. USSR, 1966, 39, 2262 (2406)

66IT   R. R. Irani and T. A. Taulli, J. Inorg. Nucl. Chem., 1966, 28, 1011

66J   L. Johansson, Acta Chem. Scand., 1966, 20, 2156

66Ja   S. Johansson, quoted in P. Paoletti, A. Vacca, and D. Arenare, J. Phys. Chem., 1966, 70, 193

66JH   N. Jordanov and I. Havezov, Z. Anorg. Allg. Chem., 1966, 347, 101

66K   C. Kiehl. Z. Phys. Chem. (Leipzig), 1966, 232, 384

66KA   P. N. Kovalenko, L. T. Azhipa, and M. M. Evstifeev, Russ. J. Inorg. Chem., 1966, 11, 1443 (2689)

66KC   R. N. Knyazeva, G. N. Chernova, and G. B. Zhukovskaya, Izv. Vyssh. Ucheb. Zaved., Khim., 1966, 9, 869

66KG   L. I. Katzin and E. Gulyas, J. Amer. Chem. Soc., 1966, 88, 5209

66KGS   W. Kraft, H. Gamsjager, and E. Schwarz-Bergkampf, Monat. Chem., 1966, 97, 1134

66KL   M. B. Kennedy and M. W. Lister, Canad. J. Chem., 1966, 44, 1709

66KS   P. N. Kovalenko and G. G. Shchemeleva, J. Appl. Chem. USSR, 1966, 39, (2440)

66KW   G. Kohler and H. Wendt, Ber. Bunsengesel. Phys. Chem., 1966, 70, 674

66L   P. J. D. Lloyd, Solvent Extr. Chem., Proc. Int. Conf., Gothenburg, Interscience Publ. (1967) 1966, 458

66LA   S. C. Lahiri and S. Aditya, J. Indian Chem. Soc., 1966, 43, 513

66LB   B. Lenarcik and A. Basinski, Rocz. Chem., 1966, 40, 165

V. BIBLIOGRAPHY

66LM   C. Luca, V. Magearu, and G. Popa, J. Electroanal. Chem., 1966, 12, 45

66LN   S. J. Lyle and S. J. Naqvi, J. Inorg. Nucl. Chem., 1966, 28, 2993

66LS   Z. L. Leshchinskaya and N. M. Selivanova, Izv. Vyssh. Ucheb. Zaved., Khim., 1966, 19, 523

66LSa   Z. L. Leshchinskaya and N. M. Selivanova, Russ. J. Inorg. Chem., 1966, 11, 143 (260)

66LV   A. Liberti and M. Vicedomini, Ric. Sci., 1966, 36, 851

66M   L. W. Marple, J. Inorg. Nucl. Chem., 1966, 28, 1319

66Ma   J. C. Morris, J. Phys. Chem., 1966, 70, 3798

66Mb   R. Matejec, Ber. Bunsengesel. Phys. Chem., 1966, 70, 703

66MA   V. M. Masalovich, P. K. Agasyan, and E. R. Nikolaeva, Russ. J. Inorg. Chem., 1966, 11, 149 (272)

66MB   W. L. Masterton and L. H. Berka, J. Phys. Chem., 1966, 70, 1924

66MD   W. U. Malik and A. Das, Indian J. Chem., 1966, 4, 203

66MG   E. C. Moreno, T. M. Gregory, and W. E. Brown, J. Res. Nat. Bur. Stand., 1966, 70A 545

66MI   R. E. Mesmer and R. R. Irani, J. Inorg. Nucl. Chem., 1966, 28, 493

66MJ   W. L. Marshall and E. V. Jones, J. Phys. Chem., 1966, 70, 4028

66MK   V. E. Mironov, F. Ya. Kulba, Yu. I. Rutkovskii, and E. I. Ignatenko, Russ. J. Inorg. Chem., 1966, 11, 955 (1786)

66MM   R. P. Mitra, H. C. Malhotra, and D. V. S. Jain, Trans. Faraday Soc., 1966, 62, 167

66MR   V. E. Mironov and Yu. I. Rutkovskii, Russ. J. Inorg. Chem., 1966, 11, 958 (1792)

66MS   W. L. Marshall and R. Slusher, J. Phys. Chem., 1966, 70, 4015

66MSY   T. V. Malkova, G. A. Shutova, and K. B. Yatsimirskii, Russ. J. Inorg. Chem., 1966, 11, 832 (1556)

66N   B. I. Nabivanets. Soviet Prog. Chem. (Ukr. Khim. Zh.), 1966, 32, 669 (886)

66NH   L. Nilsson and G. P. Haight, Jr., Acta Chem. Scand., 1966, 20, 486

66NS   G. F. Nichugovskii and V. P. Shvedov, Soviet Radiochem., 1966, 8, 110 (118)

66OA   T. Okubo and F. Aoki, J. Chem. Soc. Japan, 1966, 87, 1103

66OL   J. Oleszkiewicz and T. Lipiec, Rocz. Chem., 1966, 40, 541

66OP   Z. Orhanovic, B. Pokric, H. Furedi, and M. Branica, Croat. Chem. Acta, 1966, 38, 269

66OS   H. G. Offner and D. A. Skoog, Anal. Chem., 1966, 38, 1520

66P   J. Podlaha, Coll. Czech. Chem. Comm., 1966, 31, 7

66PD   I. V. Pyatnitskii and M. Durdyev, Soviet Prog. Chem. (Ukr. Khim. Zh.), 1966, 32,
        57 (77)

66PP   O. E. Presnyakova and R. S. Prishchepo, Russ. J. Inorg. Chem., 1966, 11, 767
        (1436)

66R   R. Radeglia, Z. Phys. Chem. (Leipzig), 1966, 231, 339

66RY   A. G. Rykov and G. N. Yakovlev, Soviet Radiochem., 1966, 8, 26 (27)

66S   J. Sobkowski, Rocz. Chem., 1966, 40, 271

66Sa   A. P. Savostin, Russ. J. Inorg. Chem., 1966, 11, 1514 (2817)

66SB   V. I. Shlenskaya and A. A. Biryukov, Russ. J. Inorg. Chem., 1966, 11, 28 (54)

66SC   R. Sabbah and G. Carpeni, J. Chim. Phys., 1966, 63, 1549

66SD   A. Swinarski, E. Danilczuk, and R. Gogolin, Rocz. Chem., 1966, 40, 737

66SG   A. Swinarski and A. Grodzicki, Rocz. Chem., 1966, 40, 373

66SH   T. Sekine and Y. Hasegawa, Bull. Chem. Soc. Japan, 1966, 39, 240

66SHa   K. Swaminathan and G. M. Harris, J. Amer. Chem. Soc., 1966, 88, 4411

66SI   A. S. Solovkin and A. I. Ivantsov, Russ. J. Inorg. Chem., 1966, 11, 1013 (1897)

66SM   V. I. Spitsyn, N. B. Mikheev, and A. Khermann, Doklady Phys. Chem., 1966, 166, 48
        (658)

66SN   I. V. Shilin and V. K. Nazarov, Soviet Radiochem., 1966, 8, 474 (514)

66SNa   V. P. Shvedov and G. F. Nichugovskii, Soviet Radiochem., 1966, 8, 62 (66)

66SNb   G. Schmid and U. Neumann, Ber. Bunsengesell. Phys. Chem., 1966, 70, 1165

66SO   G. Scibona, F. Orlandini, and P. R. Danesi, J. Inorg. Nucl. Chem. 1966, 28, 1313

66SS   Z. A. Sheka and E. I. Sinyavskaya, Russ. J. Inorg. Chem., 1966, 11, 555 (1029)

66SSH   T. Sekine, M. Sakairi, and Y. Hasegawa, Bull. Chem. Soc. Japan, 1966, 39, 2141

66SV   P. Salomaa and A. Vesala, Acta Chem. Scand., 1966, 20, 1414

66SW   G. Schwarzenbach and M. Widmer, Helv. Chim. Acta, 1966, 49, 111

66SWa   G. Schwarzenbach and D. C. Winkley, Solvent Extr. Chem., Proc. Int. Conf.,
        Gothenburg, Interscience Publ. (1967) 1966, 39

66TF   R. S. Tobias, H. N. Farrer, M. B. Hughes, and B. A. Nevett, Inorg. Chem., 1966,
        5, 2052

66TJ   J. Y. Tong and R. L. Johnson, Inorg. Chem., 1966, 5, 1902

66VL   V. M. Vdovenko, L. N. Lazarev, and Ya. S. Khvorostin, Soviet Radiochem., 1966,
        8, 613 (673)

66VS   I. M. Vasilkevich and E. A. Shilov, Soviet Prog. Chem. (Ukr. Khim. Zh.). 1966.
        32, 703 (947)

66VSB   F. Vierling, G. Schorsch, and J. Bye, Rev. Chim. Minerale, 1966, 3, 875

66VV   V. N. Vasileva and V. P. Vasilev, Izv. Vyssh. Ucheb. Zaved., Khim., 1966, 9, 185

66WD   R. M. Walters and R. W. Dodson, Solvent Extr. Chem., Proc. Int. Conf., Gothenburg,
        Interscience Publ. (1967), 1966, 71

66WM   J. I. Watters and S. Matsumoto, Inorg. Chem., 1966, 5, 361

67A   G. Anderegg, Helv. Chim. Acta, 1967, 50, 2333

67Aa   S. Ahrland, Helv. Chim. Acta, 1967, 50, 306

67Ab   I. I. Alekseeva, Russ. J. Inorg. Chem., 1967, 12, 968 (1840)

67ADR   I. A. Ammar, S. Darwish, and K. Rizk, Electrochim. Acta. 1967, 12, 647

67ADT   F. Accascina, A. D'Aprano, and R. Triolo, J. Phys. Chem., 1967, 71, 3469

67AK   R. Arnek and W. Kakolowicz, Acta Chem. Scand., 1967, 21, 1449

67AKa   R. Arnek and W. Kakolowicz, Acta Chem. Scand., 1967, 21, 2180

67AKE   L. T. Azhipa, P. N. Kovalenko, and M. M. Evstifeev, Russ. J. Inorg. Chem., 1967, 12,
        601 (1138)

67AS   M. Ardon and N. Sutin, Inorg. Chem., 1967, 6, 2268

67ASa   L. P. Andrusenko and I. A. Sheka, Russ. J. Inorg. Chem., 1967, 12, 333 (638)

67B   O. Butkevitsch, Suomen Kem., 1967, B40, 148 (see also 64SM)

67Ba   N. N. Baronova, Russ. J. Inorg. Chem., 1967, 12, 760 (1438)

67Bb   M. Barres, Rev. Chim. Minerale, 1967, 4, 803

67Bc   M. Beran, Coll. Czech. Chem. Comm., 1967, 32, 1368

67BC   L. Burlamacchi, G. Casini, O. Fagioli, and E. Tiezzi, Ric. Sci., 1967, 37, 97

67BI   H. Bilinski and N. Ingri, Acta Chem. Scand., 1967, 21, 2503

67BN   A. K. Babko, B. I. Nabivanets, and V. V. Lukachina, Russ. J. Inorg. Chem., 1967, 12,
        1568 (2965)

67BP   G. A. Boos and A. A. Popel, Russ. J. Inorg. Chem., 1967, 12, 1098 (2086)

67BS   D. Banerjea and I. P. Singh, Z. Anorg. Allg. Chem., 1967, 349, 213

67BSW   B. R. Baker, N. Sufin, and T. J. Welch, Inorg. Chem., 1967, 6, 1948

67BT   F. Bertin, G. Thomas, and J. C. Merlin, Bull. Soc. Chim. France, 1967, 2393

67C   P. Chartier, Bull. Soc. Chim. France, 1967, 2706

67CB   M. P. Collados, F. Brito, and R. Diaz Cadaviece. An. Fis. Quim., Ser. B, 1967, 63
        843

67CC   R. G. de Carvalko and G. R. Choppin, J. Inorg. Nucl. Chem., 1967, 29, 725, 737

67CE   D. W. Carlyle and J. H. Espenson, Inorg. Chem., 1967, 6, 1370

67CM   R. L. Carroll and R. E. Mesmer, Inorg. Chem., 1967, 6, 1137

67CP   B. Corain and A. J. Poe, J. Chem. Soc. (A), 1967, 1318

67CS   T. J. Conocchioli and N. Sutin, J. Amer. Chem. Soc., 1967, 89, 282

67D   P. R. Danesi, Acta Chem. Scand., 1967, 21, 143

67DE   L. Drougge, L. I. Elding, and L. Gustafson, Acta Chem. Scand., 1967, 21, 1647

67DS   R. G. Denotkina and V. B. Shevchenko, Russ. J. Inorg. Chem., 1967, 12, 1237 (2345)

67EG   W. A. Eaton, P. George, and G. I. H. Hanania, J. Phys. Chem., 1967, 71, 2016

67EH   A. J. Eve and D. N. Hume, Inorg. Chem., 1967, 6, 331

67EHP   W. J. Eilbeck, F. Holmes, G. G. Phillips, and A. E. Underhill, J. Chem. Soc. (A),
        1967, 1161

67EL   A. A. Elesin, I. A. Lebedev, E. M. Piskunov, and G. N. Yakovlev, Soviet Radiochem.,
        1967, 9, 159 (161)

67EM   A. J. Ellis and N. B. Milestone, Geochim. Cosmochim. Acta, 1967, 31, 615

67EME   A. N. Ermakov, I. N. Marov, and G. A. Evtikova, Russ. J. Inorg. Chem., 1967, 12,
        1784 (3372)

67ES   J. H. Espenson and S. G. Slocum, Inorg. Chem., 1967, 6, 906

67FD   F. H. Fisher and D. F. Davis, J. Phys. Chem., 1967, 71, 819

67FH   J. F. Fisher and J. L. Hall, Anal. Chem., 1967, 39, 1550

67FR   V. A. Fedorov, A. M. Robov, and V. E. Mironov, Russ. J. Inorg. Chem., 1967, 12,
        1750 (3307)

67G   A. Garnier, Compt. Rend. Acad. Sci. Paris, Sec. B, 1967, 265, 198

67Ga   H. J. Gardner, Aust. J. Chem., 1967, 20, 2357

67Gb   H. Gamsjager, Monat. Chem., 1967, 98, 1803

67GD   G. Gattow and M. Drager, Z. Anorg. Allg. Chem., 1967, 349, 202

67GK   Yu. I. Gryzin and K. Z. Koryttsev, Russ. J. Inorg. Chem., 1967, 12, 50 (101)

67GL   V. Gold and B. M. Lowe, J. Chem. Soc. (A), 1967, 936

67GR   H. Gamsjager, W. Rainer, and P. Schindler, Monat. Chem., 1967, 98, 1793

67GS   A. O. Gubeli and J. Ste-Marie, Canad. J. Chem., 1967, 45, 827

67GSa   A. O. Gubeli and J. Ste-Marie, Canad. J. Chem., 1967, 45, 2101

67GSb   H. Gamsjager and P. Schindler, Helv. Chim. Acta, 1967, 50, 2053

67H   H. C. Helgeson, J. Phys. Chem., 1967, 71, 3121

67HB   A. A. Humffray, A. M. Bond, and J. S. Forrest, J. Electroanal. Chem., 1967, 15, 67

V. BIBLIOGRAPHY

67HC   J. Haladjian and G. Carpeni, J. Chim. Phys., 1967, 64, 1338

67HI   G. I. H. Hanania, D. H. Irvine, W. A. Eaton, and P. George, J. Phys. Chem., 1967, 71, 2022

67HK   C. F. Hale and E. L. King, J. Phys. Chem., 1967, 71, 1779

67HL   W. J. Haffenden and G. J. Lawson, J. Inorg. Nucl. Chem., 1967, 29, 1499

67HP   J. Hala and D. Pohankova, J. Inorg. Nucl. Chem., 1967, 29, 2983

67HR   R. Haque and L. W. Reeves, J. Amer. Chem. Soc., 1967, 89, 250 (see also K. Schaumburg and C. Deverell, Ibid., 1968, 90, 2495)

67HS   M. Haeringer and J. P. Schwing, Bull. Soc. Chim. France, 1967, 708

67I   A. A. Ivakin, Russ. J. Inorg. Chem., 1967, 12, 939 (1787)

67IE   R. M. Izatt, D. Eatough, and J. J. Christensen, J. Chem. Soc. (A), 1967, 1301

67IJ   R. M. Izatt, H. D. Johnston, G. D. Watt, and J. J. Christensen, Inorg. Chem., 1967, 6, 132

67IW   R. M. Izatt, G. D. Watt, D. Eatough, and J. J. Christensen, J. Chem. Soc. (A), 1967, 1304

67JG   D. S. Jain and J. N. Gaur, Acta Chim. Acad. Sci. Hung., 1967, 51, 165

67JJ   D. V. S. Jain and C. M. Jain, J. Chem. Soc. (A), 1967, 1541

67K   Z. Kolarik, Coll. Czech. Chem. Comm., 1967, 32, 432

67Ka   K. H. Khoo, Aust. J. Chem., 1967, 20, 1287

67KG   L. N. Komissarova, V. G. Gulia, T. M. Sas, and N. A. Chernova, Russ. J. Inorg. Chem., 1967, 12, 461 (873)

67KI   I. V. Kolosov, B. N. Ivanov-Emin, L. G. Korotaeva, and Kh. Tetsu, Soviet Radiochem., 1967, 9, (473)

67KM   C. Th. Kawassiades, G. E. Manoussakis, and J. A. Tossidis, J. Inorg. Nucl. Chem., 1967, 29, 401

67KP   V. I. Kazakova and B. V. Ptitsyn, Russ. J. Inorg. Chem., 1967, 12, 323 (620)

67KPa   K. H. Khoo and M. H. Panckhurst, Aust. J. Chem., 1967, 20, 2633

67KPb   D. R. Kester and R. M. Pytkowicz, Limnol. Oceanog., 1967, 12, 243

67KR   R. T. Kolarich, V. A. Ryan, and R. P. Schuman, J. Inorg. Nucl. Chem., 1967, 29, 783

67KZ   N. P. Komar and N. T. Zung, Russ. J. Inorg. Chem., 1967, 12, 669 (1265)

67L   A. Lerman, Geochim. Cosmochim. Acta, 1967, 31, 2309

67La   M. Lucas, Bull. Soc. Chim. France, 1967, 3842

67LB   J. Lagrange and J. Bye, Bull. Soc. Chim. France, 1967, 1490

67LD   H. Loman and E. von Dalen, J. Inorg. Nucl. Chem., 1967, 29, 699

67LH   M. H. Lietzke and J. O. Hall, J. Inorg. Nucl. Chem., 1967, 29, 1249

67LK   V. I. Lobov, F. Ya. Kulba and V. E. Mironov, Russ. J. Inorg. Chem., 1967, 12, 172
        (334)

67LN   S. J. Lyle and S. J. Naqvi, J. Inorg. Nucl. Chem., 1967, 29, 2441

67M   J. Maslowska, Rocz. Chem., 1967, 41, 1857

67Ma   W. L. Marshall, J. Phys. Chem., 1967, 71, 3584

67MA   D. F. C. Morris and B. D. Andrews, Electrochim. Acta, 1967, 12, 41

67MAN   V. M. Masalovich, P. K. Agasyan, and E. R. Nikolaeva, Russ. J. Inorg. Chem., 1967,
        12, 1074 (2041)

67MB   R. E. Mesmer and C. F. Baes, Jr., Inorg. Chem., 1967, 6, 1951

67ME   A. I. Moskvin, L. N. Essen, and T. V. Bukhtiyarova, Russ. J. Inorg. Chem., 1967,
        12, 1794 (3390)

67MF   V. E. Mironov and A. V. Fokina, Russ. J. Inorg. Chem., 1967, 12, 1357 (2571)

67MFR   V. E. Mironov, A. V. Fokina, and Yu. I. Rutkovskii, Russ. J. Inorg. Chem., 1967,
        12, 1082 (2056)

67MG   S. K. Mishra and Y. K. Gupta, J. Inorg. Nucl. Chem., 1967, 29, 1643

67MK   I. F. Mavrin, F. Ya. Kulba, and V. E. Mironov, Russ. J. Inorg. Chem., 1967, 12,
        167 (324)

67MKa   I. F. Mavrin, F. Ya. Kulba, and V. E. Mironov, Russ. J. Phys. Chem., 1967, 41, 886
        (1659)

67MN   R. K. Momii and N. H. Nachtrieb, Inorg. Chem., 1967, 6, 1189

67MNU   M. G. Mushkina, M. S. Novakovskii, and N. Ch. Uen, Russ. J. Inorg. Chem., 1967, 12,
        1239 (2351)

67MP   T. Morozumi and F. A. Posey, Denki Kagaku, 1967, 35, 633

67MR   F. Maggio, V. Romano, and L. Pellerito, Ann. Chim. (Italy), 1967, 57, 191

67MRa   F. Maggio, V. Romano, and L. Pellerito, J. Electanal. Chem., 1967, 15, 227

67MS   K. Mizutani and K. Sone, Z. Anorg. Allg. Chem., 1967, 350, 216

67MSK   S. A. Merkusheva, N. A. Skorik, V. N. Kumok, and V. V. Serebrennikov, Russ. J.
        Inorg. Chem., 1967, 12, 1793 (3388)

67MSP   A. I. Moskvin, A. M. Shelyakina, and P. S. Perminov, Russ. J. Inorg. Chem., 1967,
        12, 1756 (3319)

67N   B. Noren, Acta Chem. Scand., 1967, 21, 2435, 2449

67Na   B. Noren, Acta Chem. Scand., 1967, 21, 2457

67NK   Y. Narusawa, M. Kanazawa, S. Takahashi, K. Morinaga, and K. Nakano, J. Inorg. Nucl.
        Chem., 1967, 29, 123

67NP   N. M. Nikolaev and E. D. Pastukhova, Russ. J. Inorg. Chem., 1967, 12, 796 (1514)

67NR   G. M. Nair, C. L. Rao, and G. A. Welch, Radiochim. Acta, 1967, 7, 77

67NS   J. Nassler and A. Sedlak, Coll. Czech. Chem. Comm., 1967, 32, 2405

67NT   G. H. Nancollas and K. Torrance, Inorg. Chem., 1967, 6, 1567

67OK   H. Ohtaki and H. Kato, Inorg. Chem., 1967, 6, 1935

67OW   M. Orhanovic and R. G. Wilkins, J. Amer. Chem. Soc., 1967, 89, 278

67PB   B. Pokric and M. Branica, Croat. Chem. Acta, 1967, 39, 11

67PI   G. Popa, E. Iosif, and C. Luca, Rev. Roum. Chim., 1967, 12, 169

67PM   N. A. Parpiev, I. A. Maslennikov, and Yu. A. Buslaev, Uzbek. Khim. Zh., 1967,
       No. 3, 9

67PP   A. D. Pethybridge and J. E. Prue, Trans. Faraday Soc., 1967, 63, 2019

67PS   J. Podlaha and J. Silha, Coll. Czech. Chem. Comm., 1967, 32, 3760

67R   B. N. Ryzhenko, Geokhim., 1967, 161

67RM   Yu. I. Rutkovskii and V. E. Mironov, Russ. J. Inorg. Chem., 1967, 12, 1739 (3287)

67SA   Z. A. Sheka, L. P. Andrusenko, and I. A. Sheka, Russ. J. Inorg. Chem., 1967, 12,
       36 (74)

67SB   H. Sigel, K. Becker, and D. B. McCormick, Biochim. Biophys. Acta, 1967, 148, 655

67SG   A. Swinarski and A. Grodzicki, Rocz. Chem., 1967, 41, 1205

67SI   G. Schorsch and N. Ingri, Acta Chem. Scand., 1967, 21, 2727

67SK   H. Schmid and P. Krenmayr, Monat. Chem., 1967, 98, 417, 423

67SKV Yu. I. Sannikov, E. I. Krylov, and V. M. Vinogradov, Russ. J. Inorg. Chem., 1967,
       12, 1398 (2651)

67SL   K. Sallavo and P. Lumme, Suomen Kem., 1967, B40, 155

67SS   T. Sekine, I. Sakamoto, T. Sato, T. Taira, and Y. Hasegawa, Bull. Chem. Soc. Japan,
       1967, 40, 251

67SSa T. Sekine and M. Sakairi, Bull. Chem. Soc. Japan, 1967, 40, 261

67SSb D. L. Singleton and J. H. Swinehart, Inorg. Chem., 1967, 6, 1536

67SSc Z. A. Sheka and E. I. Sinyavskaya, Russ. J. Inorg. Chem., 1967, 12, 194 (377),
       340 (650)

67SSd J. Steigman and D. Sussman, J. Amer. Chem. Soc., 1967, 89, 6400

67SV   A. A. Shidlovskii, A. A. Voskreseuskii, and E. S. Shitikov, Russ. J. Phys. Chem.,
       1967, 41, 380 (731)

67TG   J. E. Teggins, D. R. Gano, M. A. Tucker, and D. S. Martin, Jr., Inorg. Chem.,
       1967, 6, 69

67TK   M. Tanaka and I. Kojima, J. Inorg. Nucl. Chem., 1967, 29, 1769

67TP   I. V. Tananaev and S. M. Petushkova, Russ. J. Inorg. Chem., 1967, 12, 39 (81)

67VD   S. P. Vorobev, I. P. Davydov, and I. V. Shilin, Russ. J. Inorg. Chem., 1967, 12, 1129 (2142)

67VDa   S. P. Vorobev, I. P. Davydov, and I. V. Shilin, Russ. J. Inorg. Chem., 1967, 12, 1406 (2665)

67VG   V. P. Vasilev and N. K. Grechina, Russ. J. Inorg. Chem., 1967, 12, 724 (1372)

67VK   S. Varvicka and J. Koryta, Coll. Czech. Chem. Comm., 1967, 32, 2346

67VL   V. P. Vasilev and G. A. Lobanov, Russ. J. Inorg. Chem., 1967, 12, 463 (878)

67VLa   V. P. Vasilev and G. A. Lobanov, Russ. J. Phys. Chem., 1967, 41, 434 (838)

67VLb   V. P. Vasilev and G. A. Lobanov, Russ. J. Phys. Chem., 1967, 41, 1053 (1969)

67VLc   V. M. Vdovenko, L. N. Lazarev, and Ya. S. Khvorostin, Soviet Radiochem., 1967, 9, 445 (460)

67VLd   V. M. Vdovenko, L. N. Lazarev, and Ya. S. Khvorostin, Soviet Radiochem., 1967, 9, 449 (464)

67VLe   V. M. Vdovenko, L. N. Lazarev, and Ya. S. Khvorostin, Russ. J. Inorg. Chem., 1967, 12, 610 (1152)

67VM   V. P. Vasilev and P. S. Mukhina, Izv. Vyssh. Ucheb. Zaved., Khim. 1967, 10, 263

67VS   V. M. Vdovenko, O. B. Stebunov, and V. A. Shcherbakov, Russ. J. Inorg. Chem., 1967, 12, 896 (1706)

67W   R. M. Wallace, J. Phys. Chem., 1967, 71, 1271

67Wa   H. E. Wirth, J. Phys. Chem., 1967, 71, 2922

67WC   J. B. Walker and G. R. Choppin, Adv. Chem. Series, 1967, 71, 127

67WD   C. F. Wells and G. Davies, J. Chem. Soc. (A), 1967, 1858

67WM   J. I. Watters and S. Matsumoto, J. Inorg. Nucl. Chem., 1967, 29, 2955

67WW   C. H. Wu, R. J. Witonsky, P. George, and R. J. Rutman, J. Amer. Chem. Soc., 1967, 89, 1987

67YK   Yu. B. Yakovlev, F. Ya. Kulba, and V. E. Mironov, Russ. J. Inorg. Chem., 1967, 12, 1737 (3283)

67ZB   P. A. Zagorets and G. P. Bulgakova, Russ. J. Inorg. Chem., 1967, 12, 179 (347)

67ZF   I. Zsako and E. Fekete, Studia Univ. Babes-Bolyai, Ser. Chem., 1967, 45

67ZO   F. A. Zakharova and M. M. Orlova, Russ. J. Inorg. Chem., 1967, 12, 1596 (3016), 1699 (3211)

68A   R. Arnek, Acta Chem. Scand., 1968, 22, 1102

68AB   S. Ahrland and L. Brandt, Acta Chem. Scand., 1968, 22, 106

68ABa   S. Ahrland and L. Brandt, Acta Chem. Scand., 1968, 22, 1579

68AC   M. Asso and G. Carpeni, Canad. J. Chem., 1968, 46, 1795

68AD   I. A. Ammar, S. Darwish, and K. Rizk, Electrochim. Acta, 1968, 13, 797

68AL   A. Aziz and S. J. Lyle, J. Inorg. Nucl. Chem., 1968, 30, 3223

68ALN   A. Aziz, S. J. Lyle, and S. J. Naqvi, J. Inorg. Nucl. Chem., 1968, 30, 1013

68AN   V. P. Antonovich and V. A. Nazarenko, Russ. J. Inorg. Chem., 1968, 13, 940 (1805)

68AP   R. Arnek and C. C. Patel, Acta Chem. Scand., 1968, 22, 1097

68AS   R. Arnek and K. Schlyter, Acta Chem. Scand., 1968, 22, 1327

68ASa   R. Arnek and K. Schlyter, Acta Chem. Scand., 1968, 22, 1331

68ASb   R. Arnek and I. Szilard, Acta Chem. Scand., 1968, 22, 1334

68ASc   L. P. Andrusenko and I. A. Sheka, Russ. J. Inorg. Chem., 1968, 13, 1363 (2645)

68AT   E. Andalaft, R. P. T. Tomkins, and G. J. Janz, Canad. J. Chem., 1968, 46, 2959

68B   N. N. Baronova, Geokhim., 1968, 17

68Ba   R. Bury, J. Chim. Phys., 1968, 65, 1494

68BB   D. P. Baccanori, B. A. Buckman, M. M. Yevitz, and H. A. Swain, Jr., Talanta, 1968,
        15, 416

68BC   E. Bottari and L. Ciavatta, Inorg. Chim. Acta, 1968, 2, 74

68C   J. M. Cleveland, Inorg. Chem., 1968, 7, 874

68Ca   W. A. Cilley, Inorg. Chem., 1968, 7, 612

68CF   Y. C. Chiu and R. M. Fuoss, J. Phys. Chem., 1968, 72, 4123

68CFa   B. D. Costley and J. P. G. Farr, Chem. and Ind., 1968, 1435

68CG   L. Ciavatta and M. Grimaldi, J. Inorg. Nucl. Chem., 1968, 30, 197, 563

68CM   A. Chughtai, R. Marshall, and G. H. Nancollas, J. Phys. Chem., 1968, 72, 208

68CS   R. Christova and C. Stefanova, Z. Anorg. Allg. Chem., 1968, 361, 209

68D   R. W. Duerst, J. Chem. Phys., 1968, 48, 2275

68DD   R. C. Das, A. C. Dash, and J. P. Mishra, J. Inorg. Nucl. Chem., 1968, 30, 2417

68DF   A. D'Aprano and R. M. Fuoss, J. Phys. Chem., 1968, 72, 4710

68DK   N. K. Davidenko, G. A. Komashko, and K. B. Yatsimirskii, Russ. J. Inorg. Chem.,
        1968, 13, 58 (117)

68DM   P. R. Danesi, M. Magini, S. Margherita, and G. D'Alessadro, Ener. Nucl. (Milan),
        1968, 15, 333

68DP   A. R. Davis and R. A. Plane, Inorg. Chem., 1968, 7, 2565

68DS   J. Doyle, A. G. Sykes, and A. Adin, J. Chem. Soc. (A), 1968, 1314

68DT   E. N. Deichman, I. V. Tananaev, Zh. A. Ezhova, and T. N. Kuzmina, Russ. J. Inorg. Chem., 1968, 13, 23 (47)

68EP   J. H. Espenson and J. R. Pipal, Inorg. Chem., 1968, 7, 1463

68EPa   J. H. Espenson and O. J. Parker, J. Amer. Chem. Soc., 1968, 90, 3689

68F   R. Fernandez-Prini, Trans. Faraday Soc., 1968, 64, 2146

68FB   S. Fontana and F. Brito, Inorg. Chim. Acta, 1968, 2, 179

68FD   R. C. Ferguson, P. Dobud, and D. G. Tuck, J. Chem. Soc.(A), 1968, 1058

68FH   J. Feeney, R. Haque, L. W. Reeves, and C. P. Yue, Canad. J. Chem., 1968, 46, 1389

68FM   J. R. Fryer and D. F. C. Morris, Talanta, 1968, 15, 1309

68FS   V. A. Fedorov, N. P. Samsonova, and V. E. Mironov, Russ. J. Inorg. Chem., 1968, 13, 198 (382)

68G   P. Gerding, Acta Chem. Scand., 1968, 22, 1282

68Ga   R. Guillaumont, Bull. Soc. Chim. France, 1968, 162

68Gb   R. Guillaumont, Bull. Soc. Chim. France, 1968, 168

68GF   A. V. Gordievskii, E. L. Filippov, V. S. Shterman, and A. S. Krivoshein, Russ. J. Phys. Chem., 1968, 42, 1050 (1998)

68GH   R. N. Goldberg and L. G. Hepler, J. Phys. Chem., 1968, 72, 4654

68GJ   P. Gerding and I. Jonsson, Acta Chem. Scand., 1968, 22, 2247

68GJa   P. Gerding and I. Jonsson, Acta Chem. Scand., 1968, 22, 2255

68GL   V. Gold and B. M. Lowe, J. Chem. Soc. (A), 1968, 1923

68GS   A. O. Gubeli and J. Ste-Marie, Canad. J. Chem., 1968, 46, 1707

68GY   E. A. Gyunner and N. D. Yakhkind, Russ. J. Inorg. Chem., 1968, 13, 1420 (2758)

68HC   P. Hemmes, D. L. Cole, and E. M. Eyring, J. Phys. Chem., 1968, 72, 301

68HF   K. L. Hsia and R. M. Fuoss, J. Amer. Chem. Soc., 1968, 90, 3055

68HG   T. E. Haas and H. D. Gillman, Inorg. Chem., 1968, 7, 2051

68HJ   G. P. Haight, Jr., and L. Johansson, Acta Chem. Scand., 1968, 22, 961

68HM   J. Hill and A. McAuley, J. Chem. Soc.(A), 1968, 2405

68HR   C. C. Hong and W. H. Rapson, Canad. J. Chem., 1968, 46, 2053

68HRa   T. M. Hseu and G. A. Rechnitz, Anal. Chem., 1968, 40, 1054

68HS   S. Hietanen and L. G. Sillen, Acta Chem. Scand., 1968, 22, 265

68HSa K. E. Howlett and S. Sarsfield, J. Chem. Soc. (A), 1968, 683

68HSb F. M. Hall and S. J. Slater, Aust. J. Chem., 1968, 21, 2663

68HSc J. Haladjian, R. Sabbah, and P. Bianco, J. Chim. Phys., 1968, 65, 1751

68IW   R. M. Izatt, G. D. Watt, C. H. Bartholomew and J. J. Christensen, Inorg. Chem.,
       1968, 7, 2236

68IZ   B. N. Ivanov-Emin, V. A. Zaitseva, and A. M. Egorov, Russ. J. Inorg. Chem., 1968,
       13, 1368 (2655)

68J   S. S. Jorgensen, Acta Chem. Scand., 1968, 22, 335

68JD   W. Jakob and M. Dyrek, Rocz. Chem., 1968, 42, 1393

68JG   Z. Jablonski, J. Gornicki, and A. Lodzinska, Rocz. Chem., 1968, 42, 1809

68K   K. H. Khoo, J. Inorg. Nucl. Chem., 1968, 30, 2524

68Ka   K. Kleboth, Monat. Chem., 1968, 99, 1177

68KA   P. N. Kovalenko, L. T. Azhipa, and M. M. Evstifeev, J. Appl. Chem. USSR, 1968, 41,
       (198)

68KD   R. M. Kren, H. W. Dodgen, and C. J. Nyman, Inorg. Chem., 1968, 7, 446

68KK   V. N. Krylov, E. V. Komarov, and M. F. Pushlenkov, Soviet Radiochem., 1968, 10, 702
       (717)

68KKa V. N. Krylov, E. V. Komarov, and M. F. Pushlenkov, Soviet Radiochem., 1968, 10, 705
       (719)

68KKb V. N. Krylov, E. V. Komarov, and M. F. Pushlenkov, Soviet Radiochem., 1968, 10, 708
       (723)

68KM   G. B. Kolski and D. W. Margerum, Inorg. Chem., 1968, 7, 2239

68KP   D. R. Kester and R. M. Pytkowicz, Limnol. Oceanog., 1968, 13, 670

68KT   W. Kruse and D. Thusius, Inorg. Chem., 1968, 7, 464

68KTa O. I. Khotsyanovskii and V. Sh. Telyakova, Soviet Prog. Chem. (Ukr. Khim. Zh.),
       1968, 34, No. 11, 29 (1126)

68L   O. G. Levanda, Russ. J. Inorg. Chem., 1968, 13, 1707 (3311)

68La   D. Langmiur, Geochim. Cosmochim. Acta, 1968, 32, 835

68LC   E. Lanza and G. Carpeni, Electrochim. Acta, 1968, 13, 519

68LF   H. D. Ludemann and E. U. Franck, Ber. Bunsengesell. Phys. Chem., 1968, 72, 523

68LJ   H. G. Linge and A. L. Jones, Aust. J. Chem., 1968, 21, 1445, 2189

68LK   L. N. Lazarev and Ya. S. Khvorostin, Russ. J. Inorg. Chem., 1968, 13, 311 (598)

68LM   S. A. Levison and R. A. Marcus, J. Phys. Chem., 1968, 72, 358

68LMV   O. G. Levanda, I. I. Moiseev, and M. N. Vargaftik, Bull. Acad. Sci. USSR, 1968, 2237 (2368)

68LN   A. M. Lunyatskas and P. K. Norkus, Russ. J. Inorg. Chem., 1968, 13, 347 (665)

68LW   D. Levine and I. B. Wilson, Inorg. Chem., 1968, 7, 818

68M   T. Mitsuji, Bull. Chem. Soc. Japan, 1968, 41, 115

68Ma   J. Maslowska, Rocz. Chem., 1968, 42, 1819

68MB   S. Mateo Dumpierrez and F. Brito, An. Quim., 1968, 64, 115

68MD   I. N. Marov, Yu. N. Dubrov, V. K. Belyaeva, and A. N. Ermakov, Russ. J. Inorg. Chem., 1968, 13, 1107 (2140), 1262 (2445)

68MF   E. E. Mercer and D. T. Farrar, Canad. J. Chem., 1968, 46, 2679

68MG   S. K. Mishra and Y. K. Gupta, Indian J. Chem., 1968, 6, 757

68MH   D. F. C. Morris and S. D. Hammond, Electrochim. Acta, 1968, 13, 545

68MHB   W. M. McNabb, J. F. Hazel, and R. A. Baxter, J. Inorg. Nucl. Chem., 1968, 30, 1585

68ML   J. I. Morrow and J. Levy, J. Phys. Chem., 1968, 72, 885

68MM   V. E. Mironov, Yu. A. Makashev, I. Ya. Mavrina, and D. M. Markhaeva, Russ. J. Phys. Chem., 1968, 42, 1592 (2987)

68MS   O. Makitie and M. L. Savolainen, Suomen Kem., 1968, B41, 242

68MSa   J. M. Malin and J. H. Swinehart, Inorg. Chem., 1968, 7, 250

68MSb   W. U. Malik and C. L. Sharma, J. Indian Chem. Soc., 1968, 45, 29

68MT   O. N. Malyavinskaya and Ya. I. Turyan, Russ. J. Phys. Chem., 1968, 42, 144 (269)

68N   G. M. Nair, Radiochim. Acta, 1968, 10, 116

68Na   F. S. Nakayama, Soil Sci., 1968, 106, 429

68NA   V. A. Nazarenko, V. P. Antonovich, and E. M. Nevskaya, Russ. J. Inorg. Chem., 1968, 13, 825 (1574)

68NK   Yu. V. Norseyev and V. A. Khalkin, J. Inorg. Nucl. Chem., 1968, 30, 3239

68NKa   B. I. Nabivanets and E. E. Kapantsyan, Russ. J. Inorg. Chem., 1968, 13, 946 (1817)

68NM   T. W. Newton, G. E. McGrary, and W. G. Clark, J. Phys. Chem., 1968, 72, 4333

68OA   T. Okubo, F. Aoki, and R. Teraoka, Nippon Kagaku Zasshi, 1968, 89, 432

68OP   M. Orhanovic, H. N. Po, and N. Sutin, J. Amer. Chem. Soc., 1968, 90, 7224

68P   B. Prasad, J. Indian Chem. Soc., 1968, 45, 1037

68PC   R. Portanova, A. Cassol, L. Magon, and G. Tomat, Gazz. Chim. Ital., 1968, 98, 1290

68PG   M. R. Paris and Cl. Gregoire, Anal. Chim. Acta, 1968, 42, 431

68PGa   M. R. Paris and Cl. Gregoire, Anal. Chim. Acta, 1968, 42, 439

68PM   N. A. Parpiev and I. A. Maslennikov, Uzbek. Khim. Zh., 1968, No. 2, 6

68PS   Ts. V. Pevsner and I. A. Sheka, Russ. J. Inorg. Chem., 1968, 13, 1381 (2681)

68PV   B. A. Purin and I. A. Vitina, Izv. Akad. Nauk. Latv. SSR, Khim., 1968, 277, 372

68PW   B. Perlmutter-Hayman and Y. Weissmann, Israel J. Chem., 1968, 6, 17

68QM   A. S. Quist and W. L. Marshall, J. Phys. Chem., 1968, 72, 3122

68RJ   L. Rasmussen and C. K. Jorgensen, Acta Chem. Scand., 1968, 22, 2313

68RR   C. Ropars, M. Rougee, M. Momentau, and D. Lexa, J. Chim. Phys., 1968, 65, 816

68RS   C. L. Rao, C. J. Shahani, and K. A. Mathew, Inorg. Nucl. Chem. Letters, 1968, 4,
        655

68RV   R. Ripan and G. Vericeanu, Studio Univ. Babes-Bolyai, Ser. Chem., 1968, 13, 31

68S   P. Stantschett, Z. Anal. Chem., 1968, 234, 109

68SA   I. A. Sheka and L. P. Andrusenko, Russ. J. Inorg. Chem., 1968, 13, 180 (347)

68SF   U. Schedin and M. Frydman, Acta Chem. Scand., 1968, 22, 115

68SG   T. W. Swaddle and G. Guastalla, Inorg. Chem., 1968, 7, 1915

68SI   T. Sekine, H. Iwaki, M. Sakairi, F. Shimada, and M. Inarida, Bull. Chem. Soc.
        Japan, 1968, 41, 1

68SM   V. B. Spivakovskii and G. V. Makovskaya, Russ. J. Inorg. Chem., 1968, 13, 815
        (1555), 1423 (2764)

68SMR   C. J. Shahani, K. A. Mathew, C. L. Rao, and M. V. Ramaniah, Radiochim. Acta, 1968,
        10, 165

68SR   K. Srinivasan and G. A. Rechnitz, Anal. Chem., 1968, 40, 1818

68SRG   P. Schindler, M. Reinert, and H. Gamsjager, Helv. Chim. Acta, 1968, 51, 1845

68SRR   G. C. Stocco, E. Rivarola, R. Romeo, and R. Barbieri, J. Inorg. Nucl. Chem., 1968,
        30, 2409

68SS   Y. Sasaki and L. G. Sillen, Arkiv Kemi, 1968, 29, 253

68ST   T. G. Sukhova, O. N. Temkin, R. M. Flid, and T . K. Kaliya, Russ. J. Inorg. Chem.
        1968, 13, 1072 (2073)

68SW   A. Schlund and H. Wendt, Ber. Bunsengesell. Phys. Chem., 1968, 72, 652

68SY   G. A. Shutova, K. B. Yatsimirskii, and T. V. Malkova, Russ. J. Inorg. Chem., 1968,
        13, 1395 (2708)

68TH   A. H. Truesdell and P. B. Hostetler, Geochim. Cosmochim. Acta, 1968, 32, 1019

68TL   J. Tummavuori and P. Lumme, Acta Chem. Scand., 1968, 22, 2003

68TR   P. H. Tedesco, V. B. de Rumi, and J. A. Gonzalez Quintana, J. Inorg. Nucl. Chem.,
        1968, 30, 987

68TW   S. P. Tanner, J. B. Walker, and G. R. Choppin, J. Inorg. Nucl. Chem., 1968, 30,
        3067

68V   N. E. Vanderborgh, Talanta, 1968, 15, 1009

68VG   V. P. Vasilev and N. K. Grechina, Izv. Vyssh. Ucheb. Zaved., Khim., 1968, 11, 142

68VK   V. P. Vasilev and L. A. Kochergina, Russ. J. Phys. Chem., 1968, 42, 199 (373)

68VV   V. P. Vasilev and P. N. Vorobev, Izv. Vyssh. Ucheb. Zaved., Khim., 1968, 11, 971

68WM   J. I. Watters and R. Machen, J. Inorg. Nucl. Chem., 1968, 30, 2163

68WS   C. F. Wells and M. A. Salam, J. Chem. Soc. (A), 1968, 24

68WSa   C. F. Wells and M. A. Salam, J. Chem. Soc. (A), 1968, 308

68ZL   O. E. Zvyagintsev and S. B. Lyakhmanov, Russ. J. Inorg. Chem., 1968, 13, 643 (1230)

68ZP   P. Zanella, G. Plazzogna, and G. Tagliavini, Inorg. Chim. Acta, 1968, 2, 340

69A   R. Arnak, Acta Chem. Scand., 1969, 23, 1986

69AL   A. Aziz and S. J. Lyle, Anal. Chim. Acta, 1969, 47, 49

69ALa   A. Aziz and S. J. Lyle, J. Inorg. Nucl. Chem., 1969, 31, 3471

69AY   I. I. Alekseeva and K. B. Yatsimirskii, Russ. J. Inorg. Chem., 1969, 14, 221 (432)

69B   E. W. Baumann, J. Inorg. Nucl. Chem., 1969, 31, 3155

69Ba   A. M. Bond, J. Electroanal. Chem., 1969, 23, 269

69Bb   A. M. Bond, J. Electroanal. Chem., 1969, 23, 277

69Bc   N. N. Baranova, Russ. J. Inorg. Chem., 1969, 14, 1717 (3257)

69BC   S. L. Bertha and G. R. Choppin, Inorg. Chem., 1969, 8, 613

69BG   F. Becker and R. Grundmann, Z. Phys. Chem. (Frankfort), 1969, 66, 137

69BL   B. Kozlowska, F. Letowska, and J. Niemiec, Rocz. Chem., 1969, 43, 1597

69BM   Yu. A. Barbanel and N. K. Mikhailova, Soviet Radiochem., 1969, 11, 576 (595)

69BMN   A. K. Babko, E. A. Mazurenko, and B. I. Nabivanets, Russ. J. Inorg. Chem., 1969,
        14, 1091 (2079)

69BNR   E. A. Biryuk, V. A. Nazarenko, and R. V. Ravitskaya, Russ. J. Inorg. Chem.,
        1969, 14, 503 (965)

69BNT   E. A. Biryuk, V. A. Nazarenko, and L. N. Thu, Russ. J. Inorg. Chem., 1969, 14,
        373 (714)

69BP   G. R. Bruce and M. H. Panckhurst, Aust. J. Chem., 1969, 22, 469

69BPa   H. L. Bohn and M. Peech, Proc. Soil Sci. Soc. Amer., 1969, 33, 873

V. BIBLIOGRAPHY

69BS B. Behar and G. Stein, <u>Israel J. Chem.</u>, 1969, <u>7</u>, 827

69BW W. B. Baldwin and G. Wiese, <u>Arkiv Kemi</u>, 1969, <u>31</u>, 419

69BWa W. G. Baldwin and G. Wiese, <u>Arkiv Kemi</u>, 1969, <u>31</u>, 429

69C C. W. Childs, <u>J. Phys. Chem.</u>, 1969, <u>73</u>, 2956

69Ca J. M. Carpentier, <u>Bull. Soc. Chim. France</u>, 1969, 3851

69CP C. W. Childs and M. H. Panckhurst, <u>Aust. J. Chem.</u>, 1969, <u>22</u>, 911

69CPK A. K. Chuchalin, B. I. Peshchevitskii, and I. A. Kuzin, <u>Russ. J. Inorg. Chem.</u>, 1969, <u>14</u>, 937 (1785)

69CR D. L. Cole, L. D. Rich, J. D. Owen, and E. M. Eyring, <u>Inorg. Chem.</u>, 1969, <u>8</u>, 682

69DH B. Desire, M. Hussonnois, and R. Guillaumont, <u>Compt. Rend. Acad. Sci. Paris</u>, <u>Ser. C</u>, 1969, <u>269</u>, 448

69DK G. Davies and K. Kustin, <u>Inorg. Chem.</u>, 1969, <u>8</u>, 1196

69EP A. M. Erenburg and B. I. Peshchevitskii, <u>Russ. J. Inorg. Chem.</u>, 1969, <u>14</u>, 485 (932)

69ES B. Evtimova, J. P. Scharff, and M. R. Paris, <u>Bull. Soc. Chim. France</u>, 1969, 81

69F A. W. Fordham, <u>Aust. J. Chem.</u>, 1969, <u>22</u>, 1111

69FB V. A. Fedorov, I. M. Bolshakova, and V. E. Mironov, <u>Russ. J. Inorg. Chem.</u>, 1969, <u>14</u>, 805 (1538)

69FD Ya. D. Fridman and T. V. Danilova, <u>Russ. J. Inorg. Chem.</u>, 1969, <u>14</u>, 370

69FP D. P. Fay and N. Purdie, <u>J. Phys. Chem.</u>, 1969, <u>73</u>, 3462

69FR A. V. Fokina, Yu. I. Rutkovskii, and V. E. Mironov, <u>Russ. J. Inorg. Chem.</u>, 1969, <u>14</u>, 620 (1183)

69FRa V. A. Fedorov, A. M. Robov, and V. E. Mironov, <u>Russ. J. Inorg. Chem.</u>, 1969, <u>14</u>, 1432 (2720)

69FT S. Funahashi and M. Tanaka, <u>Inorg. Chem.</u>, 1969, <u>8</u>, 2159

69G P. Gerding, <u>Acta Chem. Scand.</u>, 1969, <u>23</u>, 1695

69GA H. Gamsjaeger, K. Aeberhard, and P. Schindler, <u>Helv. Chim. Acta</u>, 1969, <u>52</u>, 2315

69GK B. Gorski and H. Koch, <u>J. Inorg. Nucl. Chem.</u>, 1969, <u>31</u>, 3565

69GM R. Guillaumont, C. Ferreira de Miranda, and M. Galin, <u>Comp. Rend. Acad. Sci. Paris</u>, <u>Ser. C</u>, 1969, <u>268</u>, 140

69GN G. L. Gardner and G. H. Nancollas, <u>Anal. Chem.</u>, 1969, <u>41</u>, 202

67GP M. Gazikalovic, Z. Pavlovic, and T. Markovic, <u>Arh. Tehnol.</u>, 1967, <u>5</u>, 51

69GS H. Gamsjaeger, P. S. Schindler, and B. Kleinert, <u>Chimia</u> (Switz.), 1969, <u>23</u>, 229

69GV I. Grenthe and J. Varfeldt, <u>Acta Chem. Scand.</u>, 1969, <u>23</u>, 988

69H   G. Horn, Rodex-Rundsch., 1969, 1, 439

69HS   J. Hala and J. Smola, J. Inorg. Nucl. Chem., 1969, 31, 1133

69IE   R. M. Izatt, D. Eatough, J. J. Christensen, and C. H. Bartholomew, J. Chem. Soc.(A),
          1969, 45

69IEa   R. M. Izatt, D. Eatough, J. J. Christensen, and C. H. Bartholomew, J. Chem. Soc.(A),
          1969, 47

69IV   A. A. Ivakin and E. M. Voronova, Russ. J. Inorg. Chem., 1969, 14, 815 (1557)

69IVa   A. A. Ivakin and E. M. Voronova, Trudy Inst. Khim. Uralsk. Fil. Akad. Nauk SSSR,
          1969, No. 17, 107

69J   L. Johansson, Acta Chem. Scand., 1969, 23, 548

69K   B. Kennedy, Diss., Georgetown Univ., Wash., D.C., 1969; Diss. Abs. Int. B, 1969,
          30, 1544

69KH   C. Kappenstein and R. Hugel, Rev. Chim. Minerale, 1969, 6, 1107

69KK   V. N. Krylov and E. V. Komarov, Soviet Radiochem., 1969, 11, 94 (101) 99, (105)

69KKP   V. N. Krylov, E. V. Komarov and M. S. Pushlenkov, Soviet Radiochem., 1969, 11, 97
          (103), 237 (244), 450 (460)

69KM   K. H. Khoo and J. D. Murray, J. Inorg. Nucl. Chem., 1969, 31, 2437

69KP   D. R. Kester and R. M. Pytkowicz, Limnol. Oceanog., 1969, 14, 686

69KS   V. I. Krautsov and I. V. Simakova, Vestn. Leningrad. Univ., Fiz. Khim., 1969, 124;
          Chem. Abs., 1970, 72, 125686W

69LS   F. I. Lobanov, V. M. Savostina, L. V. Serzhenko, and V. M. Peshkova, Russ. J.
          Inorg. Chem., 1969, 14, 562 (1077)

69LV   G. A. Lobanol and V. P. Vasilev, Izv. Vyssh. Ucheb. Zaved., Khim., 1969, 12, 740

69M   J. Maslowska, Zesz. Nauk. Politech. Lodz., Chem., 1969, No. 19, 7; Chem. Abs.,
          1970, 72, 16211s

69MA   D. F. C. Morris, D. T. Anderson, S. L. Waters, and G. L. Reed, Electrochim. Acta,
          1969, 14, 643

69MB   R. E. Mesmer and C. F. Baes, Jr., Inorg. Chem., 1969, 8, 618

69MF   I. D. McKenzie and R. M. Fuoss, J. Phys. Chem., 1969, 73, 1501

69MG   M. C. Mehra and A. O. Gubeli, Radiochem. Radioanal. Letters, 1969, 2, 61

69MGa   H. Metivier and R. Guillamont, Radiochem. Radioanal. Letters, 1969, 1, 209

69MH   E. E. Mercer and J. A. Hormuth, J. Inorg. Nucl. Chem., 1969, 31, 2145

69MK   A. I. Mikhailichenko and I. E. Kurdin, Soviet Radiochem., 1969, 11, 348 (356)

69MKa   B. Marin and T. Kikindai, Comp. Rend. Acad. Sci. Paris, Ser. C, 1969, 268, 1

69MKb   O. Makitie and V. Konttinen, Acta Chem. Scand., 1969, 23, 1459

69MM   V. E. Mironov, Yu. A. Makashev, and I. Ya. Mavrina, Russ. J. Inorg. Chem., 1969,
        14, 746 (1424)

69MN   E. A. Mazurenko and B. I. Nabivanets, Russ. J. Inorg. Chem., 1969, 14, 1732 (3286)

69MNM   D. M. Mikhailova, R. N. Nacheva, and V. Ts. Mikhailova, Soviet Radiochem., 1969,
        11, 241 (247)

69MP   J. B. Macaskill and M. H. Panckhurst, Aust. J. Chem., 1969, 22, 317

69MS   D. F. C. Morris and P. J. Sturgess, Electrochim. Acta, 1969, 14, 629

69N   B. Noren, Acta Chem. Scand., 1969, 23, 379

69Na   B. Noren, Acta Chem. Scand., 1969, 23, 931

69Nb   N. M. Nikolaeva, Russ. J. Inorg. Chem., 1969, 14, 487 (936)

69NM   V. A. Nazarenko and O. V. Mandzhgaladze, Russ. J. Inorg. Chem., 1969, 14, 639
        (1219)

69NN   V. A. Nazarenko and E. M. Nevskaya, Russ. J. Inorg. Chem., 1969, 14, 1696 (3215)

69NP   V. A. Nazarenko and E. N. Poluektova, Russ. J. Inorg. Chem., 1969, 14, 105 (204)

69NPa   N. M. Nikolaeva and M. P. Primanchuk, Russ. J. Inorg. Chem., 1969, 14, 1554 (2945)

69NPS   B. P. Nikolskii, V. V. Palchevskii, and E. F. Strizhev, Vestn. Leningrad, Univ.,
        Fiz. Khim., 1969, 116; Chem. Abs., 1970, 72, 16165e

69NR   F. S. Nakayama and B. A. Rasnick, J. Inorg. Nucl. Chem., 1969, 31, 3491

69NS   G. F. Nichugovskii and V. P. Shvedov, Russ. J. Inorg. Chem., 1969, 14, 156 (299)

69PB   B. I. Peshchevitskii and V. I. Belevantsev, Izv. Sib. Otd. Akad. Nauk SSSR,
        Ser. Khim., 1969, 82; Chem. Abs., 1969, 71, 16401h

69PK   R. M. Pytkowicz and D. R. Kester, Amer. J. Sci., 1969, 267, 217

69PN   C. J. Peacock and G. Nickless, Z. Naturforsch., 1969, 24a, 245

69R   T. Ryhl, Acta Chem. Scand., 1969, 23, 2667

69RC   L. D. Rich, D. L. Cole, and E. M. Eyring, J. Phys. Chem., 1969, 73, 713

69RP   C. L. Rao and S. A. Pai, Radiochem. Acta, 1969, 12, 135

69RS   N. P. Rudenko and A. I. Sevastyanov, Russ. J. Inorg. Chem., 1969, 14, 441 (848)

69SA   I. A. Sheka and L. P. Andrusenko, Russ. J. Inorg. Chem., 1969, 14, 186 (362)

69SB   T. G. Sukhova, N. Ya. Borshch, O. N. Temkin, and R. M. Flid, Russ. J. Inorg.
        Chem., 1969, 14, 362 (694)

69SG   V. I. Sidorenko and V. I. Gordienko, J. Anal. Chem. USSR, 1969, 24, 499 (645)

69SGM   M. Shiloh, M. Givon, and Y. Marcus, J. Inorg. Nucl. Chem., 1969, 31, 1807

69SM   V. B. Spivakovskii and L. P. Moisa, Russ. J. Inorg. Chem., 1969, 14, 615 (1173)

69SMK   S. A. Sherif, N. E. Milad, and A. A. Khedr, J. Inorg. Nucl. Chem., 1969, 31, 3225

69SMT   E. V. Saksin, O. N. Malyavinskaya, and Ya. I. Turyan, Russ. J. Phys. Chem., 1969, 43, 283 (517)

69SP   G. Sahu and B. Prasad, J. Indian Chem. Soc., 1969, 46, 233

69SPa   L. Sharma and B. Prasad, J. Indian Chem. Soc., 1969, 46, 241

69SS   T. Sekine and M. Sakairi, Bull. Chem. Soc. Japan, 1969, 42, 2712

69ST   T. G. Sukhova, O. N. Temkin, and R. M. Flid, Russ. J. Inorg. Chem., 1969, 14, 483 (928)

69SW   G. Schwarzenbach and H. Wenger, Helv. Chim. Acta, 1969, 52, 644

69VB   F. H. Van Cauwelaert and H. J. Bosmans, Rev. Chim. Miner., 1969, 6, 611

69VM   V. P. Vasilev and P. S. Mukhina, Izv. Vyssh. Ucheb. Zaved., Khim., 1969, 12, 258

69VO   A. Vanni, G. Ostacoli, and E. Roletto, Ann. Chem., 1969, 59, 847

69VP   E. Verdier and J. Piro, Ann. Chim. (France) 1969, 4, 213

69VS   V. M. Vdovenko and O. B. Stebunov, Soviet Radiochem., 1969, 11, 625 (635)

69VSa   V. M. Vdovenko and O. B. Stebunov, Soviet Radiochem., 1969, 11, 630 (640)

69VV   V. P. Vasilev, P. N. Vorobev, and A. F. Belyakova, Izv. Vyssh. Ucheb. Zaved., Khim., 1969, 12, 115

69WK   J. I. Watters, S. Kalliney, and R. C. Machen, J. Inorg. Nucl. Chem., 1969, 31, 3817

69WKa   J. I. Watters, S. Kalliney, and R. C. Machen, J. Inorg. Nucl. Chem., 1969, 31, 3823

69WS   C. F. Wells and M. A.Salam, J. Inorg. Nucl. Chem., 1969, 31, 1083

69YM   L. B. Yeatts and W. L. Marshall, J. Phys. Chem., 1969, 73, 81

69ZL   O. E. Zvyagintsev and S. B. Lyakhmanov, Russ. J. Inorg. Chem., 1969, 14, 956, (1822)

70AB   J. Ascanio and F. Brito, An. Quim., 1970, 66, 617

70AL   A. Aziz and S. J. Lyle, J. Inorg. Nucl. Chem., 1970, 32, 1925

70AR   S. Ahrland and J. Rawsthorne, Acta Chem. Scand., 1970, 24, 157

70AS   A. Anagnostopoulos and P. O. Salkellaridis, J. Inorg. Nucl. Chem., 1970, 32, 1740

70AW   N. S. Al-Niaimi, A. G. Wain, and H. A. C. McKay, J. Inorg. Nucl. Chem., 1970, 32, 977

70AWa   N. S. Al-Niaimi, A. G. Wain, and H. A. C. McKay, J. Inorg. Nucl. Chem., 1970, 32, 2331

70B   A. M. Bond, J. Electroanal. Chem., 1970, 28, 433

70Ba   A. M. Bond, J. Electrochem. Soc., 1970, 117, 1145

70Bb   A. Bellomo, Talanta, 1970, 17, 1109

70Bc   E. W. Baumann, J. Inorg. Nucl. Chem., 1970, 32, 3823

70Bd   W. G. Baldwin, Arkiv Kemi, 1970, 31, 407

70BH   A. M. Bond and G. Hefter, Inorg. Chem., 1970, 9, 1021

70BO   A. M. Bond and T. A. O'Donnell, J. Electroanal. Chem., 1970, 26, 137

70BS   W. G. Baldwin and L. G. Sillen, Arkiv Kemi, 1970, 31, 391

70BT   A. M. Bond and R. J. Taylor, J. Electroanal. Chem., 1970, 28, 207

70BZ   K. A. Burkov, N. I. Zinevich, and L. S. Lilich, Izv. Vyssh. Ucheb. Zaved.,
         Khim., 1970, 13, 1250

70C   C. W. Childs, Inorg. Chem., 1970, 9, 2465

70CG   L. Ciavatta and M. Grimaldi, Inorg. Chim. Acta, 1970, 4, 312

70CGM   L. Ciavatta, M. Grimaldi, and A. Mangone, J. Inorg. Nucl. Chem., 1970, 32, 3805

70CJ   J. J. Christensen, H. D. Johnston and R. M. Izatt, J. Chem. Soc. (A), 1970, 454

70DS   V. I. Dubinskii and V. M. Shulman, Russ. J. Inorg. Chem., 1970, 15, 764 (1488)

70DSa   E. F. De Almeida Neves and L. Sant'Agostino, Anal. Chim. Acta, 1970, 49, 591

70E   B. Elgquist, J. Inorg. Nucl. Chem., 1970, 32, 937

70Ea   L. I. Elding, Acta Chem. Scand., 1970, 24, 1331, 1527

70Eb   L. I. Elding, Acta Chem. Scand., 1970, 24, 2546, 2557

70EE   M. Ebert and J. Eysseltova, J. Inorg. Nucl. Chem., 1970, 32, 967

70EL   M. Ehrenfreund and J. L. Leibenguth, Bull. Soc. Chim. France, 1970, 2494, 2498

70FC   V. A. Fedorov, G. E. Chernikova, and V. E. Mironov, Russ. J. Inorg. Chem., 1970,
         15, 1082 (2100)

70FS   V. A. Fedorov, N. P. Samsonova, and V. E. Mironov, Russ. J. Inorg. Chem., 1970,
         15, 1325 (2561)

70GF   R. Guillaumont, J. C. Franck, and R. Muxart, Radiochem. Radioanal. Letters, 1970,
         4, 73

70GG   A. W. Gardner and E. Glueckauf, Trans. Faraday Soc., 1970, 66, 1081

70GH   A. O. Gubeli, J. Hebert, P. A. Cote; and R. Taillon, Helv. Chim. Acta, 1970, 53,
         186

70GHa   A. O. Gubeli, J. Hebert, R. Taillon, and P. A. Cote, Helv. Chim. Acta, 1970, 53,
         1229

70GK   H. Gamsjaeger, W. Kraft, and P. Schindler, Helv. Chim. Acta, 1970, 53, 290

70GM   T. M. Gregory, E. C. Moreno, and W. E. Brown, J. Res. Nat. Bur. Stand., 1970,
          74A, 461

70GN   R. Ghosh and V. S. K. Nair, J. Inorg. Nucl. Chem., 1970, 32, 3041

70GNa   G. L. Gardner and G. H. Nancollas, Anal. Chem., 1970, 42, 794

70GO   I. Grenthe, H. Ots, and O. Ginstrup, Acta Chem. Scand., 1970, 24, 1067

70GS   A. M. Gorelov and A. P. Shtin, Russ. J. Inorg. Chem., 1970, 15, 1655 (3178)

70GSM   V. I. Gordienko, V. I. Sidorenko, and Yu. I. Mikhailyuk, Russ. J. Inorg. Chem.,
          1970, 15, 1241 (2397)

70GZ   A. Gunter and A. Zuberbuhler, Chimia (Switz.), 1970, 24, 340

70HK   D. S. Honig and K. Kustin, J. Inorg. Nucl. Chem., 1970, 32, 1599

70HKS   Y. Hasegawa, H. Kawashima, and T. Sekine, Bull. Chem. Soc. Japan, 1970, 43, 1718

70HR   P. Hemmes, L. D. Rich, D. L. Cole, and E. M. Eyring, J. Phys. Chem., 1970, 74,
          2859

70HS   S. Hietanen and L. G. Sillen, Arkiv Kemi , 1970, 32, 111

70HV   E. Halloff and N. G. Vannerberg, Acta Chem. Scand., 1970, 24, 55

70HW   W. J. Hamer and Y. C. Wu, J. Res. Nat. Bur. Standards, 1970, 74A, 761

70IE   B. N. Ivanov-Emin, A. M. Egorov, V. I. Romanyuk, and E. N. Siforova, Russ. J.
          Inorg. Chem., 1970, 15, 628 (1224)

70IV   A. A. Ivakin and E. M. Voronova, Tr. Inst. Khim., Akad. Nauk SSSR, Ural. Filial,
          1970, No. 17, 144; Chem. Abs., 1970, 73, 134396z

70KB   H. Krentzien and F. Brito, Ion, 1970, 30, 14

70KBM   G. N. Kozachenko, I. M. Batyaev, and V. E. Mironov, Russ. J. Inorg. Chem., 1970,
          15, 452 (888)

70KK   A. E. Klygin, N. S. Kolyada, and I. D. Smirnova, Russ. J. Inorg. Chem., 1970,
          15, 1719 (3300)

70KS   A. E. Klygin, I. D. Smirnova, and D. M. Zavrazhnova, Russ. J. Inorg. Chem., 1970,
          15, 155 (294)

70KT   V. I. Kravtsov, N. V. Titova, and G. P. Tsayun, Elektrokhim., 1970, 6, 573

70KY   F. Ya. Kulba, Yu. B. Yakovlev, and E. A. Kopylov, Russ. J. Inorg. Chem., 1970,
          15, 1088 (2112)

70L   J. W. Larson, J. Phys. Chem., 1970, 74, 3392

70La   F. Letowski, Rocz. Chem., 1970, 44, 1665

70Lb   G. M. Lafon, Geochim. Cosmochim. Acta, 1970, 34, 935

70Lc   O. Lukkari, Suomen Kem., 1970, B43, 347

70Ld   J. O. Liljenzin, Acta Chem. Scand., 1970, 24, 1655

70LK   H. Lahr and W. Knoch, Radiochim. Acta, 1970, 13, 1

70LS   B. N. Laskorin, V. F. Smirnov, and V. I. Nikonov, Russ. J. Inorg. Chem., 1970,
         15, 1724 (3310)

70M   A. I. Moskvin, Russ. J. Inorg. Chem., 1970, 15, 1756 (3368)

70MA   V. M. Mylnikova, K. V. Astakhov, and S. A. Barkov, Russ. J. Phys. Chem., 1970,
         44, 314 (560)

70MG   M. C. Mehra and A. O. Gubeli, Canad. J. Chem., 1970, 48, 3491

70MGa   M. C. Mehra and A. O. Gubeli, J. Less-Common Metals, 1970, 22, 281

70MM   V. E. Mironov, Yu. A. Makashev, I. Ya. Mavrina, and M. M. Kryshanovskii, Russ.
         J. Inorg. Chem., 1970, 15, 668 (1301)

70MS   L. P. Moisa and V. B. Spivakovskii, Russ. J. Inorg. Chem., 1970, 15, 1513 (2907)

70MSS   T. Murayama, T. Sawaki, and S. Sakuraba, Bull. Chem. Soc. Japan, 1970, 43, 2820

70NK   B. I. Nabivanets and L. V. Kalabina, Russ. J. Inorg. Chem., 1970, 15, 818 (1595)

70OE   J. D.Owen and E. M. Eyring, J. Inorg. Nucl. Chem., 1970, 32, 2217

70OP   N. A. Orlin and V. E. Plyushchev, Russ. J. Inorg. Chem., 1970, 15, 227 (439),
         229 (442)

70P   H. Persson, Acta Chem. Scand., 1970, 24, 3739

70PH   J. Prasilova and J. Havlicek, J. Inorg. Nucl. Chem., 1970, 32, 953

70PM   G. Pilloni and F. Magno, Inorg. Chim. Acta, 1970, 4, 105

70RG   W. F. Rittner, A. Gulko, and G. Schmuckler, Talanta, 1970, 17, 807

70RS   V. K. Rao, C. J. Shahani, and C. L. Rao, Radiochim. Acta, 1970, 14, 31

70SB   F. H. Sweeton and C. F. Baes, Jr., J. Chem. Thermodyn., 1970, 2, 479

70SG   P. Senise and O. E. S. Godinho, J. Inorg. Nucl. Chem., 1970, 32, 3641

70SGK   T. M. Sas, V. A. Gagarina, L. N. Komissarova, and V. G. Gulia, Russ. J. Inorg.
         Chem., 1970, 15, 644 (1255)

70SK   T. W. Swaddle and P. Kong, Canad. J. Chem., 1970, 48, 3224

70SP   L. Sharma and B. Prasad, J. Indian Chem. Soc., 1970, 47, 193

70SS   V. I. Shlenskaya, N. G. Shumkova, and A. A. Biryukov, J. Anal. Chem. USSR, 1970,
         25, 1852 (2155)

70TR   Ya. I. Turyan and O. E. Ruvinskii, J. Electroanal. Chem., 1970, 28, 381

70VA   V. P. Vasilev, S. A. Aleksandrova and L. A. Kochergina, Russ. J. Inorg. Chem.,
         1970, 15, 899 (1751)

70VAa   V. P. Vasilev, S. A. Aleksandrova and L. A. Kochergina, Russ. J. Inorg. Chem.,
         1970, 15, 1659 (3185)

70VT   N. T. Voskresenskaya and N. V. Timofeeva, Russ. J. Inorg. Chem., 1970, 15, 1352
         (2608)

70VV   V. P. Vasilev and P. N. Vorobev, Russ. J. Phys. Chem., 1970, 44, 657 (1181)

70W   J. B. Walker, J. Inorg. Nucl. Chem., 1970, 32, 2793

71AB   K. P. Anderson, E. A. Butler and E. M. Woolley, J. Phys. Chem., 1971, 75, 93

71AK   S. Ahrland and L. Kullberg, Acta Chem. Scand., 1971, 25, 3457, 3471

71AKa   S. Ahrland and L. Kullberg, Acta Chem. Scand., 1971, 25, 3677

71AKb   S. Ahrland and L. Kullberg, Acta Chem. Scand., 1971, 25, 3692

71AM   R. P. Agarwal and E. C. Moreno, Talanta, 1971, 18, 873

71AO   S. Akalin and U. Y. Ozer, J. Inorg. Nucl. Chem., 1971, 33, 4171

71B   A. M. Bond, Anal. Chim. Acta, 1971, 53, 159

71Ba   T. F. Bidleman, Anal. Chim. Acta, 1971, 56, 221

71BG   V. P. Biryukov and E. Sh. Ganelina, Russ. J. Inorg. Chem., 1971, 16, 320 (600)

71BH   A. M. Bond and G. Hefter, J. Inorg. Nucl. Chem., 1971, 33, 429

71BHa   A. M. Bond and G. Hefter, J. Electroanal. Chem., 1971, 31, 477

71BL   A. R. Bailey and J. W. Larson, J. Phys. Chem., 1971, 75, 2368

71BN   T. A. Belyavskaya and I. A. Nemirovskaya, Moscow Univ. Chem. Bull., 1971, No. 6,
         95 (745)

71BS   G. Biedermann and T. G. Spiro, Chem. Scripta, 1971, 1, 155

71BSa   L. Barcza and L. G. Sillen, Acta Chem. Scand., 1971, 25, 1250

71BSb   L. P. Barchuk and I. A. Sheka, Russ. J. Inorg. Chem., 1971, 16, 1268 (2378)

71BSB   N. Bertazzi, A. Silvestri, and R. Barbieri, J. Inorg. Nucl. Chem., 1971, 33,
         799

71BZ   K. A. Burkov, N. I. Zinevich, and L. S. Lilich, Russ. J. Inorg. Chem., 1971, 16,
         926 (1746)

71CD   R. O. Cook, A. Davies, and L. A. K. Staveley, J. Chem. Thermodyn., 1971, 3, 907

71CV   J. Cadek, J. Vesely and Z. Sulcek, Coll. Czech. Chem. Comm., 1971, 36, 3377

71D   V. I. Dubinskii, Russ. J. Inorg. Chem., 1971, 16, 607

71Da   A. D'Aprano, J. Phys. Chem., 1971, 75, 3290

71DB   N. Dezelic, H. Bilinski, and R. H. H. Wolf, J. Inorg. Nucl. Chem., 1971, 33, 791

V. BIBLIOGRAPHY

71DC   P. R. Danesi, R. Chiarizia, G. Scibona and G. D'Alessandro, _J. Inorg. Nucl._
             Chem., 1971, _33_, 3503; 1974, _36_, 2396

71DD   R. C. Das, A. C. Dash, D. Satyanarayan, and U. N. Dash, _Thermochim. Acta_, 1971,
             _2_, 435

71DG   H. Dautet and R. Guillaumont, _Radiochem. Radioanal. Letters_, 1971, _8_, 175

71EG   L. I. Elding and L. Gustafson, _Inorg. Chim. Acta_, 1971, _5_, 643

71EGa   A. J. Ellis and W. Giggenbach, _Geochim. Cosmochim. Acta_, 1971, _35_, 247

71EM   C. E. Evans and C. B. Monk, _Trans. Faraday Soc._, 1971, _67_, 2652

71FCK   V. A. Fedorov, G. E. Chernikova, T. N. Kalosh, and M. E. Mironov, _Russ. J._
             Inorg. Chem., 1971, _16_, 170 (325)

71FCM   V. A. Fedorov, G. E. Chernikova, and V. E. Mironov, _Russ. J. Inorg. Chem._, 1971,
             _16_, 489 (918)

71FK   V. A. Fedorov, T. N. Kalosh, G. E. Chernikova, and V. E. Mironov, _Russ. J. Phys._
             Chem., 1971, _45_, 106 (186)

71FKa   V. A. Fedorov, T. N. Kalosh, G. E. Chernikova, and V. E. Mironov, _Russ. J. Phys._
             Chem., 1971, _45_, 775 (1364)

71FKM   V. A. Fedorov, T. N. Kalosh, and V. E. Mironov, _Russ. J. Inorg. Chem._, 1971,
             _16_, 539 (1014)

71FKS   V. A. Fedorov, T. N. Kalosh, L. I. Shmydko, and V. E. Mironov, _Russ. J. Inorg._
             Chem., 1971, _16_, 1276 (2393)

71FR   V. A. Fedorov, A. M. Robov, I. D. Isaev, and V. E. Mironov, _Russ. J. Inorg._
             Chem., 1971, _16_, 500 (940)

71G   W. Giggenbach, _Inorg. Chem._, 1971, _10_, 1333

71GB   E. S. Ganelina and V. A. Borgoyakov, _Russ. J. Inorg. Chem._, 1971, _16_, 318 (596)

71GD   R. Guillaumont, B. Desire, and M. Galin, _Radiochem. Radioanal. Letters_, 1971,
             _8_, 189

71GF   E. A. Gyunner and A. M. Fedorenko, _Russ. J. Inorg. Chem._, 1971, _16_, 1784 (3371)

71GH   S. L. Grassino and D. N. Hume, _J. Inorg. Nucl. Chem._, 1971, _33_, 421

71HR   P. Hemmes, L. D. Rich, D. L. Cole, and E. M. Eyring, _J. Phys. Chem._, 1971, _75_,
             929

71IB   B. N. Ivanov-Emin, L. D. Borzova, A. M. Egorov, and S. G. Malyugina, _Russ. J._
             Inorg. Chem., 1971, _16_, 1474 (2766)

71IJ   R. M. Izatt H. D. Johnston, D. J. Eatough, J. W. Hansen and J. J. Christensen,
             Thermochim. Acta, 1971, _2_, 77

71K   A. S. Kereichuk, _Russ. J. Inorg. Chem._, 1971, _16_, 751 (1424)

71KB   N. N. Kozachenko and I. M. Batyaev, _Russ. J. Inorg. Chem._, 1971, _16_, 66 (125)

71KBa   H. C. Kaehler and F. Brito, An. Quim., 1971, 67, 1185

71KM   V. I. Kravtsov and L. B. Martynova, Russ. J. Inorg. Chem., 1971, 16, 457 (858)

71KMF   P. Klotz, A. Mukherji, S. Feldberg, and L. Newman, Inorg. Chem., 1971, 10, 740

71KN   P. K. Khopkar and P. Narayanankutty, J. Inorg. Nucl. Chem., 1971, 33, 495

71KP   L. N. Komissarova, N. M. Prutkova, and G. Ya. Pushkina, Russ. J. Inorg. Chem.,
         1971, 16, 954 (1798)

71KS   S. Kiciak and T. Stefanowicz, Rocz. Chem., 1971, 46, 1801

71KSa   L. O. Spreer and E. L. King, Inorg. Chem., 1971, 10, 916

71M   A. I. Moskvin, Russ. J. Inorg. Chem., 1971, 16, 405 (759)

71Ma   N. B. Milic, Acta Chem. Scand., 1971, 25, 2487

71Mb   R. E. Mesmer, Inorg. Chem., 1971, 10, 857

71MB   R. E. Mesmer and C. F. Baes, Jr., Inorg. Chem., 1971, 10, 2290

71MD   S. Mateo, A. Diaz, and F. Brito, An. Quim., 1971, 67, 1179

71MG   M. C. Mehra and A. O. Gubeli, J. Less-Common Metals, 1971, 25, 221

71MH   D. F. C. Morris, T. J. Hedger, and P. A. Watson, J. Inorg. Nucl. Chem., 1971,
         33, 2077

71MKA   Yu. A. Makashev, F. Ya. Kulba, M. I. Agaf, Yu. A. Volokhov, and V. E. Mironov,
         Russ. J. Phys. Chem., 1971, 45, 414 (735)

71MKK   I. V. Melchakova, D. G. Khadzhidemetriu, N. A. Krasnyanskaya, and V. M.
         Peshkova, Russ. J. Inorg. Chem., 1971, 16, 1054 (1981)

71MM   J. P. Manners, K. G. Morallee, and R. J. P. Williams, J. Inorg. Nucl. Chem.,
         1971, 33, 2085

71MO   K. Momoki and H. Ogawa, Anal. Chem., 1971, 43, 1664

71MS   H. Moriya and T. Sekine, Bull. Chem. Soc. Japan, 1971, 44, 3347

71N   E. F. A. Neves, J. Inorg. Nucl. Chem., 1971, 33, 571

71Na   F. S. Nakayama, J. Chem. Eng. Data, 1971, 16, 178

71Nb   F. S. Nakayama, J. Inorg. Nucl. Chem., 1971, 33, 1287

71Nc   G. Neumann, Arkiv Kemi, 1971, 32, 229

71NA   V. A. Nazarenko, V. P. Antonovich, and E. M. Nevskaya, Russ. J. Inorg. Chem.,
         1971, 16, 530 (997)

71NAa   V. A. Nazarenko, V. P. Antonovich, and E. M. Nevskaya, Russ. J. Inorg. Chem.,
         1971, 16, 1273 (2387)

71NL   B. I. Nabivanets and V. V. Lukachina, Soviet Prog. Chem. (Ukr. Khim. Zh.), 1971,
         37, No. 6, 60 (581)

710B   H. Ohtaki and G. Biedermann, Bull. Chem. Soc. Japan, 1971, 44, 1822

71P   H. Persson, Acta Chem. Scand., 1971, 25, 543

71Pa   L. Pettersson, Acta Chem. Scand., 1971, 25, 1959

71PB   B. I. Peshchevitskii and N. V. Kurbatova, Russ. J. Inorg. Chem., 1971, 16, 1007
        (1898)

71PH   J. C. Pierrard and R. Hugel, Rev. Chim. Minerale, 1971, 8, 831

71PJ   R. Paterson, S. K. Jalota, and H. S. Dunsmore, J. Chem. Soc. (A), 1971, 2116

71PK   J. C. Pierrard, C. Kappenstein, and R. Hugel, Rev. Chim. Minerale, 1971, 8, 11

71PM   G. Pilloni and F. Magno, Inorg. Chim. Acta, 1971, 5, 30

71PMP   P. R. Patel, E. C. Moreno, and J. M. Patel, J. Res. Nat. Bur. Stand., 1971, 75A,
        205

71PR   W. J. Popiel and M. S.Rustom, Chem. Ind. (London), 1971, 543

71PS   A. T. Pilipenko, I. P. Sereda, and Z. A. Semchinskaya, Russ. J. Inorg. Chem.,
        1971, 16, 1349 (2529)

71PT   T. I. Pochkaeva, N. S. Tamm, and A. V. Novoselova, Russ. J. Inorg. Chem., 1971,
        16, 113 (219)

71PW   R. A. Porter and W. J.Weber, Jr., J. Inorg. Nucl. Chem., 1971, 33, 2443

71S   U. Schedin, Acta Chem. Scand., 1971, 25, 747

71SB   I. A. Sheka and L. P. Barchuk, Russ. J. Inorg. Chem., 1971, 16, 1573 (2961)

71SK   T. Sekine, Y. Kimatsu, and M. Sakairi, Bull. Chem. Soc. Japan, 1971, 44, 1480

71SKa   A. Samotus and B. Kosowicz-Czajkowska, Rocz. Chem., 1971, 46, 1623

71SM   T. Sekine, R. Murai, and M. Iguchi, Nippon Kagaku Zasshi, 1971, 92, 412

71SS   K. Srinivasan and R. S. Subrahmanya, J. Electroanal. Chem., 1971, 31, 233

71T   J. Tummavuori, Suomen Kem., 1971, B44, 222

71Ta   J. Tummavuori, Suomen Kem., 1971, B44, 343

71Tb   J. Tummavuori, Suomen Kem., 1971, B44, 350

71TL   J. Tummavuori and P. Lumme, Suomen Kem., 1971, B44, 215

71TM   L. R. Tokareva and M. V. Mokhosoev, Russ. J. Inorg. Chem., 1971, 16, 2388 (2417)

71TR   M. M. Taqui Khan and P. R. Reddy, J. Inorg. Nucl. Chem., 1971, 33, 1427

71TS   K. Teruya, T. Seki, and I. Nakamori, Kogyo Kagaku Zasshi, 1971, 74, 142; Chem.
        Abs., 1971, 74, 103712r

71V   F. Vierling, Bull. Soc. Chim. France, 1971, 22, 25

71VP   Y. A. Volokhov, L. N. Pavlov, N. I. Eremin, and V. E. Mironov, J. Appl. Chem.
        USSR, 1971, 44, 243 (246)

71WT   J. B. Walker, C. R. Twine, and G. R. Choppin, J. Inorg. Nucl. Chem., 1971, 33,
        1813

72AB   R. Arnek and L. Barcza, Acta Chem. Scand., 1972, 26, 213

72AG   R. Ya. Aliev, M. N. Guseinov, and A. D. Kuliev, Russ. J. Phys. Chem., 1972, 46,
        1520 (2657)

72AJ   R. Arnek and S. R. Johansson, Acta Chem. Scand., 1972, 26, 2903

72AK   R. Ya. Aliev and A. D. Kuliev, Russ. J. Inorg. Chem., 1972, 17, 1639 (3118)

72B   E. W. Baumann, J. Inorg. Nucl. Chem., 1972, 34, 687

72BA   N. Bertazzi, G. Alonzo, and A. Silvestri, J. Inorg. Nucl. Chem., 1972, 34, 1943

72BB   V. I. Blokhin, T. N. Bukhtiyarova, N. N. Krot, and A. D. Gelman, Russ. J. Inorg.
        Chem., 1972, 17, 1262 (2420)

72BBa   V. I. Blokhin, T. N. Bukhtiyarova, N. N. Krot, and A. D. Gelman, Russ. J. Inorg.
        Chem., 1972, 17, 1742 (3317)

72BBM   E. A. Belousov, V. V. Bocharov, and V. E. Mironov, Russ. J. Inorg. Chem., 1972,
        17, 1717 (3265)

72BC   P. A. Baisden, G. R. Choppin, and W. F. Kinard, J. Inorg. Nucl. Chem., 1972, 34,
        2029

72BH   A. M. Bond and G. Hefter, J. Electroanal. Chem., 1972, 34, 227

72BHa   A. M. Bond and G. Hefter, J. Inorg. Nucl. Chem., 1972, 34, 603

72BP   M. C. Bonnet, R. A. Paris and R. P. Martin, Bull. Soc. Chim. France, 1972, 903

72BPa   M. C. Bonnet, R. A. Paris and R. P. Martin, Bull. Soc. Chim. France, 1972, 909

72BPb   N. G. Bogdanovich, N. I. Pechurova, L. I. Martynenko, L. S. Koltsova, and V. V.
        Piunova, Moscow Univ. Chem. Bull., 1972, 27, No. 2, 84 (236)

72BT   M. J. Burkhart and R. C. Thompson, J. Amer. Chem. Soc., 1972, 94, 2999

72CB   G. Carpeni, E. Boitard, R. Pilard, S. Poize and N. Sabiani, J. Chim. Phys., 1972,
        69, 1445

72CD   C. Collier, L. Donneau, M. Fournier, and M. Quintin, J. Chim. Phys., 1972, 69,
        945

72CG   S. Cabani and P. Gianni, Anal. Chem., 1972, 44, 253

72CK   J. J. Christensen, G. L. Kimball, H. D. Johnston, and R. M. Izatt, Thermochim.
        Acta, 1972, 4, 141

72CMP   A. Cassol, L. Magon, R. Portanova, and E. Tondello, Radiochim. Acta, 1972, 17,
        28

72CMT   A. Cassol, L. Magon, G. Tomat, and R. Portanova, Inorg. Chem., 1972, 11, 515

72CP   C. Y. Chan and M. H. Panckhurst, Aust. J. Chem., 1972, 25, 311, 317

72DD   A. D'Aprano and I. D. Donato, Electrochim. Acta, 1972, 17, 1175

72DJ   H. S. Dunsmore, S. K. Jalota, and R. Paterson, J. Chem. Soc. Faraday I, 1972,
        68, 1583

72DN   C. Dragulescu, A. Nimara, and I. Julean, Rev. Roum. Chem., 1972, 7, 1181

72DS   A. Davies and J. Staveley, J. Chem. Thermodyn., 1972, 4, 267

72E   L. I. Elding, Inorg. Chim. Acta, 1972, 6, 647

72F   D. Ferri, Acta Chem. Scand., 1972, 26, 733

72Fa   D. Ferri, Acta Chem. Scand., 1972, 26, 747

72FI   V. A. Fedorov, I. D. Isaev, A. M. Robov, A. V. Vertiprakhov, and V. E. Mironov,
        Russ. J. Inorg. Chem., 1972, 17, 495 (951)

72FKM   V. A. Fedorov, L. I. Kiprin, and V. E. Mironov, Russ. J. Inorg. Chem., 1972, 17,
        641 (1233)

72FKS   V. A. Fedorov, T. N. Kalosh, L. I. Shmydko, and V. E. Mironov, Russ. J. Inorg.
        Chem., 1972, 17, 1086 (2089)

72FR   V. A. Fedorov, A. M. Robov, T. I. Grigor, and V. E. Mironov, Russ. J. Inorg.
        Chem., 1972, 17, 990 (1909)

72FS   C. M. Frey and J. E. Stuehr, J. Amer. Chem. Soc., 1972, 94, 8898

72FSa   V. A. Fedorov, N. P. Samsonova, and V. E. Mironov, Russ. J. Inorg. Chem., 1972,
        17, 674 (1301)

72FSb   V. A. Fedorov, L. P. Shishin, and V. E. Mironov, Russ. J. Inorg. Chem., 1972,
        17, 836 (1616)

72FSL   V. A. Fedorov, L. P. Shishin, S. G. Likhacheva, A. V. Fedorova, and V. E.
        Mironov, Russ. J. Inorg. Chem., 1972, 17, 41 (79)

72GC   A. O. Gubeli and P. A. Cote, Canad. J. Chem., 1972, 50, 1144

72GR   A. O. Gubeli and J. P. Retel, Helv. Chim. Acta, 1972, 55, 1429

72H   G. Hefter, J. Electroanal. Chem., 1972, 39, 345

72HF   R. D. Hancock, N. P. Finkelstein and A. Evers, J. Inorg. Nucl. Chem. 1972, 34,
        3747

72HH   H. Hussonnois, S. Hubert, L. Aubin, R. Guillaumont, and G. Boussieres,
        Radiochem. Radionanal. Letters, 1972, 10, 231

72HP   H. D. Harmon, J. R. Peterson, and W. J. McDowell, Inorg. Nucl. Chem. Letters,
        1972, 8, 57

72HPa   H. D. Harmon, J. R. Peterson, W. J. McDowell, and C. F. Coleman, J. Inorg. Nucl.
        Chem., 1972, 34, 1381

72HPB   H. D. Harmon, J. R. Peterson, J. T. Bell, and W. J. McDowell, J. Inorg. Nucl.
        Chem., 1972, 34, 1711

72HS   C. F. Hale and F. S. Spedding, J. Phys. Chem., 1972, 76, 1887

72J   J. B. Jenson, Acta Chem. Scand., 1972, 26, 4031

72JW   R. F. Jameson and M. F. Wilson, J. Chem. Soc. Dalton, 1972, 2607

72KB   V. S. Kublanovskii and V. N. Belinskii, Russ. J. Inorg. Chem., 1972, 17, 68 (129)

72KMB   H. Kaehler, S. Mateo, and F. Brito, An. Quim., 1972, 68, 1215

72KMN   T. Kimura, K. Morinaga, and K. Nakano, Nippon Kagaku Kaishi, 1972, 664

72KO   G. Kura and S. Ohashi, J. Inorg. Nucl. Chem., 1972, 34, 3899

72L   S. Lasztity, Radiochem. Radioanal. Letters, 1972, 12, 33

72LG   V. A. Leitsin, S. D. Grekov, T. P. Sirina, and B. S. Pritsker, Russ. J. Inorg.
          Chem., 1972, 17, 687 (1325)

72LL   O. Lukkari and H. Lukkari, Suomen Kem., 1972, B45, 6

72LO   J. E. Land and C. V. Osborne, J. Less-Common Metals, 1972, 29, 147

72MB   S. Mateo and F. Brito, An. Quim., 1972, 68, 38

72MBS   R. E. Mesmer, C. F. Baes, Jr., and F. H. Sweeton, Inorg. Chem., 1972, 11, 537

72MC   W. J. McDowell and C. F. Coleman, J. Inorg. Nucl. Chem., 1972, 34, 2837

72MG   H. Metivier and R. Guillaumont, Radiochem. Radioanal. Letters, 1972, 10, 27

72MH   D. F. C. Morris, F. B. Haynes, P. A. Lewis, and E. L. Short, Electrochim.
          Acta, 1972, 17, 2017

72MR   P. G. Manning and S. Ramamoorthy, Inorg. Nucl. Chem. Letters, 1972, 8, 653

72MT   T. Murayama and A. Takayanagi, Bull. Chem. Soc. Japan, 1972, 45, 3549

72MV   O. I. Martynova, L. G. Vasina, and S. A. Pozdnyakova, Doklady Chem., 1972, 202,
          173 (1337)

72N   J. O. Nriagu, Geochim. Cosmochim. Acta, 1972, 36, 459

72Na   J. O. Nriagu, Amer. J. Sci., 1972, 272, 476

72Nb   J. O. Nriagu, Inorg. Chem., 1972, 11, 2499

72NE   L. V. Nazarova, T. M. Efremova, and S. I. Orenshtein, Russ. J. Inorg. Chem.,
          1972, 17, 186 (357)

72NS   E. F. A. Neves and P. Senise, J. Inorg. Nucl. Chem., 1972, 34, 1915

72O   H. Ots, Acta Chem. Scand., 1972, 26, 3810

72OK   H. Ohtaki and T. Kawai, Bull. Chem. Soc. Japan, 1972, 45, 1735

72OS   S. O'Cinneide, J. P. Scanlan, and M. J. Hynes, Chem. Ind. (London), 1972, 340

72P   J. C. Pierrard, Thesis, 1972, Univ. Reims (France)

72PR   J. Pouradier and J. Rigola, Compt. Rend. Acad. Sci. Paris, Sec. C, 1972, 275, 515

72R   T. Ryhl, Acta Chem. Scand., 1972, 26, 2961

72RC   R. Roulet and R. Chenaux, Helv. Chim. Acta, 1972, 55, 1959

72S   R. N. Sylva, Rev. Pure Appl. Chem., 1972, 22, 115

72SB   V. I. Shlenskaya, A. A. Biryukov, and V. M. Kadomtseva, Russ. J. Inorg. Chem.,
         1972, 17, 572 (1104)

72SC   R. Stampfli and G. R. Choppin, J. Inorg. Nucl. Chem., 1972, 34, 205

72SF   N. P. Samsonova, V. A. Fedorov, and V. E. Mironov, Russ. J. Phys. Chem., 1972,
         46, 1233 (2153)

72SK   T. Stefanowicz and S. Kiciak, Rocz. Chem., 1972, 46, 1209

72SN   P. Senise and E. F. A. Neves, J. Inorg. Nucl. Chem., 1972, 34, 1923

72TL   J. Tummavuori and P. Lumme, Suomen Kem., 1972, B45, 21

72TS   S. Tribalat and L. Schriver, Compt. Rend. Acad. Sci. Paris, Sec. C, 1972, 274,
         849

72TSa   Ya. I. Turyan and N. K. Strizhov, Russ. J. Inorg. Chem., 1972, 17, 1066 (2053)

72US   L. N. Usherenko and N. A. Skorik, Russ. J. Inorg. Chem., 1972, 17, 1533 (2918)

72V   F. Vierling, Bull. Soc. Chim. France, 1972, 2557

72Va   F. Vierling, Bull. Soc. Chim. France, 1972, 2563

72VK   C. E. Vanderzee, D. L. King and I. Wadso, J. Chem. Thermodyn., 1972, 4, 685

73A   R. G. Ainsworth, J. Chem. Soc. Faraday I, 1973, 69, 1028

73Aa   I. Ahlberg, Acta Chem. Scand., 1973, 27, 3003

73Ab   N. I. Ampelogova, Soviet Radiochem., 1973, 15, 823 (813)

73AA   R. Abu-Eittah and G. Arafa, Z. Anorg. Allg. Chem., 1973, 399, 244

73AB   J. Ascanio and F. Brito, An. Quim., 1973, 69, 177

73AK   T. Amaya, H. Kakihana, and M. Maeda, Bull. Chem. Soc. Japan, 1973, 46, 1720,
         2889

73AM   J. M. Austin and A. D. Mair, unpublished data (see 73HP)

73B   G. Bengtsson, Acta Chem. Scand., 1973, 27, 2554

73Ba   W. L. Bradford, Limnol. Oceanog., 1973, 18, 757

73BF   C. P. Bezboruah, M. Filomena, G. F. C. Camoes, A. K. Covington, and J. V. Dobson,
         J. Chem. Soc. Faraday I, 1973, 69, 949

73BI   O. Budevsky, F. Ingman, and D. H. Liem, Acta Chem. Scand., 1973, 27, 1277

73BL   K. A. Burkov, L. S. Lilich, N. D. Ngo, and A. Yu. Smirnov, Russ. J. Inorg. Chem.,
           1973, 18, 797 (1513)

73BN   E. A. Biryuk and V. A. Nazarenko, Russ. J. Inorg. Chem., 1973, 18, 1576 (2964)

73BR   M. Barres, J. P. Redoute, R. Romanetti, H. Tachoire, and C. Zahra, Compt. Rend.
           Acad. Sci. Paris, Sec. C, 1973, 276, 363

73BS   P. Benes and H. Selecka, Radiochem. Radioanal. Letters, 1973, 13, 339

73CB   G. R. Choppin and S. L. Bertha, J. Inorg. Nucl. Chem., 1973, 35, 1309

73CD   A. K. Covington, J. V. Dobson and K. V. Srinivasan, J. Chem. Soc. Faraday I,
           1973, 69, 94

73CDa   R. Chiarizia, . R. Danesi, and G. Scibona, J. Inorg. Nucl. Chem., 1973, 35,
           3595

73CG   L. Ciavatta and M. Glimaldi, Gazz. Chim. Ital., 1973, 103, 731

73CH   C. L. Christ, P. B. Hostetler, and R. M. Siebert, Amer. J. Sci., 1973, 273, 65

73CP   Y. Couturier and C. Petitfaux, Bull. Soc. Chim. France, 1973, 439

73CZ   D. M. Czakis-Sulikowska and A. Zajdler, Rocz. Chem., 1973, 47, · 3

73DH   D. Dyrssen and I. Hansson, Marine Chem., 1973, 1, 137

73FA   G. Ferroni, G. Antonetti, R. Romanetti and J. Galea, Bull. Soc. Chim. France,
           1973, 3269

73FC   V. A. Fedorov, G. E. Chernikova, M. A. Kuznechikhina, and V. E. Mironov, Russ.
           J. Inorg. Chem., 1973, 18, 337 (645)

73FP   M. M. Farrow and N. Purdie, J. Soln. Chem., 1973, 2, 503

73FR   V. A. Fedorov, A. M. Robov, I. I. Shmydko, and V. E. Mironov, Russ. J. Inorg.
           Chem., 1973, 18, 180 (342)

73FS   V. A. Fedorov, I. I. Shmydko, A. M. Robov, L. S. Simaeva, V. A. Kukhtina, and
           V. E. Mironov, Russ. J. Inorg. Chem., 1973, 18, 673 (1274)

73G   V. I. Gordienko, J. Gen. Chem. USSR, 1973, 43, 2109 (2117)

73GG   J. Y. Guennec and R. Guillaumont, Radiochem. Radioanal. Letters, 1973, 13, 33

73GS   A. Gulko and G. Schmuckler, J. Inorg. Nucl. Chem., 1973, 35, 603

73GT   V. P. Goncharik, L. P. Tikhonova, and K. B. Yatsimirskii, Russ. J. Inorg.
           Chem., 1973, 18, 658 (1248)

73HH   M. H. Hutchinson and W. C. E. Higginson, J. Chem. Soc. Dalton, 1973, 1247

73HHa   J. Havel and E. Hogfeldt, Acta Chem. Scand., 1973, 27, 3323

73HHb   M. Hussonnois, S. Hubert, L. Brillard, and R. Guillaumont, Radiochem. Radioanal.
           Letters, 1973, 15, 47

73HP   G. R. Hedwig and H. K. J. Powell, J. Chem. Soc. Dalton, 1973, 798

73IV   A. A. Ivakin and E. M. Voronova, Russ. J. Inorg. Chem., 1973, 18, 465 (885)

73IVa   A. A. Ivakin and E. M. Voronova, Russ. J. Inorg. Chem., 1973, 18, 956 (1809)

73J   L. Johansson, Acta Chem. Scand., 1973, 27, 1832

73KB   Yu. A. Kozlov, V. V. Blokhin, V. V. Shurukhin, and V. E. Mironov, Russ. J. Phys.
            Chem., 1973, 47, 1343 (2386)

73KK   F. Ya. Kulba, E. A. Kopylon, and Yu. B. Yakovlev, Russ. J. Inorg. Chem., 1973,
            18, 38 (76)

73KM   H. C. Kaehler, S. Mateo, and F. Brito, An. Quim., 1973, 69, 1269, 1273

73KN   N. P. Komar, V. A. Naumenko, E. M. Sokolskaya, and T. G. Shapovalova, Russ. J.
            Phys. Chem., 1973, 47, 1588 (2832)

73KO   T. Kawai, H. Otsuka, and H. Ohtaki, Bull. Chem. Soc. Japan, 1973, 46, 3753

73KP   N. N. Kozachenko, N. A. Panteleeva, V. S. Netsvetaeva and I. M. Batyaev, Russ.
            J. Inorg. Chem., 1973, 18, 938

73KR   E. G. Krunchak, A. G. Rodichev, Ya. S. Khvorostin, B. S. Krumgalz, V. G.
            Krunchak, and Yu. I. Yusova, Russ. J. Inorg. Chem., 1973, 18, 1519 (2859)

73KT   V. I. Kravtsov, E. G. Tsventarnyi, and A. N. Kochetkova, Russ. J. Inorg. Chem.,
            1973, 18, 1060 (1998)

73L   Z. Libus, Inorg. Chem., 1973, 12, 2972

73LG   J. Lipkowski and Z. Galus, J. Electroanal. Chem., 1973, 48, 337

73LR   S. Libich and D. L. Rabenstein, Anal. Chem., 1973, 45, 118

73MC   C. Mehrbach, C. H. Culberson, J. E. Hawley, and R. M. Rytkowicz, Limnol. Oceanog.,
            1973, 18, 897

73MS   U. von Meyenburg, O. Siroky, and G. Schwarzenbach, Helv. Chim. Acta, 1973, 56,
            1099

73MSa   T. P. Makarova, A. V. Stepanov, and B. I. Shestakov, Russ. J. Inorg. Chem., 1973,
            18, 783 (1485)

73N   J. O. Nriagu, Geochim. Cosmochim. Acta, 1973, 37, 2357

73Na   B. Noren, Acta Chem. Scand., 1973, 27, 1369

73NM   T. Nozaki, T. Mise, and K. Torii, Nippon Kagaku Kaishi, 1973, 2030

73NP   B. P. Nikolskii, V. V. Palchevskii and F. TyPang, Doklady Chem., 1973, 209, 253
            (624)

73NS   V. A. Nazarenko, G. G. Shitareva, and E. N. Poluektova, Russ. J. Inorg. Chem.,
            1973, 18, 609 (1155)

73OO   L. L. Olson and C. R. O'Melia, J. Inorg. Nucl. Chem., 1973, 35, 1977

73P   H. K. J. Powell, J. Chem. Soc. Dalton, 1973, 1947

73Pa   V. P. Poddymov, Russ. J. Phys. Chem., 1973, 47, 1063 (1883)

73PE   I. Z. Pevzner, N. I. Eremin, N. N. Knyazeva, Ya. B. Rozen, and V. E. Mironov,
           Russ. J. Inorg. Chem., 1973, 18, 596 (1129)

73PR   S. K. Patil and V. V. Ramakrishna, Radiochim. Acta, 1973, 19, 27

73PRa   S. K. Patil and V. V. Ramakrishna, J. Inorg. Nucl. Chem., 1973, 35, 3333

73PS   B. Perlmutter-Hayman and F. Secco, Israel J. Chem., 1973, 11, 623

73R   P. H. Rieger, Aust. J. Chem., 1973, 26, 1173

73RM   S. Ramamoorthy and P. G. Manning, J. Inorg. Nucl. Chem., 1973, 35, 1279

73RR   F. Rodante, F. Fallo, and P. Fiordiponti, Thermochim. Acta, 1973, 6, 369

73RS   C. Ruzycki, Yu. B. Solovev, and V. E. Mironov, Russ. J. Inorg. Chem., 1973, 18,
           28 (57)

73RT   R. S. Ramakrishna and R. Thuraisingham, J. Inorg. Nucl. Chem., 1973, 35, 2809

73S   A. V. Stepanov, Russ. J. Inorg. Chem., 1973, 18, 194 (371)

73SB   A. Szymaszek and J. Biernat, Monat. Chem., 1973, 104, 74

73SP   H. K. Sinha and B. Prasad, J. Indian Chem. Soc., 1973, 50, 177

73SS   M. Sieprawski, J. Said and R. Cohen-Adad, J. Chim. Phys., 1973, 70, 1417

73SZ   V. I. Sidorenko, E. F. Zhuravlev, V. I. Gordienko, and N. P. Grineva, Russ. J.
           Inorg. Chem., 1973, 18, 670 (1270)

73T   J. Tummavuori, Suomen Kem., 1973, B46, 97

73TR   M. M. Taqui Khan and P. R. Reddy, J. Inorg. Nucl. Chem., 1973, 35, 179

73V   F. Vierling, Ann. Chim., (France) 1973, 8, 53

73VA   V. P. Vasilev and S. A. Aleksandrova, Russ. J. Inorg. Chem., 1973, 18, 1089,
           (2055)

73VAK   V. P. Vasilev, S. A. Aleksandrova, and L. A. Kochergina, Russ. J. Inorg. Chem.,
           1973, 18, 1549 (2912)

73VK   V. P. Vasilev and E. V. Kozlovskii, Russ. J. Inorg. Chem., 1973, 18, 1544 (2902)

74A   G. Anderegg, Helv. Chim. Acta, 1974, 57, 1340

74Aa   R. Aruga, J. Inorg. Nucl. Chem., 1974, 36, 3779

74AA   S. Ahrland, E. Avsar, and L. Kullberg, Acta Chem. Scand., 1974, A28, 855

74AB   S. Ahrland and J. O. Bovin, Acta Chem. Scand., 1974, A28, 1089

74AC   K. P. Anderson, A. L. Cummings, J. L. Bills, and K. J. Walker, Jr., J. Inorg.
           Nucl. Chem., 1974, 36, 1837

74AM   Z. Amjad and A. McAuley, J. Chem. Soc. Dalton, 1974, 2521

V. BIBLIOGRAPHY

74AS   R. Arnek and Y. Sasaki, Acta Chem. Scand., 1974, A28, 20

74BA   F. Brito, J. Ascanio and M. Franceschetto, An. Quim., 1974, 70, 465

74BB   E. A. Belousov, V. V. Bocharov, and V. E. Mironov, Russ. J. Inorg. Chem., 1974, 19, 99 (186)

74BC   A. J. Barker and A. B. Clarke, J. Inorg. Nucl. Chem., 1974, 36, 921

74BL   J. W. Bixler and T. M. Larson, J. Inorg. Nucl. Chem., 1974, 36, 224

74BLL   G. Biedermann, J. Lagrange, and P. Lagrange, Chem. Scripta, 1974, 5, 153

74EM   I. Eliezer and A. Moreno, J. Chem. Eng. Data, 1974, 19, 226

74F   A. M. Fedorenko, Russ. J. Inorg. Chem., 1974, 19, 841 (1543)

74Fa   G. Ferroni, Bull. Soc. Chim. France, 1974, 2698

74FG   A. M. Fedorenko and E. A. Gyunner, Russ. J. Inorg. Chem., 1974, 19, 1397 (2560)

74FGA   G. Ferroni, J. Galea, G. Antonetti, and R. Romanetti, Bull. Soc. Chim. France, 1974, 2695

74FK   V. A. Fedorov, L. I. Kiprin, N. S. Shchekina, N. P. Samsonova, M. Ya. Kutuzova, and V. E. Mironov, Russ. J. Inorg. Chem., 1974, 19, 474 (872)

74FKa   V. A. Fedorov, T. N. Kalosh, and L. I. Shmydko, Russ. J. Inorg. Chem., 1974, 19, 991 (1820)

74FKb   L. N. Filatova and T. N. Kurdyumova, Russ. J. Inorg. Chem., 1974, 19, 1746 (2190)

75FP   M. M. Farrow, N. Puride, and W. D. White, J. Soln. Chem., 1974, 3, 395

74FR   V. A. Fedorov, A. M. Robov, I. I. Shmydko, V. V. Zadnovskii, and V. E. Mironov, Russ. J. Inorg. Chem., 1974, 19, 830 (1523)

74FRa   V. A. Fedorov, A. M. Robov, I. I. Shmydko, N. A. Vorontsova, and V. E. Mironov, Russ. J. Inorg. Chem., 1974, 19, 950 (1746)

74FRI   V. A. Fedorov, A. M. Robov, I. D. Isaev, and A. A. Alekseeva, Russ. J. Inorg. Chem., 1974, 19, 798 (1466)

74FRP   V. A. Fedorov, A. M. Robov, V. P. Plekhanov, V. V. Kudruk, M. A. Kuznechikhina, and G. E. Chernikova, Russ. J. Inorg. Chem., 1974, 19, 666 (1225)

74FS   Ya. D. Fridman, D. S. Sarbaev, and T. V. Danilova, Russ. J. Inorg. Chem., 1974 19, 471 (867)

74G   S. Gobom, Acta Chem. Scand., 1974, A28, 1180

74HH   J. Havel and E. Hogfeldt, Chem. Scripta, 1974, 5, 164

74HI   G. I. H. Hanania and S. A. Israelian, J. Soln. Chem., 1974, 3, 57

74IG   A. A. Ivakin and V. A. Gurevich, Russ. J. Inorg. Chem., 1974, 19, 1655 (3027)

74IGG   A. A. Ivakin, V. A. Gurevich, and M. P. Glazyrin, Russ. J. Inorg. Chem., 1974, 19, 1309 (2397)

74IK   A. A. Ivakin, L. D. Kurbatova, and E. M. Voronova, Russ. J. Inorg. Chem., 1974,
        19, 387 (714)

74IS   E. M. Ivashkovich and M. I. Skoblei, Russ. J. Inorg. Chem., 1974, 19, 411 (760)

74J   L. Johansson, Coord. Chem. Rev., 1974, 12, 241

74JJ   D. V. S. Jain and C. M. Jain, Indian J. Chem., 1974, 12, 178; D. V. S. Jain,
        Ibid., 1970, 8, 945

75JL   R. L. Jacobson and D. Langmuir, Geochim. Cosmochim. Acta, 1974, 38, 301

74K   L. Kullberg, Acta Chem. Scand., 1974, A28, 829

74Ka   L. Kullberg, Acta Chem. Scand., 1974, A28, 897

74Kb   L. Kullberg, Acta Chem. Scand., 1974, A28, 979

74Kc   Kabir-Ud-Din, Z. Phys. Chem. (Frankfort), 1974, 88, 316

74KC   W. F. Kinard and G. R. Choppin, J. Inorg. Nucl. Chem., 1974, 36, 1131

74KH   C. Kappenstein and R. Hugel, J. Inorg. Nucl. Chem., 1974, 36, 1821

74KI   H. Kakihana and S. Ishiguro, Bull. Chem. Soc. Japan, 1974, 47, 1665

74KM   P. K. Khopkar and J. N. Mathur, J. Inorg. Nucl. Chem., 1974, 36, 3819

74KN   N. P. Komar, V. A. Naumenko, and T. A. Karpova, Russ. J. Phys. Chem., 1974, 48,
        954 (1613)

74KO   G. Kura, S. Ohashi, and S. Kura, J. Inorg. Nucl. Chem., 1974, 36, 1605

74KY   F. Ya. Kulba, Yu. B. Yakovlev, and D. A. Zenchenko, Russ. J. Inorg. Chem., 1974,
        19, 502 (923)

74LP   J. K. Lawrence and J. E. Prue, J. Soln. Chem., 1974, 3, 553

74M   D. F. C. Morris, Radiochem. Radioanal. Letters, 1974, 19, 141

74MB   R. E. Mesmer and C. F. Baes, Jr., J. Soln. Chem., 1974, 3, 307

74MC   C. Musikas, F. Couffin, and M. Marteau, J. Chim. Phys., 1974, 71, 641

74MG   Yu. I. Mikhailyuk and V. I. Gordienko, Russ. J. Inorg. Chem., 1974, 19, 1114
        (2033)

74MS   H. Moriya and T. Sekine, Bull. Chem. Soc. Japan, 1974, 47, 747

74MSB   Yu. A. Makashev, M. I. Shalaevskaya, V. V. Blokhin, and V. E. Mironov, Russ.
        J. Phys. Chem., 1974, 48, 1219 (2066)

74MV   O. I. Martynova, L. G. Vasina, and S. A. Pozdnyakova, Doklady Chem., 1974, 217,
        552 (1080)

74MVa   O. I. Martynova, L. G. Vasina, S. A. Pozdnyakova, and V. A. Kishnevskii,
        Doklady Phys. Chem., 1974, 217, 730 (862)

74NB   V. A. Nazarenko and E. A. Biryuk, Russ. J. Inorg. Chem., 1974, 19, 341 (632)

74NBa   N. D. Ngo and K. A. Burkov, Russ. J. Inorg. Chem., 1974, 19, 680 (1249)

74NK   B. I. Nabivanets, E. E. Kapantsyan, and E. N. Oganesyan, Russ. J. Inorg. Chem.,
         1974, 19, 394 (729)

74P   H. Persson, Acta Chem. Scand., 1974, A28, 885

74Pa   H. K. J. Powell, J. Chem. Soc. Dalton, 1974, 1108

74Pb   L. Pellerito, J. Electroanal. Chem., 1974, 54, 405

74RB   Ts. Ruzhitski, V. V. Blokhin, and V. E. Mironov, Russ. J. Phys. Chem., 1974, 48,
         282 (480)

74RL   E. J. Reardon and D. Langmuir, Amer. J. Sci., 1974, 274, 599

74RM   S. Ramamoorthy and P. G. Manning, Inorg. Nucl. Chem. Letters, 1974, 10, 623

74RN   D. R. Rosseinsky, M. J. Nicol, K. Kite, and R. J. Hill, J. Chem. Soc. Faraday I,
         1974, 70, 2232

74RO   D. L. Rabenstein, R. Ozubko, S. Libich, C. A. Evans, M. T. Fairhurst, and C.
         Suvanprakorn, J. Coord. Chem., 1974, 3, 263

74SB   L. H. Skibsted and J. Bjerrum, Acta Chem. Scand., 1974, A28, 740

74SM   F. H. Sweeton, R. E. Mesmer, and C. F. Baes, Jr., J. Soln. Chem., 1974, 3, 3

74SP   V. I. Shlenskaya, G. V. Pichugina, V. P. Khvostova, and I. P. Alimarin, Bull.
         Acad. Sci. USSR, Chem. Sci., 1974, 23, 240 (268)

74SS   P. H. Santschi and P. W. Schindler, J. Chem. Soc. Dalton, 1974, 181

74T   D. J. Turner, J. Chem. Soc. Faraday I, 1974, 70, 1346

74TM   Ya. I. Turyan, L. M. Makarova, and V. N. Sirko, Russ. J. Inorg. Chem., 1974, 19,
         969 (1778)

74TR   M. M. Taqui Khan and P. R. Reddy, J. Inorg. Nucl. Chem., 1974, 36, 607

74VK   V. P. Vasilev and E. V. Kozlovskii, Russ. J. Inorg. Chem., 1974, 19, 807 (1481),
         971 (1781)

74VKL   V. P. Vasilev, V. E. Kalinina, and A. I. Lytkin, Russ. J. Inorg. Chem., 1974, 19,
         989 (1815)

74VZ   G. Velinov, P. Zirolov, P. Tchakarova and O. Budevsky, Talanta, 1974, 21, 163

76AA   S. Ahrland and E. Avsar, Acta Chem. Scand., (in press)

76BM   C. F. Baes, Jr. and R. E. Mesmer, The Hydrolysis of Cations, John Wiley, Inc.-
         Interscience, 1976 (in press)

LIGAND FORMULA INDEX

Order of elements: C,H,O,N, others in alphabetical order.

AsF_6^-	74		HO^-	1
Br^-	115		$HOBr$	134
Br_6Ir^-	135		$HOCl$	134
			HOI	134
$CHON$	28		HO_2N	47
CHN	26		HO_2Cl	134
$CHNS$	29		HO_3I	126
CH_2O_3	37		$HO_4NF_2S_2$	135
CH_2N_2	135		HO_4S^-	79
CH_2S_3	131		HO_4Se^-	93
CH_2S_4	131		HO_4Tc	135
CH_2Se_3	131		HF	96
$CNSe^-$	35		HN_3	45
$C_2N_3^-$	36		H_2O_2	75
$C_4H_2O_4Fe$	135		$H_2O_2N_2$	53, 135
$C_4N_3^-$	36		$H_2O_3N_2$	135
C_5HO_5Mn	135		H_2O_3FP	132
$C_6H_4N_6Fe$	21		H_2O_3S	78
$C_6N_6Co^{3-}$	24		$H_2O_3S_2$	86
$C_6N_6Fe^{3-}$	22		H_2O_3Se	91
$C_8H_3N_8W$	135		H_2O_4Cr	17
$C_8H_4N_8W$	135		H_2O_4Mn	135
			H_2O_4Mo	18
Cl^-	104		$H_2O_4S_2$	135
Cl_6Ir^{3-}	135		H_2O_4W	19
			H_2O_5S	133
F_6P^-	74		$H_2O_8S_2$	89

LIGAND NAME INDEX

Ammonia, 40

Antimonic acid, 133

Arsenic acid, 133

Arsenous acid, 132

Boric acid, 25

Bromate ion, 121

Bromide ion, 115

Carbonic acid, 37

Chlorate ion, 113

Chloride ion, 104

Chlorous acid, 134

Chromic acid, 17

Cyanamide, 135

Cyanic acid, 28

Cyclo-tri-μ-imidotris(dioxophosphate), 72

Dicyanimide ion, 36

Diimidotriphosphoric acid, 71

Hexabromoiridate (III) ion, 135

Hexachloroiridate (III) ion, 135

Hexacyanocobaltate (III) ion, 24

Hexacyanoferrate (III) ion, 22

Hexafluoroarsenate ion, 74

Hexafluorophosphate ion, 74

Hexametaphosphoric acid, 70

Hydrazine, 43

Hydrazoic acid, 45

Hydrocyanic acid, 26

Hydrofluoric acid, 96

Hydrogen amidophosphate, 132

Hydrogen amidosulfate, 88

Hydrogen antimonate, 133

Hydrogen arsenate, 133

Hydrogen arsenite, 132

Hydrogen azide, 45

Hydrogen borate, 25

Hydrogen carbonate, 37

Hydrogen chlorite, 134

Hydrogen chromate, 17

Hydrogen cyanate, 28

Hydrogen cyanide, 26

Hydrogen diamidophosphate, 132

Hydrogen diamidothiophosphate, 132

Hydrogen diimidotriphosphate, 71

Hydrogen diphosphate, 59

Hydrogen diphosphate (III,V), 72

Hydrogen μ-disulfidohexaoxodiphosphate, 135

Hydrogen dithionite, 135

Hydrogen fluoride, 96

Hydrogen fluorophosphate, 132

Hydrogen fluorotriphosphate, 135

Hydrogen germanate, 131

Hydrogen hexacontaphosphate, 135

Hydrogen hexacyanoferrate (II), 21

Hydrogen hexametaphosphate, 70

Hydrogen hexaphosphate, 135

Hydrogen hydroxylamidosulfate, 133

Hydrogen hypobromite, 134

Hydrogen hypochlorite, 134

Hydrogen hypoiodite, 134

Hydrogen hyponitrite, 53